TECHNOLOGICAL INFRASTRUCTURE POLICY

Economics of Science, Technology and Innovation

VOLUME 7

The titles published in this series are listed at the end of this volume.

THE JERUSALEM INSTITUTE FOR ISRAEL STUDIES

TECHNOLOGICAL INFRASTRUCTURE POLICY

An International Perspective

Edited by

MORRIS TEUBAL

Hebrew University, Israel,
and
Head of the Industrial Policy Group,
The Jerusalem Institute for Israel Studies

DOMINIQUE FORAY

Sr. Researcher,
Centre National de la Recherche Scientifique (CNRS),
and
École Centrale Paris

MOSHE JUSTMAN

Chair,
Department of Economics,
Ben Gurion University,
Israel

and

EHUD ZUSCOVITCH

Department of Economics,
Ben Gurion University,
Israel
and
Centre National de la Recherche Scientifique (CNRS)

KLUWER ACADEMIC PUBLISHERS
DORDRECHT / BOSTON / LONDON

A C.I.P. Catalogue record for this book is available from the Library of Congress.

ISBN 0-7923-3835-9

Published by Kluwer Academic Publishers,
P.O. Box 17, 3300 AA Dordrecht, The Netherlands.

Kluwer Academic Publishers incorporates
the publishing programmes of
D. Reidel, Martinus Nijhoff, Dr W. Junk and MTP Press.

Sold and distributed in the U.S.A. and Canada
by Kluwer Academic Publishers,
101 Philip Drive, Norwell, MA 02061, U.S.A.

In all other countries, sold and distributed
by Kluwer Academic Publishers Group,
P.O. Box 322, 3300 AH Dordrecht, The Netherlands.

Printed on acid-free paper

Printed in the Netherlands

Table of Contents

Foreword

This collaborative volume examines the notion of Technological Infrastructure (TI) and the form and function of Technological Infrastructure Policy (TIP) from a variety of perspectives within the general context of the Economics of Innovation and Technology Policy fields. It is comprised of an introductory chapter and three sections. The introductory chapter and Section 1 offer a conceptual discussion of TI, and introduce the distinction between basic TI aimed at absorbing and diffusing externally-developed technology, and advanced TI involving the development of generic, cutting-edge technologies through the cooperative efforts of business firms, often in collaboration with public-sector institutions. Section 2 then elaborates on advanced TI while Section 3 focuses on basic TI.

The book is an outgrowth of two earlier initiatives of the Industrial Development Policy Group (IDPG) of the Jerusalem Institute for Israel Studies. The first was an IDPG volume in Hebrew, published in 1993, and edited by Justman, Teubal, and Zuscovitch, entitled *Technological Infrastructure Policy for Renewed Growth.* It aimed at contributing to the design of a new policy effort in Israel supporting collaborative precompetitive research and development. The second was an International Symposium on *Liberalization and Technological Infrastructrue* in which most of the contributors of this volume participated. The long-time collaboration between Justman, Teubal, and Zuscovitch within IDPG, and D. Foray's insights into its subject matter and contribution to its design resulted in the present volume.

This book would not have been possible without the long-term commitment of the Jerusalem Institute for Israel Studies to the activities of IDPG and in particular to its support of the latter's efforts to apply a conceptual framework to deal with the concrete industrial and technology policy issues of the day. We much appreciate the continued encouragement of Ora Ahimeir and Professor Abraham (Rami) Friedman, directors of the Jerusalem Institute for Israel Studies. Our heartfelt thanks to the individual authors for their contributions and for putting up with our editorial whims. The volume benefited greatly from the skilled and studied hand of our dedicated copy editor, Larry Herman. During the last months of intense and often hectic activity Tsiona Hizkiahu has provided us with valuable administrative assistance.

Finally, special thanks to Esti Boehm whose effort and skill in organizing our work and in designing and producing this volume were as invaluable as they were previously in her capacity as coordinator of IDPG and organizer of the International Symposium.

The Editors

About the Authors

Professor **Patrick COHENDET** is at BETA, Université Louis Pasteur, Strasbourg, France. His current fields of research are evolutionary theory of the firm (concepts of flexibility, coherence, organisational learning); theory of industrial organisation (networks and distribution of knowledge, transfer of technology and knowledge), and evaluation of research and technology policy.

Paul A. DAVID is Senior Research Fellow, All Souls College, University of Oxford; Professor of Economics, and Quondam William Robertson Coe Professor of American Economic History, Stanford University; and Visiting Research Professor of the Economics of Science and Technology, Rijksuniversiteit Limburg (MERIT). His research interests are path-dependence in technological, institutional and demographic phenomena; the economics of science and technology, and the organization of science and engineering research; interdependence of learning processes and diffusion of innovations; standards and standardization in past and present; and coevolution of information network technologies and organizations.

Charles EDQUIST is Professor at the Department of Technology and Social Change at Linkoping University, Linkoping, Sweden. His research interests include the economics of innovation, innovation systems, and technology policy. He is currently coordinator of a Research Network on Conceptual and Theoretical Aspects of Systems of Innovation.

Professor Dr. **Dominique FORAY** is currently Senior Researcher at the Centre National de la Recherche Scientifique (CNRS) and Professor of Economics at the Ecole Centrale Paris. He served as a coordinator of the 1994-1996 CNRS Program "Management of innovation, public policy for science and technology, and knowledge appropriation". He is a permanent consultant at the OECD (Division for Science, Technology and Industry; the Director's Office) where he is leading the Program "National System of Innovation". He received his Ph.D. in 1984 and his *habilitation* in 1992 from the University Lumiere in Lyon. His research interests include the economics of science and technology, the economics of production and distribution of knowledge and the analysis of path-dependent processes of economic change.

Peter C. GRINDLEY is a Senior Economist at the Law and Economics Consulting Group, Berkeley, California. He is a consultant specializing in the economics of innovation, intellectual property, and competition policy. He has current research interests in the relationship between intellectual property, licensing and competitiveness, and the effect of interconnection on innovation in telecommunications and other network industries.

Moshe JUSTMAN, Chair, Department of Economics, Ben Gurion University, Ber-Sheba, Israel. His current research interests are: industrial development and technological change; the political economy of growth; and regional development.

Dan KAUFMANN is a researcher at the Jerusalem Institute for Israel Studies. He received his M.A. degree on Innovation Management and Technology Policy from the University of Limburg (MERIT), Maastricht, The Netherlands. His research interests include economics of innovation, generic research, managing R&D consortia, and technology policy design.

David C. MOWERY is Professor of Business and Public Policy at the Walter A. Haas School of Business, University of California at Berkeley, and a Research Associate of the Canadian Institute for Advanced Research. He has published a number of papers and books on the economics of technological change and on technology policy.

Margaret SHARP, is Senior Fellow, Science Policy Research Unit, University of Sussex, England. She began researching and writing about biotechnology as a new industrial activity in the early 1980s. She has recently been working on the way in which Europe's major pharmaceutical/chemical multinationals developed their interests in this area.

Philip SHAPIRA is Associate Professor, School of Public Policy, Georgia Institute of Technology, Atlanta, Ga., USA. He teaches and conducts research on industrial restructuring, economic and regional development, and technology policy in the United States, Japan, and Europe. Dr. Shapira is currently directing studies of manufacturing-technology diffusion and the effectiveness of industrial modernization initiatives, and serves as an adviser and consultant to agencies in the United States and elsewhere involved in policy analysis, economic development, and technology transfer.

Brian S. SILVERMAN is Assistant Professor of Strategy on the Faculty of Management, University of Toronto. His current research interests include technology strategy, technology policy, and the exploitation of intellectual capital. He has studied technology transfer and commercialization issues at SEMATECH and the U.S. National Laboratories.

Professor Dr. **W. Edward STEINMUELLER** is Professor of the Economics of Technical Change, MERIT (Maastricht Economic Research Institute on Innovation and Technology), Rijkuniversiteit, Limburg, The Netherlands. He is internationally known for research on the integrated circuit, software, telecommunications, and other information-technology industries in which he has a continuing interest. His current research interests also include industrial, technology, and science policy.

Gregory TASSEY is Senior Economist for the National Institute of Standards and Technology. He is engaged in analyses of the economics of high-tech industries; conducts strategic planning, economic impact evaluations, and economic policy assessments for NIST; participates in government-wide and joint industry-government growth policy development. He received his B.A. in physics from Western Maryland College and his Ph.D. in economics from George Washington University. His publications include 20 articles and several books on technology policy and economic growth. The most recent book (1992) is *Technology Infrastructure and Competitive Position*. He is currently writing a book on the economics of industrial growth policy.

Morris TEUBAL is Professor of Economics at The Hebrew University, Jerusalem, Israel, and Head of the Industrial Development Policy Group, The Jerusalem Institute for Israel Studies. His research includes the economics of innovation, processes of growth, national systems of innovation, and technology policy. Current policy-related activity includes both advice to governments and attempts at consolidating a technology policy framework following evolutionary theory principles.

Graham VICKERY is at the Directorate for Science, Technology and Industry, OECD, Paris, where he is Project Director for OECD technology and sectoral projects covering new technology, industrial performance, flexibility of manufacturing and work organisation, and industrial globalisation. He is currently designing and implementing projects on employment-unemployment issues and new growth industries, including the environment industry.

Dr. **A. Tamar YINNON**, The Fritz Haber Center for Molecular Dynamics, Department of Theoretical Chemistry, The Hebrew University, Jerusalem. She is also member of the Industrial Development Policy Group at the Jerusalem Institute for Israel Studies, Jerusalem. Her research on the economics of technological change is currently focused on the dynamics of evolutionary economic processes, while that in chemical physics involves characterization of disordered surfaces of solids.

Dr. **Ehud ZUSCOVITCH**, Department of Economics, Ben Gurion University of the Negev, and the French National Centre for Scientific Research (CNRS). Dr. Zuscovitch's research is in the area of economics of technological change and his current research interests are evolutionary economics, knowledge and information-intensive economic regimes, and technological learning in complex systems.

An Introduction to Technological Infrastructure and Technological Infrastructure Policy

The Editors

Technological infrastructure policy (TIP) is increasingly coming to the forefront of policy discussions, both in the specific context of *technological policy*, and more generally in regards to growth-promoting policies in advanced and developing economies. This has been accompanied by an expansion of the conventional notion of *infrastructure* that has occurred since the mid-eighties. In particular, the structuralist perspective on economic growth and development has emphasized the importance of *structural change* in growth while incorporating a wider notion of infrastructure that includes both universal and *structural change-specific* components (Justman and Teubal 1990, 1991; Guerrieri 1992; Lipsey and Bekar 1995). In addition to the traditional infrastructure elements of physical capital, human capital, and institutions, this approach also incorporates a separate *technological infrastructure* (TI) component. This broader view of the nature and enhanced role of infrastructure in economic growth is consistent with the policy and growth experience of the successful newly industrialyzing countries of Asia (Justman, Teubal, Zuscovitch 1993). It is also increasingly being recognized as being of wider relevance by both policy makers and academic economists. However, a complete acknowledgement of the role played by a distinct technological-infrastructure component has not yet taken place within large segments of the macroeconomic and advanced-neoclassical communities of scholars. We trust that this volume will make a contribution to this process.

Enhancing the understanding of the increasing importance of TI and of TIP is the main objective of this book. We view this process as inevitable for several reasons. First, the increasing importance of innovation and technological development in national economic performance is gaining widespread recognition. Second, the specific policy roles of generic and precompetitive research and of institutional change are increasingly becoming the focus of industrial policy in many countries, as for example, throughout the EU. Finally, scholarship has made significant theoretical and empirical

M. Teubal et al. (eds.), Technological Infrastructure Policy, 1–17.

contributions to this process by consolidating our understanding of TIP; this book is a continuation of several of these recent studies.

The pioneering work of Tassey (1991, 1992) helped identify the crucial role of TI as well as many of the implications for TIP. Early work of Justman and Teubal (1986, 1993), further developed in this volume, integrates salient features and categories of TI and TIP into a structuralist perspective of economic growth. Other traditions such as David and Foray's 1994 (further developed in this volume) and the National Systems of Innovation (NSI) framework of analysis (Nelson 1993, Lundvall 1992) also assign important roles to TI, TIP and associated institutions. They are components of existing NSI and active agents of change of such systems. Finally, Romer's (1993) recent proposal for the creation of an institutional framework for actively promoting "self-organizing industry boards" embodies a policy framework oriented towards the development of institutional tools which enable firms to solve collective-action problems raised by the production of industry-specific public goods. This fits into a modified *basic TIP framework* as proposed in Chapter 1 and is an example of a structural change-specific infrastructure.[1]

This introduction uses many of the arguments and conclusions developed in this body of scholarship in presenting a concise overview of three important TI- and TIP-related issues. The topics, all of which are covered in detail in a number of chapters of this book, include:

- the role and importance of TI in economic growth;
- alternative definitions and components of TI and TIP; and
- TI and technology markets.

1. The Role of TI in Economic Growth

1.1 TI Within the Structuralist Perspective

Examining the link between technological infrastructure development and economic growth according to the structuralist perspective requires focusing on the conditions for successful *structural change* (SC) (Justman and Teubal 1990, 1991). This perspective goes beyond identifying physical-capital accumulation as the critical determinant of growth, by proposing three

[1]The role played by technology institutes in mid-tech and low-tech industries also has been the subject of systematic research in a number of countries. Goldman's (1994) investigation of the subject has since been followed by a steady stream of similar studies.

additional decisive requirements for growth: infrastructure (conventional, human capital, institutional, and technological); the development of firm-based capabilities; and the emergence of an efficient pattern of inter- and intrasectoral links (see Box).

Conditions for Successful Structural Change (in addition to directly productive physical capital investment)	
	Examples
1. Infrastructure (I) • Conventional • Human Capital Infrastructure (HCI) • Institutional Infrastructure • Technological Infrastructure (TI) – portfolio of specific, "collective" capabilities	 • power, water, and roads • minimum pool of electronics engineers • network of technology institutes; functioning patent system • microelectronics capability within a technology institute
2. Firm-Based Technological Capabilities (FBC)	• production and investment capabilities in process industries • innovation capabilities in emerging high-tech firms
3. Intra- and Intersectoral Links	• user-producer relations • clusters • industrial districts

While conventional infrastructure may be necessary for spearheading physical-capital accumulation, other infrastructure components – human capital, technological, and institutional – facilitate and stimulate the accumulation of firm-based technological capabilities, the success of which is the acid-test of successful structural change. The availability of ample pools of basic skills at all levels may be *sine qua non* since firm-based capabilities must build upon them. For example, under certain conditions a minimum pool of electronics engineers may be required for guaranteeing a capability for effectively using and adopting available microelectronics technology; similarly, both this factor and the involvement of consultants and a technology center may be preconditions for successful diffusion of the technology to SMEs of emerging industrial branches. Also, a minimum pool of mechanical, metallurgical, and chemical engineers may be a requirement for creating production and investment capabilities within infant steel-industry firms. Needless to say, this notion of infrastructure (i) transcends conventional notions of universal, structure-independent infrastructure (e.g., power and primary education) and (ii) includes a less-recognized element which we term *technological infrastructure* (TI).

TI in turn should be distinguished both from the human-capital component of the infrastructure of structural change and from the network of science and technology organizations and institutions which exist beyond the boundaries of individual firms. Only a part of the human-capital infrastructure will be used in activities which generate the set of *collective capabilities* comprising TI. The remainder will be utilized as a critical factor in generating firm-based capabilities, through a process of mutually reinforcing collective learning. Moreover, a particular network of science and technology institutions may house alternative TIs, that is, alternative sets of specific – although broad and generic – capabilities. These considerations mean that technological infrastructure policy should also consider selective aspects of the human capital and institutional infrastructures while ignoring others.

Coevolution of TI and Firm-Based Capabilities

There are two extreme models of short-term – long-term, complementary – substitution relationships between TI and firm-based capabilities (FBC), each of which has clear implications for policy.

Model 1: TI-Led Structural Change: New TI is initially required for structural change to enable firms to acquire their own capabilities. This process is a result, *inter alia*, of technology transfers from the technology institutions housing the capabilities. Thus, TI and FBCs are initially complementary but eventually become substitutes. There is a significant likelihood that initially TIP is relevant to stimulate TIs, and then subsequently for transferring technology to the private sector. Examples of this pattern include the use of Japanese Engineering Research Associations (ERAs) of the '50s and '60s to transfer capabilities to auto-parts firms (Sigurdson 1986), and the beginnings of Taiwan's microelectronics industry in the early 1980s (Hou and Gee 1993).

Model 2: TI-Assisted Structural Change: Firms initially acquire capabilities directly, with or without government support, as the result of classical firm-based support of innovation and technology adoption, and not as the result of TIP. The acquired capabilities enable firms to accomplish significant economic achievements. Further growth, however, necessitates TI to complement these capabilities. At that point technological infrastructure policies may not be relevant since market forces by themselves may be effective at initiating, organizing, and financing the collective undertaking. Examples of this pattern include the establishment of textile test centers by the Japanese textile industry, and numerous cases in Korea. However, this model may still assign an important role for TIP, especially when strong user involvement and cooperation are required. This is particularly true when a culture of cooperation among firms does not exist, and when the uncertainties and transactions costs associated with TI-related cooperation are high (see chapters in this volume by Justman and Teubal; Grindley, Mowery, and Silverman; Sharp; Kaufmann and Yinnon). Government's role under these conditions would involve, among other things, generating an appropriate institutional framework for (and catalytic support of) TI generation by market forces.

The actual patterns of coevolution between TI and FBC are, in all likelihood, much more complicated than that described by either model. A more realistic approach would be to characterized the phenomenon as also involving coevolution between technology and institutions (Nelson 1994). Under this approach the role of TIP in promoting structural change should be to assure self-reinforcing patterns of growth and interaction between TI and FBCs, and to stimulate the enabling and catalytic role of institutional change in this process.

1.2 Tassey

Tassey reacts to Ergas's (1987) view that country specialization in one of the stages of the technology cycle (early, growth, and mature phases) is consistent with strong economic performance. According to this *single-stage strategy* view, the US could specialize in the early phase of the cycle and adopt a policy of *shifting* – that is, developing new sectors – while Germany, by specializing in the mature phase, could adopt a strategy of *deepening* –

increasing productivity and exploiting market niches in existing, mature sectors and technologies.

In contrast to this view, Tassey proposes a *multistage strategy* for the U.S.; he argues that a country cannot and should not specialize exclusively in the early stages of new technologies, but rather, should be involved simultaneously in all stages. His proposition is based on three central arguments. First he observes that shortened technology life cycles reduce the value added that can be expected to be extracted from any one stage. Next he argues for the desirability of diversification in order to reduce risks. And finally he warns that a problem particular to specializing in the early phase of technologies is that it may be difficult to extract sufficient value added since much of the economic rent from the new technology might end up in the hands of owners of complementary assets (Teece 1988), such as Japanese firms possessing superior manufacturing capabilities. Thus, the appropriate strategy for large countries (U.S., Japan), or for regional blocks of countries (EC), is to generate a portfolio of technologies where substantial numbers of emerging, strategic, and mature technologies are continuously under development and exploitation.

Since an increasingly important requirement for competitivity is speed, individual firms will – to a greater degree than in the past – require access to external sources of technology (see Willinger and Zuscovitch 1988). TI, which is a major source of external technology, will therefore have to expand its functions and increase both the speed and flexibility of its interactions with the private sector.

Tassey's (1991) more specific views on the role of TI are based on the following three postulates:

- *Technology does not automatically move from the laboratory to the marketplace*; nor does it readily become widely used within an economic sector. This is contrary to the view which became prevalent in the eighties that the only role for government is to fund basic science but not technology, and that both the process of innovation and transfer of technology from public laboratories, and the diffusion of new technologies should be performed by market forces without government support;
- Since typical civilian technology is a complex entity consisting of a number of proprietary and nonproprietary elements, *a variety of private and public institutions must make critical contributions at the appropriate times*. This clearly implies that there is an important role to be played by public institutions, even after the basic research or science stage;

- *Private and public institutions that produce technology must be integrated with business and economic institutions.* This suggests the need for consortia that include public laboratories together with private firms, a course of action not opened to these laboratories in the U.S. before the Technology Transfer Act of 1986.

TIP implications

The logical conclusion from Tassey's premises is an active government policy which supports generic research and technological infrastructure. This strong support for a policy role that goes beyond basic research insinuates an intermediate stage which precedes the entrepreneurial function of applied research and development. Beyond the simple need for additional resources for TI, it also calls for new mechanisms and institutions through which TI can be effectively developed and implemented. These would:

- enable and stimulate consortia between government laboratories and private firms;
- lead to an explicit diffusion policy or technology-transfer component at government laboratories; and
- recognize the multiplicity of public-private collaboration patterns.

Moreover, these assumptions highlight the need for enhanced support of *emerging technologies*, including a combination of modest precompetitive support for many technologies and more massive (and selective) support for a smaller group of technologies having strategic value. Concerning the former, Tassey emphasizes that small amounts of support may have important leverage effects on the scope and timing of private complementary investments. The latter group involves system technologies which have broad downstream market impacts and which create large upstream demands. Finally, and no less important, Tassey's framework signals the critical role of *infratechnologies*, an aspect that he further develops in his chapter in this volume.

1.3 The Role of TI in the Distribution of Knowledge

The particular characteristics of science and technology knowledge have important implications for TI. First, knowledge possesses one of the essential properties of a public good; it is *nonrival*, meaning that it is infinitely expansible without loss of its intrinsic qualities (David 1993). Second, the process of knowledge generation in science and technology is *cumulative* and *integrative*;[2] consequently, knowledge is at once both an output and the critical input in the process of its own creation. As a result of the nonrival nature of knowledge, the efficient utilization of existing knowledge requires expansive distribution. At the same time, because of its cumulative and integrative character, the efficient generation of new knowledge in science and technology is highly dependent upon the accessibility of the existing stock of knowledge. These properties underlie David and Foray's principal conclusion regarding technological innovation in the knowledge-based economy: the efficient distribution of knowledge and the assurance of widespread access to information are *sine qua non* conditions for increasing the quantity of innovative opportunities in technologically advanced economies.[3]

The nature of technological change has evolved as a result of the technological revolution and the process of globalization. Both have transformed the innovation process and greatly increased the significance of capabilities for accessing the existing knowledge base. The ability to generate valuable innovations has become less dependant upon the discovery of new technological principles, and more reliant on the capacity to systematically exploit the existing stock of knowledge by constructing new uses and devising fresh combinations. This new model of innovation is less characterized by "leaps" in technology than by routine uses of the technological base, involving systematic access to state-of-the-art techniques.[4]

Industry can facilitate this new form of innovation by introducing procedures for the dissemination of information regarding the stock of

[2]Cumulative forms of knowledge are those in which today's advances lay the basis for tomorrow's, which in turn lay the basis for the next round. The integrative aspect of the production of knowledge means that new knowledge is selectively applied and integrated into existing systems to create new systems.

[3]Conventional theory posits that widespread diffusion of knowledge generates static efficiency gains that are, to some degree, mitigated by losses in dynamic efficiency. The gains in static efficiency arise from the wider application of innovations in the production of goods and services. But dynamic efficiency suffers to the extent that innovators perceive that they will be unable to fully exploit potential economic rents, thus undermining incentives to create new knowledge. However, this dichotomy fails to account for the positive dynamic implications attributable to the cumulative and integrative nature of the creation of science and technology knowledge. Thus, the widespread diffusion of this form of knowledge is likely to yield mutually reinforcing static and dynamic efficiency outcomes.

[4]This view of innovation has led to new innovation-related data-gathering activities. For a discussion of the objectives, targets, and methodology of the new innovation surveys see OECD 1992.

technologies available, enabling individual innovators to draw upon the work of others. It cannot be expected, however, that the conditions for efficient distribution and utilization of knowledge will arise automatically from the interplay of market forces. In effect, making the knowledge stock itself more socially useful will, in many instances, call for different policy measures than those designed to make it more economically valuable to private parties. Indeed, in the short run these two goals are likely to be in conflict.

This is where technological infrastructure plays a crucial role, by improving the efficiency of knowledge distribution, otherwise known as *knowledge distribution power*. TI can provide mechanisms for the transfer and transformation of existing and new knowledge, making the stock of knowledge more socially useful. TIP can be designed to improve these mechanisms, help create capabilities to exploit existing knowledge, and facilitate the accessibility of the stock of knowledge. Because many new products, processes, and ideas arise from novel and often unanticipated combinations of knowledge, it is particularly important that:

- the distribution of knowledge be widespread and timely;
- access to new findings be easy and inexpensive;
- wasteful duplication of research efforts be minimized; and
- information be put into the hands of a more diverse population of researchers.

David and Foray's framework is significant both because of their insights into existing growth-promoting mechanisms and capabilities that contribute to knowledge distributive power, and because their analysis points to additional components of TI (see next section). They emphasize that the new mode of knowledge generation – based on the recombination and reuse of known practices – raises information and search problems due to the increasing costs of storing, retrieving, evaluating, and using knowledge. They also identify important institutional problems requiring carefully designed TIP; for example, the existing legal framework for protecting intellectual property rights tends to impede the accessibility of the existing stock of information.

A number of specific policies and institutions designed to strengthen the distributive power of knowledge are identified by David and Foray, and other contributors to this volume. These include:

- mechanisms and incentives supporting coordination among private agents and local sharing of knowledge (see David and Foray 1994; Grindley, Mowery, and Silverman in this volume);

- mechanisms and incentives for public support of technology transfer between government laboratories and the business sector (see Tassey 1991, 1992, and in this volume);
- the support of transfer sciences and of industry-university centers housing the associated capabilities;
- sectoral infrastructures capable of providing *industry-specific public goods* resulting both from the transfer and absorption of technology developed elsewhere and from the sectoral diffusion of generic technologies (see Romer 1993; and chapters in this volume by Justman and Teubal; Steinmueller; and Sharp).
- specific incentives, mechanisms and organizations associated with the transfer and diffusion of technology and business practices to SMEs (see Justman 1995; and chapters in this volume by Cohendet; Shapira; and Vickery).

The common theme in all of these approaches is the enhanced technological infrastructure policy role in a knowledge-based economy. A substantial portion of the TI must respond to the needs associated with accessing and distributing knowledge. The formulation and the implementation of TIP should therefore support generation of firm-based capabilities for accessing and utilizating knowledge.

2. Alternative Definitions and Components of TI and TIP

In the first chapter of this volume, Justman and Teubal define technological infrastructure as "the set of specific industry-relevant capabilities which have been supplied collectively and which are intended for several applications in two or more firms or user organizations." As with firm-based capabilities, TI is embodied in a variety of forms: human capital (formal education, tacit capabilities resulting from both training and experience), physical capital (instrumentation), knowledge (design methodologies), and organization (e.g., networks linking users of technological services with a technological center supplying them).

The TI concept presented by Justman and Teubal in this volume is broader than the so-called *classical innovation infrastructure* they described earlier (1986), but still restricted to technology-related activities that fall between basic and applied research. Here a *basic TI* component plays the role of providing technological services and other routine activity, while an *advanced TI* component involves generic research in cutting-edge technologies. In this respect, Justman and Teubal's notion of TI coincides with the multiphase TI concept proposed by Tassey. This notion of TI is clearly differentiated from large segments of the science infrastructure even though there are elements of overlap between the two. It also excludes the institutional infrastructure supporting the capabilities comprising such infrastructure. Despite the fact that a given set of collective organizations such as technology institutes could house alternative sets of generic multiuse capabilities there is no unambiguous answer to the question whether it is better to define TI as including or excluding such organizations; neither is it clear whether or not to define TI even more broadly by also including the institutional aspects governing the transfer of results from these organizations to individual firms.

Tassey (1991) gives a clearly affirmative response to these questions by characterizing TI to include:

- infratechnologies,
- generic technologies,
- technical information,
- research and test facilities,
- information relevant for strategic planning,
- forums for joint industry-government planning and collaboration, and
- assignment of property rights.

Note the broad nature of his definition, as he includes both organizational and institutional aspects into TI. This also follows from his analysis of the functional role of TI, wherein he calls for it to take account of both funding agencies and the mechanisms for the transfer of research results. Finally, Tassey makes it very clear that an analysis of TIP requires consideration of both the U.S. laboratory research system and associated technology transfer mechanisms.

Thus, Tassey – as well as others, including David and Foray (1994 and in this volume) – takes a strong position for including organizations, institutional mechanisms, and even policy-setting mechanisms in TI, in addition to the universally accepted collective technological capabilities.[5] Going even further, Tassey's TI concept includes nontechnical information and capabilities, such as those required for strategic planning. This raises the question of whether

[5]Note that whatever definition of TI is adopted, organizational and institutional aspects always lie within the purview of TIP.

also to include *techoeconomic capabilities* or *technology-needs coupling capabilities* in the definition of TI; these are the capabilities for establishing intermediation between existing TI and specific applications at the firm, industry, and sector levels. For example, for purposes of transferring and diffusing a new technology into the business sector, a technological center may develop capabilities to configure specific services for different types of users. It is clear that a wider definition of TI such as this could be useful in a number of important TIP contexts.

We have previously discussed David and Foray's knowledge-based economy framework and its implications for both the nature and objectives of TI. Among the TI components that particularly enhance the knowledge distributive power of the national system of innovation, they include:

- a specific *information infrastructure* (e.g., protocols, format standards, codified knowledge);
- institutional structures governing the transfer of knowledge across organizations; and
- at least part of the knowledge generated by university research on the transfer sciences.

Having come this far, a final extension of the TI concept would be the inclusion of firm-based capabilities. It is well known that many such capabilities involve generic components and are potentially available to other firms and users, as firms seek to exploit the nonrival nature of knowledge, or via unintended spillovers (see Steinmueller in this volume). Moreover, globalization is changing the nature of markets, leading to the opening up of many private-sector laboratories which previously catered exclusively to the in-house needs of firms (for example, technological services from Fiat's laboratories are increasingly being offered to third parties).[6] Similarly, the restructuring of government laboratories is part of a shift in focus from single-user (government) to multi-user (market-oriented) capabilities, a process that resembles market building in connection with basic TI (see Teubal and Justman in this volume). Finally, the nature of certain technologies and market structures in some countries can lead to a situation in which the industry and

[6]The process of globalization and the information technology revolution raise additional issues concerning national TI since users may increasingly exploit capabilities available elsewhere. These processes are increasingly transforming some of the outputs of these capabilities into tradeable goods. This shift and the associated "marketization" of certain classes of knowledge and advisory services may actually reduce national TI requirements. It also has implications for the measurement of TI.

firm are identical, meaning that, by definition, sectoral and firm-based capabilities are one and the same (see Edquist in this volume for examples). Consequently, a subset of firm-based capabilities might have to be included in any practical definition (and eventually, measurement) of the growth-promoting TI of a country.

Defining TI is not only an academic exercise. The various (and conceptually alternative) components of TI call for appropriate and often targeted TI policies. How TIP is configured will naturally depend on how narrow or expansive a notion of infrastructure one takes. Effectively designed policy also requires comprehensive and systematically gathered information on TI; ironically, despite the prominent role that knowledge plays in TI theory, empirical knowledge of TI itself remains a sparse commodity.[7]

3. TI and Markets

It is commonplace to think that markets for TI or for its outputs and applications generally do not exist or may be imperfect (Justman and Teubal). This is consistent with the traditional view of TI which is characterized by the public-good nature of the capability and by economies of scope in generating it. Typical examples are the technological capabilities housed in a public institution and available in principle to all national users. We may add that the traditional view of TI is largely a static one derived from extrapolating to TI the context surrounding some conventional physical infrastructures.

However, paradigmatic changes in NIS are creating new markets in areas where they previously didn't exist including technology markets for outputs and applications of TI. Once such outputs and applications are transacted in the market, will the underlying generic capability still be considered part of TI? The answer is, probably not.

[7]The fact that no systematic data on TI exists reflects the conceptual complexity and the still-indistinct nature of the concept. Partial and fragmentary information exists on selected components of TI: on consortia of different countries (for Japan see Wakasugi 1988; for the U.S. see chapters by Tassey and by Grindley, Mowery, and Silverman in this volume); on research and technology organizations in Europe (Sprint 1994); and on university-industry joint research centers in the U.S. (Cohen, Florida, and Coe 1994). Unfortunately, even the budget for NIST in the U.S. is not disaggregated among the various categories of TI (generic technology, infratechnology, and technology), nor are the budgets of other U.S. government agencies (private communication from G. Tassey).

Consider the process of diffusion throughout the business sector of a previously unavailable technological service (e.g., a more precise measurement technique used in production and quality control) flowing from a technological capability transfered from abroad and absorbed in a sectoral technological center (an example of basic TI). Initially this capability qualifies as TI since it is intended for (and is potentially beneficial to) numerous users and uses; and its generation was subject to economies of scope.

The process of diffusion of the service and, eventually, of the underlying capability to provide the service among a set of private laboratories and consultants[8] essentially means the transformation of a *collective capability* (i.e., TI) into a set of *private capabilities.*[9] Very likely, the process of *privatization* of TI will be more complete and faster whenever diffusion is accompanied by the building of a market for the novel service.

Consequently, the contribution of initially collectively generated technological capabilities to TI may decline through time, even in the absence of competitive technologies (and the associated obsolescence factor). In fact this possibility is valid also for advanced TI since successful cutting-edge generic technologies which are initially part of advanced TI may eventually become part of basic TI during the diffusion phase. We are confronting an apparent paradox: given the fact that basic TIP promotes explicit market-building policies, it accelerates the process of transforming TI into enterprise-based capabilities.[10] However, if considered in the context of correcting market failure it will be recognized that this process is fundamentally a successful and sustainable remedy.

[8]This is an appropiate policy for a technology center attempting to maximize social (as opposed to private) profitability; see Justman and Teubal (this volume).

[9]This is the process described by the TI-led model of structural change (see 1.1 above). During the process of diffusion both the value of the externalities flowing from the service and the indivisibilities from capability replication decline, although this need not be a monotonic process.

[10]The market-building objective of basic TIP complements the main objective of stimulating the generation of new TI, both basic and advanced, since "marketization" liberates public resources which may be reinvested in new capabilities whose emergence is blocked by significant market failures.

Concluding Remarks

This introduction analyzed what we regard as some of the central conceptual issues surrounding TI and TIP. The chapters that follow further develop some of them while also introducing others. A tentative conclusion seems to be that despite recent advance, much work clearly remains to be done. We hope this book succeeds in stimulating additional research in these increasingly important areas.

References

Cohen, W.R., R. Florida, and W.R. Coe. 1994. University-industry research centres in the United States. Carnegie Mellon University. Typescript.

David, P. 1993. Knowledge, property and the system dynamics of technical change. World Bank Annual Conference on Development Economics. Washington D.C.

David, P. and D. Foray. 1994. *Accessing and Expanding the Science and Technology Knowledge-Base: A Framework for Interpreting Available Quantitative Measures*. OECD-DSTI-CSPT.

Ergas, H. 1987. Does technology policy matter? In *Technology and Global Industry: Companies and Nations in the World Economy*, edited by B. Guile and H. Brooks. Washington, D.C.: National Academy Press.

Goldman, M. 1994. Technology institutions: When are they useful? World Bank. Typescript.

Guerrieri, P. 1992. Technology and trade competition: the changing positions of the United States, Japan and Germany. In *Linking Trade and Technology Policies*, edited by M. Caldwell Harris and G. Moore. National Academy of Engineering. Washington D.C.: National Academy Press.

Hou, C. and S. Gee. 1993. National systems supporting technical advance in industry: The case of Taiwan. Chapter 12 in Nelson 1993.

Justman, M. 1995. *Modernization and Growth in the Small Business Sector – Lessons from Public Policy* (in Hebrew). Industrial Development Policy Group. The Jerusalem Institute for Israel Studies.

Justman, M. and M. Teubal. 1986. Innovation policy in an open economy: A normative approach to strategic and tactical issues. *Research Policy* 15: 121-138.

—. 1990. The structuralist approach to economic growth and development: Conceptual foundations and policy implications. In *Science and Technology: Lessons for Development Policy*, edited by R. Evenson and G. Ranis. Boulder, CO: Westview Press.

—. 1991. The Structuralist perspective on the role of technology in economic growth and development. *World Development* 19 (9): 1167-1183.

—. 1993. Strategic technological policy or new industrial infrastructure: developing capabilities and building markets (in Hebrew). Chapter 4 in *Technological Infrastructure Policy for Renewed Growth*, edited by M. Justman, M. Teubal, and E. Zuscovitch. The Jerusalem Institute for Israel Studies. Forthcoming in *The European Periphery Facing the Next Century*, edited by S. Metcalfe and J. Vences. Kluwer.

Justman, M., M. Teubal, and E. Zuscovitch, eds. 1993. *Technology Infrastructure Policy for Renewed Growth* (in Hebrew). The Jerusalem Institute.

Lipsey, R. and C. Bekar. 1995. A structuralist view of technical change and economic growth. Working Paper No. 45. CIAR Program in Economic Growth and Policy.

Lundvall, B.A., ed. 1993. *National Systems of Innovation: Towards a Theory of Innovation and Interactive Learning*. Pinter Publishers.

Nelson, R., ed. 1993. *National Systems of Innovation*. Oxford University Press.

—. 1994. Recent evolutionary theorising about economic change. Typescript.

OECD. 1992. *OECD Proposed Guidelines for Collecting and Interpreting Technology Innovation Data* (OSLO MANUAL). Paris: OECD.

Ouchi, W. 1989. The new joint R&D. *Proceedings of the IEEE* 77: 1318-1326.

Romer, P. 1993. Implementing a national technology strategy with self-organizing industry investment boards. *Brookings Papers: Microeconomics* 2: 345-399.

Sigurdson, J. 1986. *Industry and State Partnership in Japan: The VLSI Circuits Program*. Discussion Paper No. 186, Research Policy Institute, Lund, Sweden.

Sprint. 1994. *The Future of Research and Technology Organisations in Europe*. Proceedings of a Sprint Conference, 16-17 November 1993. Brussels.

Tassey, G. 1991. The functions of technology infrastructure in a competitive economy. *Research Policy* 20: 345-361.

—. 1992. *Technology Infrastructure and Competitive Position.* Norwell, Mass: Kluwer Academic Publishers.

Teece, D.J. 1986. Profiting from technological innovation: Implications for integration, collaboration, licensing and public policy. *Research Policy* 15: 285-305.

Teubal, M. 1995. R&D and technology policy at NICs as learning processes. IDPG: The Jerusalem Institute (forthcoming in *World Development*).

Wakasugi, R. 1988. A consideration of innovative organization: Joint R&D of Japanese firms. Paper presented at the International Schumpeter Conference.

Willinger, M. and E. Zuscovitch. 1988. Towards the economics of information-intensive production systems. In *Technical Change and Economic Theory*, edited by Dossi et al. London: Pinter.

World Bank. 1993. *The East Asian Miracle: Economic Growth and Public Policy*. Oxford University Press.

I. Underpinnings

Technological Infrastructure Policy (TIP): Creating Capabilities and Building Markets

Moshe Justman and Morris Teubal*

Accelerated technological development and the globalization of trade and investment have changed the nature of competition in world markets, increasing the importance of technological capabilities as a source of competitive advantage. These changes have raised new needs and new opportunities for collective action in support of individual firms' efforts to acquire the necessary capabilities. We refer to the public goods that are the object of such collective efforts as *technological infrastructure*, and to the policies aimed at promoting their creation or emergence as *technological infrastructure policy*.

The role of *conventional* infrastructure in industrial development is well understood today and informs a coherent and detailed theory of public intervention that can identify where government should be involved (e.g., "natural monopolies"), what modes of intervention are appropriate (e.g., rate-of-return regulation of public utilities), and what implementation problems might be encountered in practice (e.g., overinvestment in fixed assets). *Technological* infrastructure is less tangible, answers needs that are less clearly articulated, and at the same time is more differentiated and specific, serving narrower constituencies. Where conventional infrastructure is aimed at meeting well-defined needs through known methods within existing markets, technological infrastructure often involves the articulation of new needs that can only be met through the generation of new capabilities within markets that have yet to be created. Policies for fostering the timely emergence of technological infrastructure must therefore address issues that arise from the

*Reprinted from *Research Policy* 24 (1995) 259-281, with kind permission from Elsevier Science B.V., Amsterdam, The Netherlands. We are grateful to members of the Industrial Development Policy Group (IDPG) of the Jerusalem Institute for Israel Studies and to researchers at SPRU (University of Sussex) where a preliminary version of this paper was presented. We are particularly grateful to M. Bell, M. Dodgeson, M. Hobday, D. Kaufmann, K. Pavitt, M. Sharp, B. Toren, E. Zuscovitch, T. Yinnon and to a referee for his comments.

An earlier draft of this paper entitled "Strategic Technology Policy: Capability Creation and Market Building", appeared as Chapter 4 of *Technology Infrastructure Policy for Renewed Growth*, M. Justman, M. Teubal and E. Zuscovitch (eds.), Industrial Development Policy Group, The Jerusalem Institute for Israel Studies, 1993 (in Hebrew).

M. Teubal et al. (eds.), Technological Infrastructure Policy, 21–58.

indivisibility of infrastructure and its interdependence with private investment in firm-based capabilities, in an unfamiliar context of limited appropriability, incomplete information, strong differentiation, and rapid change.

This calls for new approaches to technology policy that place greater emphasis on institutional innovation as opposed to conventional price-based measures that compensate for the external benefits of technological change and view the role of the public sector as mostly catalytic. Broad-based subsidies or tax credits cannot resolve the complex dimensions of coordination and public choice that arise in building technological infrastructure. Hence the need for separate consideration of a specific category of policies that are distinct in their rationale and in their administrative requirements from earlier modes of intervention, requiring new forms of interaction between the public and private sector, new modes of analysis, and new lines of coordination and control within government.[1]

In this paper we suggest a conceptual framework for these new policies, drawing on a growing body of theory and practical experience on the nature of technological infrastructure and the challenges it raises. We begin by clarifying what we mean by technological infrastructure (TI), before going on to discuss the distinct structure and features of technological infrastructure policy (TIP).

1. Technological Infrastructure

The role of TI in growth is best understood in the context of a structuralist perspective, as one of several preconditions for structural change (Justman and Teubal 1990, 1991). These also include the existence of the necessary

- conventional infrastructure (transportation, communications, power, etc.),
- human capital (e.g., a critical mass of electronics engineers),
- institutional infrastructure (e.g., a patent system, market for high-risk stocks);
- firm-based capabilities in production, investment and innovation (cf. Westphal et al. 1984; Dahlman et al. 1985);

[1] These changes also have far-reaching implications for relations among firms, especially the new importance of networks as an intermediate form of industrial organization. See Teubal, Yinnon and Zuscovitch 1991; and Zuscovith and Justman 1993.

- and the resolution of the implicit interdependencies of investment decisions on which structural change depends.[2]

In this context the distinct nature of *technological* infrastructure emerges more clearly: a set of collectively supplied, specific, industry-relevant capabilities, intended for several applications in two or more firms or user organizations. They are embodied in human capital (both formal education and experience), and include also elements of physical capital (such as instrumentation) and knowledge. Hence the distinction between TI and other forms of infrastructure: more differentiated and less tangible than conventional infrastructure, TI cannot be established through public works programs; transcending human-capital infrastructure in its orientation to specific, systemic needs which generally cannot be generated through the education system or through broad-based support for science; centered in capabilities rather than knowledge, its acquisition requires sustained effort; and TI's essential requirement of industry-relevance sets it apart from science infrastructure.[3] Moreover, though TI complements and is dependent on firm-based capabilities, it is a collective good that is not addressed by traditional support for industrial R&D targeted at single firms. Finally, our emphasis on the link between TI and structural change leads us to exclude the *routine* provision of technological services.[4]

An excellent example of activity leading to multiuser/multiuse capabilities embodied in design data and a design methodology is the propeller tests conducted at Stanford University by W. Durand and E. Leslie during 1916-1926 (Rosenberg and Nelson 1992). Extensive experimental testing was required because of the absence of a body of scientific knowledge that would permit a more direct determination of optimal propeller design, compatible both with the power-output requirements of the engine and the flight requirements of

[2]See Hirschman (1958) and Chenery (1959) for seminal analyses of the interdependence between investment in infrastructure and in directly productive activities; see also Justman (1993) for a formal dynamic model.

[3]Though scientific *results* are not generally part of TI, scientific *capabilities* often are. See Pavitt (1993) on the usefulness of basic research for technological practitioners.

[4]Our definition of TI is similar in spirit to Tassey's (1991) though he defines TI in terms of knowledge rather than capabilities. That his meaning is similar to ours can be seen from his specification of the components of TI: infratechnologies, generic technologies, research and testing facilities, technical information and other elements. It is important to distinguish between the capability itself and the output it generates. The former is generic (multiuse and multiuser) while the latter may be specific, e.g., a series of efforts at solving very specific problems facing a number of firms.

the airframe. The experiments relied extensively on wind-tunnel testing. The outcome not only enabled improvement in propeller design as such, but also led to a better ability to match the propeller to the engine and the airframe. Moreover, it also increased the reliability of certain techniques utilized in aircraft design. The improvements in aeronautical research and in design methods that resulted from the Stanford experiments made an important contribution to the maturing of the American aircraft industry – a maturity crowned by the success of the DC-3.

The need for TI has increased greatly in recent years, but some examples date back a hundred years and more. Early forms of TI, each with its modern counterparts, include applied research and extension systems for agriculture and related sectors with little or no firm-based R&D capability; and measures and standards institutes and capabilities, e.g., for grading of quality, etc. More recent examples of TIP include creation of Engineering Research Associations (ERAs), established in the UK after World War I, and in Japan after World War II, including related capabilities to support modernization and enhance the competitiveness of medium and low-tech industries; and, more recently, the stimulation of growth-oriented consortia for developing frontier industrial technologies, such as Japan's pioneering VLSI (Very Large Scale Integration) program of 1976-1979 (Sigurdson 1986).

These examples point to a number of common features that characterize TI in its various forms. First, it has a *generic* quality: the capabilities involved are intended for different uses by a number of firms, not directly geared to individual innovations or firm-specific R&D projects. Generic research is an important component of some technological infrastructures; but not all technological infrastructure involves research, and not all scientific research generates technological infrastructure.[5] Second, it usually requires a

[5]"Generic industrial research describes research activities of an applied nature that are of broad relevance to some entire industry or sector – as opposed to some specific product, device or firm" (Rosenberg and Nelson 1992). According to Tassey (1991), *generic technological research* "identifies and characterizes relevant performance attributes; and demonstrates - up to a laboratory prototype - how these attributes will be bundled together as an eventual product (operating characteristics, product architecture, etc.)". It is important to distinguish generic technological research from TI capabilities, which while also generic (multiple users/uses) are frequently based on problem solving experience rather than on generic research. For example part of the TI of the United States after the Civil War was located at universities which devoted themselves to regionally localized problem solving in industry and in agriculture. The focus of US university activity after World War II shifted to generic research in applied science and engineering, only a part of which should be regarded as TI. (Some universities continued to provide some TI-related services through their engineering and agriculture schools.)

multidisciplinary effort that combines scientific and engineering skills. Thus development of TI in biotechnology requires the combined effort of microbiologists, biochemists, and bioengineers; and TI initiatives in semiconductors typically place great importance on the development of production capabilities in conjunction with design capabilities.

A third salient feature of TI is its *indirect* economic value or "precompetitiveness" and the potential absence of a market for its output. While TI attempts to bridge the gap between scientific research and commercial application, further investments in specific innovations may be necessary before a return on investments in TI can be realized.[6] The lack of direct economic value makes it less likely that individual firms have adequate incentive to undertake such activities singly (we shall have more to say about this in the following sections) while increasing the chances for successful cooperation among competing firms that need to protect their trade secrets from each other. Nonetheless, application-specific activities often figure prominently in TI programs because they help focus capability development on actual user needs and offer private firms added incentives for participation. Precompetitiveness raises a problem of implementation in forming effective consortia; in practice, precompetitive research of indirect value may be difficult to distinguish from irrelevant research of no value at all.[7]

Economies of scope are a fourth feature of technological infrastructure which distinguish it from conventional infrastructure such as roads, power, or

[6]This leads us to the place of university research in the applied sciences and engineering in generating TI. It is worth illustrating this in connection with the engineering discipline 'Chemical engineering'. Chemical engineering emerged in the early decades of this century in American universities (e.g., MIT) because of the enormous distance between the discovery of a new chemical entity and its production on a commercial scale. Chemical process development is not simple scaling up of laboratory processes and requires an idiosyncratic methodology which is based on the concept of 'unit operations' (Rosenberg and Nelson 1992). Part of this activity which is carried out in universities should in fact be considered as contributing to TI. However, the generic activity involved in translating a laboratory chemical process into a full-scale commercial one need not all be performed at universities since a lot of engineering data and other very practical effort is required, and since the engineering disciplines themselves are becoming more and more basic. Therefore, there certainly is room for pilot plant development or other effort associated with specific processes to be executed at a Technology Center; an example is the chemical pilot plant established at the LANFI Technology Centers in Mexico during the seventies. See also Section 1.5 below.

[7]In Japan's VLSI in 1976-79, over 85% of the 200 million dollars support was devoted to the design and production of specific chips by individual consortia members (Sigurdson 1986). In ESPRIT 1, "application-specific" projects (in contrast to "precompetitive R&D" and "standardization") accounted for 23.1% of the total, increasing to 34.6% in ESPRIT 2 (Mytelka 1991).

water. Whereas the latter exhibit economies of scale in the production and supply of a standard commodity, the critical mass associated with technological infrastructure derives from the need to provide a spectrum of linked but specialized and distinct capabilities on which firms in the industry draw in a variety of patterns.

The production and cost functions of potential directly productive activities (DPAs) in the industry depend on the availability of these capabilities. Table 1a describes a very simple additively separable cost structure, in which selected capabilities (C1-C5), when available, can lower the unit cost of output for each DPA (α, β, γ); we call it a *dependency matrix*.[8] It also includes information on the market conditions for the output of each DPA, and the fixed cost of establishing each of the capabilities. DPA α requires the services of C1, C2, and C4. Without any of them, the average cost of a unit of DPA α output ($300) exceeds the price it commands on the market ($200). To be viable it must have C1 and either C2 or C4, but might need all three, depending on their prices (e.g., DPA α breaks even if it has free access to C1 services, or if it has access to all three for a combined fee of no more than $90 per unit of output). Similarly, DPA β uses C3, C4, and C5; and γ uses C2 and C3. At the same time, each capability is commercially viable if it can generate the revenues necessary to cover its fixed costs. Capability C1, involving a fixed cost of $3,000 and used only in producting α output, is viable if its services can be sold for at least $30 per unit of α (as overall α demand is 100 units). In this way Table 1a defines the interdependencies between potential DPAs and capabilities. It is fully articulated, and illustrates the possibility of a low-level equilibrium trap. The industry is profitable as a whole: with all capabilities available, the combined revenues of the three DPAs ($64,000) more than cover variable production costs ($33,000) and the fixed cost of establishing the five capabilities ($29,000); and if none of the DPAs or capabilities exists initially then it is not immediately profitable to establish any one of them singly. However, the coordination problems this raises are no different, in principle, than those normally associated with interdependent investment decisions.

[8]Exogenous growth models in which the diversity of inputs lowers the cost of production generally follow the Spence-Dixit-Stiglitz formulation of monopolistic competition with identical firms, symmetric inputs, and constant elasticity of substitution between inputs (e.g., Romer 1990). Table 1 highlights differences between firms and the asymmetry of inputs which, in conditions of incomplete information, are a key source of difficulty in building supply and demand for TI services, and creating missing capabilities. However this simple formulation abstracts from important issues of superadditivity and the distribution of monopoly quasi-rents.

Table 1: A Dependency Matrix

a. fully articulated

DPA	unit price	basic average cost	quantity of est. demand	capability's impact on average cost				
				C1	C2	C3	C4	C5
α	$200	$300	100	-$100	-$50	0	-$40	0
β	$200	$400	160	0	0	-$120	-$80	-$100
γ	$100	$150	120	0	-$60	- $40	0	0
Fixed cost of capability:				$3,000	$8,000	$6,000	$5,000	$7,000

b. with incomplete information

DPA	unit price	basic average cost	quantity of est. demand	capability's impact on average cost				
				C1	C2	C3	C4	C5
α	(A)	$300	?	X	(A)	0	(A)	0
β	(B)	(B)	$160	0	0	-$120	?	-$100
γ	$100	?	?	?	-$60	(C)	0	(C)
Fixed cost of capability:				(A,B)	$8,000	$6,000	(B)	?

The special character of TI derives from lack of information regarding the dependency matrix: some entries may not be known at all, some may only be known to a vague approximation, and some may only be privately known to individual firms. This is described in Table 1b, where we assume the existence of three firms – A, B, and C – with a potential to engage, respectively, in the DPAs α, β, and γ. (The information structure of the dependency matrix cannot be discussed without reference to the pattern of ownership in the industry). Question marks denote unknown entries, Xs denote entries known only to be non-zero, and parentheses denote private information (of the firms denoted in the parentheses). We assume that there are no active consultants offering the technological services associated with any of the capabilities, and that there is no private knowledge outside the firms – all external knowledge is common. Some entries are unknown because firms have yet to define their needs in terms of the unavailable capabilities, possibly because these capabilities are not yet in a usable form. Under these circumstances a collective effort may be needed not merely to *coordinate* but also to *articulate* supply and demand for the relevant technological capabilities. This has a static dimension – finding out which elements of infrastructure should be developed, who should cooperate with whom, and in what sequence; and a dynamic aspect – shaping capabilities to better meet firms' needs and redefining firms' needs in terms of newly available capabilities.

The differentiation of TI and the vagueness of its dependency matrix set it apart from conventional infrastructure. Conventional infrastructure has far fewer dimensions, and once its location is determined, may even be unidimensional (e.g., the generating capacity of a power station or the width of a road). Moreover, its dependency matrix is not only much simpler in structure but also much better known (e.g., the electric power requirements of different industries can be closely approximated from published sources). Hence the decision to establish such infrastructure can be taken unilaterally by a single agent – whether an entrepreneurial firm or a government agency – acting on this knowledge. It does not need to consult its customer base because it can anticipate its needs with sufficient accuracy, and it has sufficient concrete knowledge of the feasibility and cost of its meeting those needs. This is a key difference from TI where the determination of needs and capabilities requires the concerted efforts of many agents, each with its own perspective, its private information, and its powers of discrimination. Hence the need to define the role of government in building TI in different terms from its role in establishing conventional infrastructure: it is primarily a catalyst of cooperation in the private sector.

We distinguish in what follows between two extreme types of TI: "basic" and "advanced".[9] *Basic* (or sectoral) TI typically serves small and medium

[9]Our two types of technological infrastructure parallel Tassey's (1991) multi-stage analysis of the changing role of TI along the development cycle of a technology: providing the technical basis for fundamental research in the emerging stage of new technologies (corresponding to our advanced TI); and supporting product improvement, quality control and other activities associated with production and marketing in the growth and mature stages of the technology (our basic TI). We also agree with Tassey on the importance of including the promotion of both types of infrastructure within the purview of government policy, although the rationale for each type will differ. Finally, we should mention that Tassey deals with a third component of TIs which he calls "infratechnologies" and which in fact underlies both types of TI covered in our paper. Infratechnologies include novel methods of measurement, agreed research procedures, properly analyzed scientific and engineering data (e.g., on materials), etc. Much research on new technologies cannot be undertaken efficiently without these infratechnologies. Infratechnologies also provide the technical basis for implementing, testing, and other quality assurance procedures (part of our basic TIs) and are also related to the issue of standards.

enterprises (SMEs) in a low- to mid-tech activity (e.g., plastics products), providing them with technological services (often through a sectoral technology center), such as

- design,
- information on new production technologies,
- testing and analysis, and
- solutions to environmental or ecological problems.

Basic TI comprises routine or conventional capabilities that are generally available in other countries and supports the effort (mostly engineering) needed for their domestic or local absorption. For basic TI the general structure of the dependency matrix is known or can be deduced in large measure from the prior experience of other countries, even if precise numerical values can only be obtained through actual experience; hence separate control of DPAs and capabilities is feasible. Sufficient information is available for DPA firms to redefine their needs in terms of new, unavailable capabilities, and for expert consultants (in the TCs or working independently) to develop technological services that are tailored to the yet unarticulated needs of local DPA firms. We refer to this simultaneous creation of supply and demand for new technological services as market building, and we expand on it below.

Advanced (or functional) TI serves high-tech, leading-edge industries, providing necessary R&D inputs to the specific innovations or development projects of user firms. The necessary capabilities are not initially available anywhere and must be developed. This does not allow a separation of user-need determination from capability creation. Hence the need for a concerted, user-led effort by DPA firms (e.g., working through a consortia), to apply significant R&D effort in developing capabilities that can meet their vague, currently undefined needs. It is frequently more specific than basic TI, serving a narrower constituency defined in terms of a function (e.g., superconductivity) rather than an existing industrial sector. For advanced TI, little of the structure of the dependency matrix is known at all and most of what knowledge exists initially is in the private domain of individual firms.

Though basic TI is generally identified with low- to mid-tech sectors, it can also serve firms in industries that are usually classified as high-tech, such as electronics, providing them with access to improved – but *generally available* – capabilities. By the same token advanced TI can serve a low-tech industry. The key distinction between basic and advanced TI lies in whether the capabilities exist and are obtainable from external sources,[10] or must be created. This leads to a key distinction regarding the potential role of government: in establishing basic TI government can play an active intermediary role that complements the efforts of the private sector to import technology (cf. Braunling 1993); with regard to establishing advanced TI government is at most a facilitator of essentially private efforts to create new technology.

The distinction between basic TI and advanced TI also relates to the "degree of user need or market determinateness" (Teubal et al. 1976; Teubal 1979), i.e., the degree to which user needs are specified, or specifiable, in terms of the services offered by the new infrastructure. While users of conventional infrastructure generally have well-defined demand curves (strong need determinateness), potential users of basic TI may not be aware of the existence of new capabilities, and users of advanced TI may be incapable of defining their needs without actively participating in capability creation (low or weak need determinateness). These differences have important implications for the degree of user involvement required, the existence of markets for TI outputs, and the respective roles of private entrepreneurs and government in building TI.[11]

Key differences between conventional infrastructure, basic TI, and advanced TI are summarized in Table 2. The distinction between both types of TI informs much of the subsequent discussion of the nature of TI and implications for TIP.

[10]Obtainability is crucial. When Japan initiated its VLSI program, the capabilities it sought to develop existed in large part in IBM, but were not accessible to other firms.

[11]A fully articulated dependency matrix (Table 1a above) necessarily implies strong need determinateness and even clear demands for TIs. User-need determination, therefore, is a useful concept when such a matrix is not known, both in connection with the relevant capability categories and in relation to the particular configuration within each capability category. It has been defined as the process by which users gradually translate general needs first into product classes, and then into product functions and features. The higher the degree of user-need determination concerning TI, the greater the likelihood of a good fit or coupling between the targeted capabilities and the needs of their users.

Table 2: Differences between Types of Infrastructure

	Conventional	**Basic TI**	**Advanced TI**
Nature of output	Production inputs	Technological services	R&D inputs
Activity supported	Production	Diffusion	Innovation
Focus	Geographic	Sectoral	Functional
User-base structure	Indefinite	Many SMEs	Select few
Differentiation of output	Little	Some	Very high
Definition of need	Complete	Within reach	Inarticulate
User involvement in need determination	Unnecessary	Moderate	Intensive
Market for outputs	Exists	Does not exist, but feasible	May not be feasible
Possibility of independent entrepreneur	Yes	initially unlikely	unlikely
Typical entrepreneurial organization	Government	Industry association	Consortium of users
Government's role	Investor, regulator	Knowledgeable catalyst	Catalyst, broker
Policy focus	Capacity, pricing	Market building	Capability creation

2. Basic Technological Infrastructure

At initial stages in the development of a traditional industry there may be neither supply nor demand for essential skills; a cooperative effort may be necessary to articulate the needs of local industry and to elicit a mutual commitment to a path of progressive growth that no individual firm can follow singly. The role of basic or sectoral TI is then to mediate between the technological needs of the industry and potential sources of supply. In the first instance, it promotes static efficiency by providing information and advising local industry regarding the availability of foreign technology. But it also has a role to play in promoting dynamic efficiency through a twofold action: stimulating local demand for foreign technology by helping local industry redefine its needs in terms of the possibilities that the new technology offers, i.e., *user-need determination*;[12] and increasing the effective supply of technology inputs by stimulating investment in adapting them to local needs and promoting local sources of supply. The static dimension focuses on stimulating demand, for instance, through "awareness" campaigns that have played a central role in diffusion programs in the past. The dynamic dimension at the center of TI focuses on need determination and on building new, commercial sources of supply for new technologies.

Typical examples of sectoral TI, referred to briefly above, demonstrate that it implies a need for collective behavior of some kind rather than simply accessing world technology on a firm-by-firm basis. Indeed, in many cases it may be a precondition for accessing world technology, especially for SMEs. Moreover, the above examples make it clear that basic TI may involve little, if any, R&D, and often deals with the "bread and butter" problems facing firms. *Quality control, testing and analysis capabilities* such as those provided by Engineering Research Associations (ERAs) in Japan to the auto parts industry from the mid 1950s have involved the application of generic techniques such as measurement and statistical analysis that are inherently standardized (Sigurdson 1986). The current focus of such activities might be the acquisition of TQM and ISO 9000 capabilities to support expansion to a more demanding customer base. *Product design capabilities* for industries such as textiles, plastics, metalworking, furniture, and footwear involve the assimilation of common design elements from abroad and their application to the specific market needs of individual firms. Design centers serving regional groupings

[12]User-need determination refers here to the needs of industry firms for services that derive from the new capabilities.

of SMEs in traditional sectors have played an important role in Italy's recent industrial development (Sabel 1988; Malerba 1993), and in the successful development of design capabilities for Israel's clothing industry. It is hoped that a reorganized plastics technology center, removed from academic control and under the new auspices of an industry association, will do the same for Israel's plastics industry (Yinnon et al. 1993). *Capabilities for identifying, selecting and absorbing novel production technologies, equipment, and raw materials* are becoming increasingly important given the exponential growth in the options available. SMEs – which are particularly good at providing specialized products for niche markets – find it difficult to scan the technology horizon effectively and select what they need. There are external demonstration effects as well as economies of scale in joint learning. The Manufacturing Technology Centers (MTCs) in the United States are aimed at helping firms acquire such capabilities (Shapira 1992). *Capabilities for solving contamination and ecological problems* facing firms in a particular sector or region are constantly gaining importance. External effects are inherent in such problems and provide ample motivation for concerted action by a consortium of firms or by the local industry association. At the very least, there are advantages from a common effort in tapping foreign sources of knowledge, and a second stage of assessing this information and arriving at a collective solution may also be necessary.

In each of these cases, there may be some firms that have the capability both to define their needs in terms of the foreign technology and to access it directly without the intermediation of a specialized, multifirm technological infrastructure. However, many small- and medium-sized enterprises in mid- and low-tech sectors lack such capabilities. Enabling these firms to access foreign technology (including developing their awareness of its relevance to their activities) may necessitate both establishing local capabilities and building a market for the services flowing from them.

3. Market Building in Basic Technological Infrastructures

Market building is a dynamic approach to the transfer of technology. It proceeds in stages: first, the local market for imported technological inputs must be developed; next a derived market for local linking or intermediation services emerges; finally, these stimulate the creation of a market for local substitutes for foreign technology, when the domestic economy is able to develop a competitive advantage in an increasingly mature foreign technology. Though this process can proceed autonomously, there are circumstances when public intervention may be beneficial or even necessary.[13]

Consider business software as an example of such infrastructure services. Once there exists a core of basic software capabilities relevant to the needs of a target group of users, the process of building markets might begin with an import agent acquiring distribution rights to a specific software package and advertising its availability to potential users. This might lead to additional investment, either by the import agent or by other parties in translating operation manuals, translating the software, modifying the software to accommodate special local needs, etc. At the same time, potential users might be modifying their operations to take advantage of the new software, e.g., computerizing their accounts-payable files or building an automated warehouse. This might lead, in turn, to other agents importing other software packages that enhance the utility of the first package, further increasing the user base. Eventually, the local user base might be large enough to support locally produced business software for domestic use, and in some cases this might even lead to the exportation of domestically-produced software.

Generally speaking it is useful to separate the demand-building aspects from the supply-building aspects of market building. To this we briefly turn our attention.

Building demand: This aspect of market building is based on a distinction between general needs and a clear demand for well-defined goods or product characteristics. Some firms may "need" new technological services but still will not demand them, either because they are not aware of their existence or incorrectly believe that conventional solutions are available. (In the NIC liberalization context, many firms still believe that their competitiveness may be maintained by continuing to "demand" hardware solutions, failing to

[13]NIST's formal introduction of institutional mechanisms for technology transfer, and the added emphasis on this activity in the U.S. government laboratory system (an effect of the 1988 Omnibus Trade and Competitiveness Act; see Tassey 1991) relates to *market building*. Market building is a much broader concept than technology transfer, however, since it explicitly considers several supply and several demand factors over and above supporting regional manufacturing centers, providing technology extension services, and serving as a clearing house for shared experience. We believe that it also provides a more useful basis for understanding desired policies. The market building approach includes, for example, the explicit promotion of alternative sources of supply of new technology in the private sector and of private consultants providing technoeconomic advisory services to prospective purchasers of new technology. This in turn has implications for the management and organization of Sectoral Technological Centers.

recognize that changing circumstances call for new technology and capabilities; these capabilities are "needed", but not in demand).

Building demand involves two different aspects: generating *awareness* and *user-need determination.* Awareness programs were a common component in the diffusion policies, fashionable in Europe during the 1980s in areas such as information technology. Awareness programs may take the form of mobile exhibits, which demonstrate new technological services to reluctant users, transforming them into convinced demanders. In user-need determination a critical process is learning by interaction within user-producer networks (Lundvall 1985, 1991; Teubal, Yinnon, and Zuscovitch 1991). This is a collective learning process involving numerous externalities which may best fit within an organized network. In the context of sectoral TIs providing novel services to firms, the network PBX (the agency in charge of building and developing the network) could be the sectoral technology center (TC) which, after absorbing foreign technology, builds a market for the services flowing from it.[14]

Sectoral TCs should explicitly consider the social benefits derived from market building rather than focussing exclusively on private benefits. This implies not only promoting demand by building a network of users such that each user benefits both from its experience with the new service and from the experience of other users, but also building supply by creating a set of agents who may compete with it in providing the service and who may even make its provision by the TC totally redundant.

Building Supply: This involves at least three processes: learning-by-doing, training consultants, and spinning off TC personnel. Over time, suppliers of new technological services build up their knowledge of the purely technical aspects of the technology while at the same time they develop insights regarding adaptation of the new technology to the need of potential users. When the initial supplier is a collective organization like a technology center then it is

[14]The concept "network PBX" was introduced in Teubal et al. (1991) to describe the critical node in user-producer networks which is the locus of collective learning about adaptation of a radical new capital good to the needs of users. In that context, the "network PBX" is usually associated with the innovator, whereas in this context, it could be the sectoral technological center (which is responsible for collectively absorbing and diffusing the new technology). In both contexts, the "network PBX" is in charge of both network creation and network development.

likely also to engage in training consultants and in spinning off TC personnel.[15] In fact, a technology center policy whose objective is maximizing social (i.e., economy-wide) benefits must explicitly adopt these policies.

4. Market Failure in Basic Technological Infrastructures

Frequently, market building can proceed without public intervention. Nonetheless, there are a number of reasons why a purely market-driven process might not succeed when it was needed, indicating a potential role for public policy:

Uncertainty and externalities among early users learning about the application of the new technology. New users may hesitate to adopt the new technology because they are uncertain about its potential for meeting local needs, and information that could reduce this uncertainty may be costly to obtain and difficult to appropriate. However, if a group of potential users can arrange a coordinated effort to study common aspects of the new technology based on a free sharing of information, learning costs can be reduced, risks shared, and external effects internalized. This need not occur spontaneously,[16] and even if it does it may not eliminate all externalities, especially when there are many users.

Codification and standardization. By this we mean transforming individual experience about the adaptability of the new technology to local conditions into a codified body of knowledge that allows distinct user types and product types to be identified and effectively linked. Teubal and Zuscovitch (1991) refer to this as "general discriminating capabilities" and argue that it requires an explicit allocation of resources, over and above tacit experience. This activity may have enormous social value in promoting rapid diffusion of the

[15]The Stuttgart Institute of Microelectronics (SIM) offers a good example of spinning off a mature capability once the market for its services has developed. Specialized services that were initially offered directly led to the establishment of an external consulting firm that now provides these services on a commercial basis (private communication, and Hofflinger 1989). An even more dramatic example of spinning off technological *capabil*ities absorbed from abroad occurred in Taiwan in the early 1980's, in connection with microelectronics technology. C-MOS capabilities absorbed during the late 1970s at the Industrial Technology Research Institute Electronic Research & Service Organization (ITRI-ERSO) were later transferred, embodied in the form of human capital, to Taiwan's first custom chip firm, UCM (private communication, and Hou and San 1993).

[16]Teubal, Yinnon, and Zuscovitch (1991) refer to this as market failure in network creation; see also below.

new technology, but individual entrepreneurs may be reluctant to invest the necessary resources because of the difficulty in appropriating the benefits from such an effort. This is especially true for small-scale entrepreneurs who may fear that any early effort on their part to expand the market through standardization will be co-opted by larger established firms with stronger complementary assets. A cooperative effort may be needed, and government support in imposing a particular standard may be necessary.[17]

Network externalities. These arise when late adopters of a new technology derive inappropriable benefits from the prior existence of a large user base. Telephone networks are the obvious example, but such effects can arise in a variety of contexts; for example, late adopters of widely used computer hardware may benefit from a pre-existing supply of specialized software, maintenance and repair technicians, and expert consulting services. Such external effects will benefit the entrepreneur supplier of the new technology – e.g., a local distributor of computer hardware – who may indeed want to subsidize early adopters in some way. But without some measure of coordination this may be a very risky investment, especially if there are a number of competing entrepreneurs. If such private efforts fall short of inducing the socially desirable rate of diffusion, some measure of neutral public support for early adopters of the new technology may be warranted.

Expert consulting services. A new technology in the early stages of its diffusion may require the support of a pool of local consultants able to advise users about which products to select and how to make the best use of their selection. There may be market failure in the training of such a pool. Early experts in the technology may find that it is in their best interests to limit the spread of knowledge by offering consulting services themselves and extracting a high price for the transmission of information, rather than work for a wider dissemination of knowledge in line with social welfare. A dynamically efficient market must, of course, pay a premium for the early acquisition of valuable knowledge. Nonetheless, in some cases adequate compensation may be compatible with a wider dissemination of knowledge than is achieved in practice. A technology center *not run on strict profit lines* can generate a pool of independent consultants who will cut into its own revenues.

[17]This problem may not arise when there is a recognized leader in the market willing to set (open) standards for its suppliers, as IBM did in personal computing, and Bombardier in snowmobiles. And when such a leader exists its active participation is essential, as was demonstrated by an unsuccessful effort, in the 1970s, to set standards for data base design that bypassed IBM's dominant IMS architecture.

If private entrepreneurship does not provide the necessary capabilities when they are needed there is a range of policy responses that can be implemented to fill the gap. The least intrusive – and least costly – involve the collection and dissemination of information. Industry studies can be commissioned to map the dependency matrix: the technological needs of the industry, the available capabilities, and the structure of the linkages between them. The approach outlined in the present paper has served as a general conceptual framework for a series of industry studies in Israel,[18] but the methodology for such studies is still evolving. The information that they present may be sufficient to trigger an effective response in the private sector.

A more direct course of action is for government, in cooperation with the private sector, to take an active role in establishing technology centers that can act as catalysts to building a market for needed technology services.

5. Sectoral Technology Centers (TCs)

The enhanced expected importance of sectoral technology centers – especially in view of the enhanced role assigned to SMEs by flexible manufacturing technologies and as a result of stiffer competition from imports (due to trade liberalization) – together with dissatisfaction with their past performance, have motivated several countries to restructure the operations and management of existing centers. Mexico and Israel are two concrete examples. In the 1960s and 1970s, some of the institutes in both countries had a clear supply-push orientation, in line with the linear model of innovation which prevailed at that time. The first task, therefore, was to enhance their demand-pull orientation, which meant first and foremost, ensuring efficient supply of technological services, information, and consulting services to the private sector.

Some Mexican technological centers initiated such a reorientation during the mid-1980s. An outstanding example is CIQA – Centro de Investigaciones en Quimica Aplicada. The research agenda of CIQA during 1984/85 was characterized by curiosity-oriented research, publishability as the criterion for project evaluation and approval, and few projects involving industrial applications. Moreover, no pattern of interaction with industry existed, and lack of trust characterized the attitude of firms towards anything connected

[18]They are collected in Justman, Teubal, and Zuscovitch (1993): Yinnon et al. (1993) on plastic products relates to basic TI; Shappir (1993) on microelectronics and Toren et al. (1993) on software relate to advanced TI; and Wachs (1993) on chemicals relates to both.

with the government. The first task in the reorganization was to impart a clear industry focus to the activities of the institute, a process implying first, a shift from elastomers to plastics (including processing aspects such as injection and extrusion), and second, a new focus on polymer additives. The restructuring involved hiring a new director for the center who focused on efficiently providing services and advice to the private sector. This had the gradual effect of generating what we could call a user-producer network. A similar process has been taking place in the Plastics and Rubber Technology Center in Israel since its restructuring in 1991, involving a change in ownership from a university to an industry association. The new industry-oriented outlook enabled the Center to tap the enormous latent demand of plastic firms for services and commissioned R&D in the area of compounding, additives, etc. The effect was an increase in revenues by several hundred percent during the first year following the change.[19]

The enhanced demand-pull orientation described above also involved greater (private) profit orientation; in fact one of the objectives was to reduce the deficits of existing sectoral technological centers. However, the emphasis on higher private profits may be taken too far since it might lead to ignoring investments in new technological capabilities – the basis for future services.[20] It is clear that in taking account of capability creation, an explicit distinction should be made between social profitability and private (institute) profitability. An important stumbling block to adopting an appropriate mix of "service provision" and "capability development" is the difficulty of measuring social profitability, even *ex post*. These issues will remain central ones in relation to sectorial TIs in the years to come.

We have also seen how TCs can act as catalysts for market building in connection with new technology services. Successful examples of such centers are typically controlled by industry firms that have an interest in the center's activities, but government has a role to play by virtue of its central role in education, land use, physical infrastructure, standardization, etc. TCs build new sources of supply for new technologies by initially "importing" missing capabilities and providing services to interested firms at subsidized rates,

[19]The change in CIQA was not easy – some changes in personnel were required though it was stated that the main problem was the change in mentality and overall outlook (private communication, and Teubal 1991). Yinnon et al. (1993) describe the problematic situation of the Plastics Institute before the recent restructuring.

[20]Of course, this will also depend on accounting procedures, but these are generally biased against the intangible components of such investments, which normally appear as current expenses rather than as investments.

thus enabling the TC to learn the technology's potential value through commercial interaction with actual users; by sponsoring awareness and demonstration programs; by training consultants; and by eventually spinning off commercial consulting services that are absorbed by the private sector. Even though it operates on a commercial basis, the TC is not profitable at an early stage – or possibly at any stage; its lack of private profitability is the rationale for a collective effort. An initial government contribution towards covering the center's deficit may be an essential catalytic factor in starting a self-fueling process of capability generation.[21]

It is, however, one test of the TCs success that the private sector, recognizing its value, is eventually prepared to assume this burden. Public financial support for such a center can be justified only as long as the services it offers *complement* and do not compete with those which the private sector can offer. When the center becomes a locus of expertise in an *established* technology it becomes possible to spin off its know-how in the form of private consultancies that provide the same services on a purely commercial basis. Phasing out public financial support for the TC is essential for redirecting its resources to new challenges. Successful TCs conceived along these lines have been established in numerous locations, often operating under the aegis of local, state, or regional authorities, and playing an important part in regional development initiatives. In some cases, these are new institutions established through recent initiatives (e.g., the MTCs in the United States), and in others they are older institutions converted to a new purpose.

6. Advanced TI: Capability Creation and the Need for Cooperation

Where basic sectoral TI *mediates* between the technology needs of domestic users and potential (often foreign) sources of supply, advanced functional TI also involves the *development* of advanced capabilities, often harnessing the results of scientific research, both domestic and foreign, to industrial use. This goes beyond the dissemination of scientific knowledge, requiring also the development of enabling engineering capabilities. Creating these capabilities is a necessary first stage in the process of establishing advanced TI.

[21]The more government acts on a thorough understanding of technological needs, and the more credible is its undertaking to make good any undersubscription of the technology center's deficit, the less likely it is to be called upon to do so.

The best-known example of functional TI is the development in Japan, in the late 70s, of design and production capabilities for one-megabit DRAMs through the MITI-orchestrated VLSI project. They enabled the participating electronics firms – NEC, Toshiba, Fujitsu, Mitsubishi and Hitachi – to launch a series of innovative semiconductor devices in the decade that followed. The specific capabilities were associated with crystal technology (how to avoid the bending of silicon crystals); fine processing technology (the electron beam delineator); testing and evaluation technologies; and design capabilities (Sigurdson 1986, 46).[22] Since then, and due in no small measure to the success of the Japanese model, similarly conceived programs have been implemented by other countries or groups of countries; these include the Alvey program in the UK, Esprit and Jessi in the EEC, and the MCC and Sematech programs in the US (Arnold and Guy 1986, 1987; Guy and Arnold 1991).

The VLSI program involved establishing a temporary joint laboratory for the project. Each of its project teams included researchers from all the participating companies (although some teams were dominated by a single company). Members of MITI's Electrotechnical Laboratory also participated and one of them played a crucial role in the overall direction of the research program. A major problem was ensuring the *bona fide* participation and collaboration of the various firms. This was achieved through a number of measures: generous financing was assured for complementary firm-based R&D; the development of design capabilities – which involved aspects very close to the specific chip designs of each one of the fiercely competing participating firms – was moved from the joint lab to the individual labs of each firm; and an explicit effort was made to create an *esprit de corps*. Overall, the project was considered a success, although the competitive advantage of Japanese firms in DRAMs during the mid-1980s is attributable to other factors as well, including a related program sponsored by the Ministry of Posts and Telecommunications (Sigurdson 1986). The VLSI program is an excellent example of an initial TIP program involving horizontal (rather than vertical) collaboration and taking the organizational form of a temporary consortium with a "Joint R&D Laboratory".

A major benefit of the VLSI program was its having paved the way for subsequent collaboration among the participating firms. Cooperation offers

[22]Functional TI for biotechnology might include specific capabilities in fermentation technology, genetic engineering, biosensors, protein engineering, and downstream processing (Senker 1991).

the obvious advantage of distributing development costs over a broad base of users as well as allowing a division of labor in developing the new capabilities that takes advantage of the respective strengths of the participating firms. But its most important contribution may lie in facilitating *user-need determination* and *coordinating future supply and demand.* Fundamental uncertainty regarding the new technology, and the absence of a market for the services it makes possible, may *preclude the separation of user-need determination from the capability development process.*[23] In this case it is essential that potential industry firms work together to identify their needs as they develop these capabilities. Moreover, *a collective specification effort may be more efficient than a sequence of bilateral interactions*. This is especially important if the critical mass for a viable infrastructure is large, requiring the support of many firms with varied needs.

The development of cutting-edge technological infrastructure is subject to an enormous amount of *target uncertainty*, much more so than for regular product innovation. Initially, nobody knows the critical parameters of the new technology or how they are related to the desired performance dimensions of future products, and potential users cannot specify their needs in terms of the new technology. At this point needs will either be defined very generally without reference to a particular technology or product class, or in relation to existing technologies and products. User-led development of new capabilities implies a partial fusing of the process of need determination with that of establishing development targets, enabling users to exercise direct influence on the specification of capabilities that are developed. This greatly simplifies the use-experience feedback process: when users are also producers we have – almost by definition – automatic reception of feedback signals and improved translation of user needs regarding product types into directions for technological development.[24]

Additional advantages derive from user *coordination* in developing the technology, and are distinct from the advantages of user-involvement in the

[23]Problems of early ascertainment of desired product type may be one reason why in some areas – such as scientific instruments – users are frequently the innovators (von Hippel 1976). See also Lundvall (1985) on "user-producer interaction".

[24]The fact that a set of users collaborate in developing the technology also enables direct interaction among themselves. Neither direct nor indirect user interaction will normally occur in a neoclassical market. If the supplier of a novel capital good manages to organize a user-producer network around his innovation, then we might expect indirect interaction, that is, mediation between experience with the technology by user i and utilization of this information by user j (see Teubal et al. 1991). Direct user interaction would seem to be a specific advantage of user-consortia, although it may also arise within user-producer networks at a later stage.

development process. Direct interaction among users, rather than through the intermediation of a single developing firm, enables user coordination in setting development targets. Direct user coordination within the project allows development targets to be set on the basis of simultaneous confrontation of various user needs, and horizontal exchanges of information on desirable technical parameters and the tradeoffs between them. These simultaneous exchanges are potentially more rapid and efficient than sequential exchanges between a single developer and a group of potential users even if the developer is also a user. Because the main users operate within the project rather than being separate from it, more direct trials of alternative prototypes take place within the project, susequently diminishing the need for actual sales. Moreover, these trials and relative assessments are more likely to take place simultaneously rather than sequentially.

It follows that the advantages of user cooperation in capability creation relate to information gathering and processing, and not necessarily to transactions costs. Indeed, transactions costs associated with user cooperation for advanced TI development will probably be higher than those incurred in a neoclassical market, and in some cases may block the emergence of "user-based cooperative backward integration" for developing the new technology. However, improved information, and especially a *shared* information base should facilitate transactions among users by removing subjective differences of evaluation as a potential source of conflict. The informational advantages of this form of organization (which is strictly neither market nor hierarchy) also hold vis-a-vis Lundvall's "organized market'" (Lundvall 1985, 1991) which is similar in this respect to the user-producer networks surrounding a radical new product innovation (Teubal, Yinnon, and Zuscovitch 1991).[25]

[25]These considerations further confirm that to the effects on transactions cost of alternative systems of governance we should add the effects on innovation (see also Imai and Baba 1988; Willinger and Zuscovitch 1988) which contribute to the advantages of a network form of organization in terms of incentives, capabilities, and information flows for systemic innovations. Note that unlike atomistic vertical integration in connection with regular inputs, "collective" vertical integration in connection with radical innovations is likely to enhance transactions costs (see above). Our framework of analysis is also consistent with Teece (1992) for whom strategic alliances are "an attractive organizational form for an environment characterized by rapid innovation and by geographical and organizational dispersion in the sources of know-how..." since they efficiently enable the type of non-market coordination required for emerging generic and systemic technologies. While a strategic alliance could involve a precompetitive joint capability-development component, its emphasis would rather be on activities near the market (see Arnold, Dodgeson, and Guy 1991). Both strategic alliances and precompetitive consortia are intermediate forms of organization, lying between the neoclassical market and a fully integrated organization.

A number of organizational features of the VLSI program enhanced its ability to exploit the advantages of user-led capability-creation by promoting interaction between the participating firms, including:

- establishing a temporary joint lab with each research team including members from all participating firms;
- explicit policies for mixing researchers from different companies and encouraging the exchange of information;
- undertaking parallel R&D in the critical fine-processing-equipment area with each subteam dominated by a different user firm; and
- simultaneous absolute and relative evaluation of all designs.[26]

The complexity of the processes described above, compounding the usual complexity of R&D, required elaborate organization design and learning. Without these, and without appropriate supportive institutions, North's "hospitable environment for cooperative solutions to complex exchange..." would not have been achieved (1990). This indicates some of the scope for government involvement in TIP.

7. Functional TIP for Advanced Technologies

As with sectoral TI, the new capabilities of functional TI can sometimes be created through the autonomous workings of the market economy. Notwithstanding the advantages of cooperation between industry firms discussed in the preceding section, advanced capabilities could be achieved through the efforts of an individual firm developing the know-how it needs in-house, and then spinning it off as an independent business unit at a later stage. Rosenberg (1963) describes just such a process in his account of technological convergence and vertical disintegration in the U.S. machine tool industry in its early stages. Alternatively, an entrepreneur from outside the industry might undertake to develop the necessary capabilities and sell the industry the services it needs. However, there are common reasons that might undermine the feasibility of either solution, or substantially delay its realization.[27]

[26]See Sigurdson 1986, especially pp. 45-60. Note that functional TI linking university research with users of new technology may also take the organizational form of a permanent laboratory. An interesting example is the Swedish program in support of powder technology which recently supported the creation or expansion of three such laboratories (Granberg 1993).

[27]The autonomous changes in the machine tool industry described by Rosenberg took seventy years to evolve.

Individual firms in the industry may be reluctant to invest in capability creation because its results are difficult to appropriate. Nelson, Pech, and Kalachek (1967) point out that the results of generic research are often unpatentable; Ouchi (1989) has used the term "leaky technologies" in describing the development of generic capabilities. Moreover, because these activities are so far removed from commercialization, the uncertainty that surrounds them may be an added deterrent to investment in creating generic capabilities. A firm outside the industry faces additional barriers: it often lacks sufficient knowledge of industry conditions to enable it to identify commercially viable elements of infrastructure, and it may be daunted by its poor bargaining position vis-a-vis industry firms with superior "complementary assets" in technological know-how, product lines, marketing skills, distribution networks, etc. (Teece 1988).

Alternately, two or more firms could form a commercial joint venture for the same purpose; there are numerous such examples.[28] Cooperative efforts at capability development on a commercial basis need not deal only with the inappropriability and uncertainty of innovation itself, but must also surmount the difficulties of cooperation. Adversarial relations between potential industry participants may deter any one of them from placing itself in the difficult position of relying on a competitor for infrastructure services. And while it may be clear that cooperation among firms is likely to be beneficial, the early vagueness of needs and possibilities may undermine their ability to form business partnerships for infrastructure development on a commercial basis. These difficulties may be overwhelming, especially in socioeconomic contexts where cooperation is generally infrequent.[29]

Since widespread cooperation is a relatively recent phenomenon, government has a role to play in defining its form and institutional context in addition to its role in ensuring that such cooperation does not subvert desirable competition among the same firms in their product markets. Moreover, early efforts at cooperation in capability development generate knowledge and experience about the conditions for successful cooperation which are a special type of externality of high potential value to subsequent cooperation efforts. Valuable lessons may be learned even from failed efforts – indeed, they are

[28]Zuscovitch and Shahar (1993) provide a comprehensive discussion.

[29]In the 1980s, macroeconomic instability created such a climate in Israel, reducing the fund of goodwill in the economy (Teubal 1993). Cf. also Gerschenkron (1962) on the extreme absence of trust in the Russian economy at the turn of the century and the need it raised for government to play a more active role in industrial development.

often the most instructive – including the type of partners and partnerships that offer the best chances of success; the writing of contracts regarding the disposition of intellectual property and the transfer of knowledge among partners; and needed changes in the legal and institutional framework.

The potential importance of learning from early consortia is demonstrated by a comparison of the Esprit and Alvey programs (Quintas and Guy 1991). While a standard and very simple contract exists for Esprit programs, with very few disputes surrounding it, no such norm was established for the Alvey programs and firms were left to negotiate the terms of agreement in each case. But the negotiating process "negatively affected 48% of projects, according to industrial participants," reducing firms' goodwill and readiness to collaborate with each other. Moreover, in Alvey, "the most commonly cited factor negatively affecting progress, mentioned by 56% of participants, was 'changes affecting collaborations'," indicating the importance of careful prior screening of potential consortia participants.

Learning from early collaborative efforts should manifest itself in a significant reduction in transactions costs in subsequent efforts. Preliminary evidence on Israel's experience with its one-year old Magnet Program for advanced TI suggests that the transactions costs associated with such early efforts are indeed substantial. They include the entrepreneurial activities of consortia initiators, time invested by other participants in setting the research agenda and the details of the contract, direct legal costs, and costs associated with a possible need for enabling legislation that modifies the existing institutional framework. The high level of transactions costs sets a threshold for the size of eligible collaborative projects that discriminates against small projects, and possibly small firms (except where small firms play a pivotal role in a consortia dominated by larger firms). Experience with these early projects and consequent changes in the overall institutional framework will hopefully reduce the threshold size of collaborative projects, and also make possible consortia comprised exclusively of small firms.

Government can play a fruitful role in this process by encouraging and supporting private initiatives in this regard while such efforts are experimental. It can participate in the systematic collection of information, in developing multidisciplinary skills that can help foster cooperation in building technological infrastructure, and in modifying the institutional framework where necessary. In return it can require firms to share with others what they have learned about the process of building a cooperative effort of this type. It can serve as a clearinghouse *cum* arbiter *cum* guarantor of collaborative agreements undertaken within the industry to take concerted action.[30] The need for government initiative and significant government involvement in triggering cooperation on an agreed project configuration is more likely to arise when significant levels of cutting-edge technological capabilities are required.

Finally, user cooperation and coordination will generally be facilitated by expectations that other users are willing to invest in the capabilities they need, or by the knowledge that the government has targeted the industry, the infrastructure, or the technology. Whenever TIP necessitates such targeting, a clearly stated government "vision" or at least a credible announcement may be critical. For example, the government may declare its determination to create a significant level of infrastructure no matter what level of industry financing and participation is forthcoming; indeed, in some extreme cases user coordination may be unthinkable without some confidence that a critical mass of infrastructure will materialize. This will significantly reduce the uncertainty facing individual users intent on investing in the new capabilities by reducing their dependence on the actions of other users. A clearly stated vision may have a significant effect in inducing individual user participation even when the level of explicit government commitment is low.

8. The Technological Infrastructure Policy Framework: A Tentative Summary

Growth-oriented TIP is an area of technological policy which has received increased emphasis during the last decade. Rapid technological change stemming from the revolution in information technology, and political changes that have spurred a dramatic trend towards economic liberalization have signalled the need for new market-oriented approaches to technology policy that can address widespread and recurring crises of industrial renewal. With greater frequency, economies arrive at nodes of structural change where it is evident that a critical mass of new capabilities is needed to pave the way for further growth, and these nodes require a *strategic* policy effort that is inherently different from *tactical* corrections of market failure in periods of routine

[30]This type of intervention which characterized MITI's involvement in Japanese industry, also played a role in the renewal of Pittsburgh's central business district in the last twenty years.

growth.[31] This need has elicited growth-oriented policy responses that place new emphasis on helping the business sector develop the technological infrastructure it needs. Much experience has been gained from these efforts, and technological infrastructure policy is gradually establishing itself as a separate category with distinctive characteristics which set it aside from established policy areas such as science policy, price-based incentives for R&D, or mission-oriented policies that promote defense and other noneconomic goals (Ergas 1986).[32]

However, despite the widening application of TIP in practice, we still lack a well-formed, systematic, analytical framework for formulating, implementing, and evaluating such policies. The type of project-oriented "anatomy of market failure" which served as a framework for the previous generation of technology policies – for instance, broad-based subsidies and tax advantages for R&D – cannot by itself provide an analytical basis for this new generation of policies; they are defined by a different set of issues. We have outlined in this paper the beginnings of such a framework, which we summarize here briefly.

The starting point is to recognize that the requirements for TIP are shaped by the nature of TI: its indivisibility as "infrastructure" and its differentiation because it is "technological". Indivisibility imparts a strategic dimension to TIP dictating explicit attention to structural change, which in turn requires a capacity to make a discrete choice among alternative growth paths. Pinpointing market failures, while necessary, is very likely to be insufficient since they are pervasive at nodes of structural change (Nelson 1987). TI is a form of public good, and as such raises issues of public choice which must be resolved. A consensual vision of long-run economic – and noneconomic – objectives, based on an understanding of the underlying tradeoffs may be more useful in this regard than a narrowly-defined, partial-equilibrium, cost-benefit analysis of individual projects.

The differentiation of TI, which distinguishes it from conventional infrastructure, implies a tradeoff between *neutrality* and cost, which places a premium on the government's powers of discrimination. Consider, by way of

[31]Tassey (1991) cites shortening product life cycles as increasing the importance of technological infrastructure. For a distinction between the strategic and tactical dimensions of technological policy see Justman and Teubal (1986, 1990). We implicitly assume TIP is implemented to achieve strategic goals, though some versions of TIP may be employed tactically.

[32]For a graphic summary of the evolution of S&T policies in Europe see Dodgeson and Rothwell (1992, Table 2).

contrast, a neutral broad-based subsidy for firm-based R&D; it can be administered with little specific knowledge and what capabilities it creates are an indirect result of its subsidizing the derived demand for their services. But complete neutrality and a limited budget may dictate low subsidization rates, with the doubly undesirable result of directing funds where they are not needed, and not spending enough money to make a difference where intervention can be effective.[33]

The TIP approach involves some measure of selectivity, since it is often aimed in its implementation at specific sectors and functional areas. Moreover, within those sectors and functional areas specific activities must be targeted (e.g., ISO 9000 quality control, or the electron beam delineator) requiring specific knowledge of the underlying dependency matrix on which to base a choice of cooperative efforts worth pursuing. This implies a level of targeting that is not within the ordinary purview of government, but allows it an important role as a *catalyst* and *broker* of cooperative efforts in the business sector. Such a catalytic role requires capabilities and activities in government which are largely absent when administering broad-based neutral R&D subsidies: multidisciplinary skills involving knowledge of technology, economics and management, as well as public policy skills. These skills are more likely to be present if there is a history of direct government support to individual firms which provides government with a vantage point for assessing sectoral needs.

TIP also requires extensive policy coordination within government to align TI with other elements of infrastructure: physical infrastructure and human capital development. This may require, for example, mechanisms for interministerial coordination and efficient patterns of division of labor between the Ministries of Industry, Defense, Eduction and Science. TIP must also explicitly deal with organizational issues such as how to efficiently promote consortia formation, what are appropriate forms of industry-university cooperation, and to what extent should technology centers behave as profit maximizers. These are issues which have received little attention in the past and require new capabilities within government. It is clear that not every government can acquire them, and therefore the practical scope for TIP may be effectively much more limited than what it would seem a priori.

In contrast to the related and better known concepts of innovation and scientific research, successful approaches to the development of capabilities

[33]In Israel, for example, this quandary led to a revision of subsidy guidelines that has reduced support for larger established firms, favoring small firms and new start-ups.

have been neither pure supply-push (which characterizes curiosity-oriented research) nor pure demand-pull (which characterizes most innovations), but rather hybrids of the two; they must satisfy an *industry-relevance criterion*, which cannot be based, in general, on existing demand (since the market for the activities flowing from the capabilities is not yet developed) but on undetermined needs. The absence of markets also explains the necessity of generating alternative mechanisms of linking needs to capabilities. These distinctive characteristics separate technological infrastructure policy from traditional policies for industrial R&D and from traditional science policy.

Finally, our analysis of TIP distinguishes between its market-building and capability-creation aspects, roughly corresponding to basic (or sectoral) TIP and advanced (or functional) TIP.[34] In promoting basic TI, the capabilities which enable the provision of these services exist elsewhere but need to be imported, adapted, and absorbed in the local economy. This involves stimulating demand for these services through awareness programs and user-need determination; independent sources of supply are developed through learning-by-doing, training consultants, and spinning off independent consulting services. In promoting advanced TI, the user-need determination aspect of market building cannot be separated from capability creation, implying a need for user-led cooperation. Such cooperation represents a new form of industrial organization which still lacks a well-defined institutional framework.

A summary of the essential characteristics of TIP is shown in Table 3.

9. Concluding Remarks: Notes on Implementation

Like any new policy area, strategic technology policy requires especially careful consideration of policy formulation and implementation issues, and many aspects of implementation can only be fully examined within a specific institutional context. Nonetheless, a number of broader principles seem to cut across most of the current spectrum of applications.

[34]The general process of market building transcends the scope of this paper; we limit our focus here to creating markets for technological services.

Table 3: Essential Characteristics of Tip

1.	TIP is conceived and defined in the broad context of structural change.
2.	It involves a discrete choice of public goods.
3.	This implies a measure of selectivity, and a tradeoff between *neutrality* and cost.
4.	Market failure analysis is not enough; a vision of the future may be more useful.
5.	Institutional change and organizational aspects are critical, including the promotion of networks.
6.	Multidisciplinary skills and extensive coordination in government are required.
7.	TIP involves explicit promotion of user cooperation in defining and establishing TI.
8.	Market building plays a central role in basic TIP; it involves stimulating supply and demand for technological services.
9.	Advanced TIP involves stimulating user-led capability creation.

"Market failure" analysis does not provide a sufficient foundation for implementing strategic technology policy. Nelson's (1987) exegesis of the general limitations of market failure as a policy guide is very much applicable to the role of government in developing technological infrastructure. The term market failure carries the presumption that government intervention is a last resort. This is clearly appropriate in most routine contexts but not in establishing infrastructure where there is no reason to assume that the market works well at all. Where market failure is ubiquitous such an analysis cannot serve as an effective screening device for identifying suitable projects for intervention. At nodes of structural change, projects must be chosen on the basis of comparative effectiveness; it is not enough to identify socially profitable projects which the unhindered market will not bring about. Rather than focus our attention on what the market can or cannot accomplish, we must also ask: "(i) How much government infrastructure is needed for profit incentives and markets to work well?... (ii) What are the things that government can do well?... (iii) Can an appropriate government structure be defined and if so be established?" (Nelson 1987). Moreover, the need for institutional change must be considered explicitly, including changes in antitrust laws, patent laws, and university and government laboratory regulations concerning participation of in-house researchers in TI projects.

TIP is still in an experimental stage. Therefore, a linear, planning approach to policy, in which formulation clearly precedes implementation is not appropriate here; an evolutionary approach that emphasizes sequential experimentation is needed. Initial projects should be chosen not only for their intrinsic value, but also for the new information they generate on technological infrastructure, for the capabilities they develop, and for their demonstration effect (cf. Nelson and Winter 1982). In this early stage, the boundary between policy formulation and implementation is fuzzy, and a systematic learning effort should be incorporated in TIP programs.

The government's role is catalytic. Strategic technology policy has little to do with subsidization of R&D, certainly not long-term subsidization. It is about stimulating private-sector initiatives for cooperative infrastructure development, with a view to endogenizing collaborative TI development in the economy, with little or no government subsidization.

Its first goal is to generate a set of "reasonable" projects. The government's first objective in promoting strategic technology policy is to trigger an endogenous process of generating reasonable cooperative technology programs with the active participation and financing of the private sector. Past experience with previous cycles of innovative policy initiatives, such as Israel's industrial R&D support program in the early seventies, suggests that initially, the budget constraint may not be binding; there will be very few suitable candidates that meet the requirements of a *bona fide* cooperative technology infrastructure project, and funding needs will be easily met. In this stage government may have to take a more active role in gathering information (surveying local firms, mapping global trends) and in building coalitions in the private sector. This may involve some mild targeting, directing government initiatives at projects that are likely to contribute most to the general momentum – projects that address key industry bottlenecks and promise strong learning and demonstration effects. But the overall policy approach should be "open-minded" in its industrial orientation if not strictly neutral. As TIP matures, and the proportion of "routine TI projects" which market forces can implement independently increases, budgets can be diverted to stimulating more complex projects that entail great uncertainty.

In later stages, greater selectivity may be justified, implying a need for an explicit technology strategy and a broad vision of the future. Once the endogenous process has been triggered and greater demands are placed on the public sector in supporting private initiatives for infrastructure development, it may be necessary to exercise greater selectivity in allocating government support. This is an inherently political process; its subversion to the interests of narrow constituencies within government and in the private sector is a possibility that must be addressed. Consequently, adequate safeguards against such misuse are a prerequisite of any such policy. This requires that the government's strategy be explicit (Justman and Teubal 1990), and grounded in a clearly defined, consensual vision of the country's future path of social and economic development (cf. Johnson 1982). Explicitness in strategy formulation and a consensual vision of the future provide a coherent framework for evaluating competing policy measures on the basis of widely held goals.

Successful TIP will require institutional innovation. This cannot be over-emphasized. The machinery of government must be able to provide *an effective interface between political forces and professional analysis.* Central banks perform this function in determining interest rates. In Japan, MITI fills this function in formulating and implementing its industrial technology policy. Other countries will find other organizational solutions: an independent Technology Council acting outside the normal channels of government on the strength of the personal standing of its leading members and the professional replutation of its technical staff; a tripartite commission comprising representatives of government, labor, and industry acting as a steering committee for a professional staff seconded on a similar basis; or an interministerial committee, within government, with a professional staff that is part of the civil service. Institutional innovation implements and contributes to the process of "learning from cooperation".

Whatever the organizational form adopted, ***it should serve as a clearing house for information and learning***. Whatever the form of institutional innovation, it has an important role to play in infrastructure development: in scanning the horizon and suggesting an agenda for future infrastructure projects; as a locus of evaluation capabilities and coordination services; and not least, as a source of proposals for a broad vision of the nation's future. To meet these objectives it must adopt a systematic approach to information-gathering, learning, and experimentation, requiring both detailed knowledge gained from in-depth case studies, and a broad strategic understanding of the substance of generic capability development. Systematic knowledge implies that information should be codified and related to broader contexts and existing theory, and integrated in an educational effort aimed at creating a capable professional staff. Such an effort could be undertaken in cooperation with an interdisciplinary university graduate program combining business and economics training on the one hand, and science and engineering on the other. There are quite a number of such programs in existence today.

Government itself requires substantial capability in order to trigger action in others. Ultimately, the success of TIP will depend on the business sector, on its initiative, on its forward-mindedness, on its ability to work together for common goals. But triggering a process that culminates in such success requires an awareness on the part of government of the role of technological infrastructure in industrial development, and the capability to act on this awareness. It demands that some of the discriminating capabilities that the private sector requires to build TI – including both knowledge of existing needs and capabilities and experience in the mechanics of cooperation – be present, in some measure, within the government itself.

References

Arnold, E. and K. Guy. 1986. *Parallel Convergence: National Strategies in Information Technology*. London: F. Pinter.

—. 1987. *Information Technology and Emerging Growth Areas.* Science Policy Research Unit.

Arnold, E., K. Guy, and M. Dodgeson. 1992. *Linking success: Making the most of collaborative R&D.* Report to NEDC Electronics Industry Sector Group. London: IEE.

Braunling, G. 1993. The role of innovation infrastructure in economic development. Paper presented at the International Conference on "The New Industrial Policy – Liberalization and Technological Infrastructure." Jerusalem.

Chenery, H.B. 1959. Interdependence of investment decisions. In *The Allocation of Economic Resources*, edited by M. Abramovitz.

Dahlman, C., B. Ross-Larsen , and L. Westphal. 1985. *Managing Technological Development: Lessons from the Newly Industrialized Countries.* Washington, D.C.: The World Bank.

Dodgeson, M. and R. Rothwell. 1992. European technology policy evolution. *Technovation* 12 (4).

Ergas, H. 1987. Does technology policy matter? In *Technology and Global Industry: Companies and Nations in the World Economy*, edited by B. Guile and H. Brooks, Washington, D.C.: National Academy Press.

Gerschenkron, A. 1962. Economic backwardness in economic perspective. In *Economic Backwardness in Economic Perspective: A Book of Essays*. Cambridge, MA: Harvard University Press.

Granberg, A. 1993. Mapping the cognitive and institutional structures of an evolving advanced-materials field – the case of powder technology. Typescript.

Guy, K. and E. Arnold. 1991. IT diffusion policies. Science Policy Research Unit, University of Sussex. Typescript.

Hirschman, A.O. 1958. *The Strategy of Economic Development*. New Haven, CT: Yale University Press.

Hofflinger, B. 1989. Cooperative research on application-specific microelectronics. *Proceedings of the IEEE* 77 (9): 1390-1395.

Imai, K. and Baba. 1988. Systemic innovation and cross-border networks: Transcending markets and hierarchies to create a new techno-economic system. In *Technology and Productivity – The Challenge for Economic Policy*. Paris: OECD.

Johnson, C. 1982. *MITI and the Japanese Miracle*, Stanford: Stanford University Press.

Justman, M., M. Teubal, and E. Zuscovitch, eds. 1993. *Technological Infrastructure Policy for Renewed Growth* (Hebrew). The Jerusalem Institute for Israel Studies.

Justman, M. 1993. *Infrastructure and Growth*. The Jerusalem Institute for Israel Studies.

Justman, M. and M. Teubal. 1991. The structuralist perspective on the role of technology in economic growth and development. *World Development* 19 (9): 1167-1183.

——. 1986. Innovation policy in an open economy: A normative approach to strategic and tactical issues. *Research Policy* 15: 121-138.

——. 1990. The structuralist approach to economic growth and development: conceptual foundations and policy implications. In *Science and Technology: Lessons for Development Policy*, edited by R. Evenson and G. Ranis, Boulder, CO: Westview Press.

Lundvall. B.A. 1991. Innovation, the organized market and the productivity slowdown. Paper presented at OECD's International Seminar on *Science, Technology and Economic Growth*. Paris, France.

---. 1985. *User-Producer Interaction*, Aarlborg. Denmark: Aarlborg University Press.

Malerba, F. 1993. Italy: The national system of innovation. In *National Systems of Innovation*, edited by R. Nelson and N. Rosenberg. Oxford University Press.

Mytelka, L.K. 1991. States, strategic alliances and international oligopolies: The European Espirt program. In *Strategic Partnerships in the World Economy*, edited by L.K. Mytelka. London: Pinter.

Nelson, R.R. 1987. Roles of government in a mixed economy. *Journal of Policy Analysis and Management* 6 (4): 541-557.

Nelson, R., M. Peck, and E. Kalachek. 1967. *Technology, Economic Growth and Public Policy*. Washington D.C.: Brookings Institute.

Nelson, R. and S. Winter. 1982. *An Evolutionary Theory of Economic Change*. Harvard University Press.

North, C.D. 1990. *Institutions, Institutional Change and Economic Performance*. Cambridge University Press.

Ouchi, W. 1989. The new joint labs. *Proceedings of the IEEE* 77 (9): 1318-1326.

Pavitt, K.L.R. 1993. Why British basic research matters to Britian. Science Policy Research Unit, University of Sussex. Typescript.

Quintas, P. and K. Guy. 1991. Collaborative, pre-competitive R&D and the firm: Some lessons from the UK Alvey Programme. Typescript.

Romer, P.M. 1990. Endogenous technological change. *Journal of Political Economy* 98: S71-102.

Rosenberg, N. 1963. Technological change in the U.S. machine tool industry, 1840-1910. *Journal of Economic History* 23: 414-443.

Rosenberg, N. and R.R. Nelson. 1992. American universities and technical advance in industry. Typescript.

Senker, J. 1991. Evaluating the funding of strategic science: Some lessons from British experience. *Research Policy* 20 (1): 29-44.

Shappir, Y. 1993. National infrastructure for microelectronics in Israel. In Justman et al. 1993.

Shapira, P., J.D. Roessner, and R. Barke. 1992. Federal-state collaboration in industrial modernization. School of Public Policy, Georgia Institute of Technology.

Sigurdson, J. 1986. Industry and state partnership in Japan: The VLSI Circuits Program. Discusssion Paper No. 168, Research Policy Institute. Lund, Sweden.

Tassey, G. 1991. The functions of technology infrastructure in a competitive economy. *Research Policy* 20: 345-361.

Teece, D.J. 1986. Profiting from technological innovation: Implications for integration, collaboration, licensing and public policy. *Research Policy* 15: 285-305.

—. 1992. Competition, cooperation and innovation – organizational arrangements for regimes of rapid technological progress. *Journal of Economic Behaviour and Organization* 18: 1-25.

Teubal, M. 1979. On user needs and need determination. In *Industrial Innovation: Technology, Policy, Diffusion*, edited by M. Baker. London: Macmillan.

—. 1993. The innovation system of Israel: Description, analysis and outstanding issues. In *National Systems of Innovation*, edited by R. Nelson. Oxford University Press.

Teubal, M., N. Arnon, and M. Trachtenberg. 1976. Performance in innovation in the Israeli electronics industry. *Research Policy* 6: 354-379.

Teubal, M., T. Yinnon, and E. Zuscovitch. 1991. Networks and market creation. *Research Policy* 20: 381.

Teubal, M. and E. Zuscovitch. 1991. Networks and evolutionary differentiation. The Jerusalem Institute for Israel Studies. Typescript.

Toren, B., Z. Adelman, and R. Goldring. 1993. Software: An individualistic industry with collective needs. In Justman et al. 1993.

von Hippel, E. 1976. The dominant role of users in the scientific instrument innovation process. *Research Policy* 5: 212-239.

Wachs, R. 1993. Infrastructure for the chemical industry. in Justman et al. 1993.

Westphal, L.E., L. Kim, and C.J. Dahlman. 1984. *Reflections on Korea's Acquisition of Technological Capability*. Washington D.C.: The World Bank.

Willinger, M. and E. Zuscovitch. 1988. Towards the economics of information-intensive production systems. In *Technical Change and Economic Theory*, edited by Dossi et al. London: Pinter.

Yinnon, T., H. Kalansky, Y. Ben Barak, and H. Urbach. 1993. A proposal for the organization in the plastics industry in Israel. In Justman et al. 1993.

Zuscovitch, E. and M. Justman. 1993. Networks, sustainable differentiation and economic development. Forthcoming in *Networks in Action*, edited by D. Batten, J. Casti, and R. Thord.

Zuscovitch, E. and M. Schahar. 1993. Industrial and technological joint ventures. In Justman et al. 1993.

Infratechnologies and Economic Growth

Gregory Tassey

Economic infrastructure is widely recognized as an essential element of economic growth. The economics literature has pointed to the critical role of infrastructure (roads, railroads, canals, etc.) during the latter part of the 1800s in expanding and unifying regional markets within the U.S. economy. By allowing realization of comparative advantage among regions of the domestic economy, by permitting attainment of economies of scale, and by reducing transactions costs within markets, infrastructure accelerated the industrial revolution and thereby leveraged the United States into a position of world economic leadership.

In this century investment in infrastructure has evolved substantially by drawing upon advances in technology. Governments in industrialized nations have made vast investments in advanced infrastructure, such as communications, not only to accelerate the pace of economic activity but to make it more efficient as well. This latter benefit of investment in infrastructure, *efficiency*, or more conventially, *productivity*, has not been widely cited and certainly not sufficiently appreciated.

Only recently have economists begun to systematically study the roles and economic impacts of infrastructure. The initial results have surprised even supporters of the importance of this source of economic growth. Analyzing the substantial drop in productivity growth in the U.S. economy between the period 1950-1970 and the following 15-year period (1970-1985), David Aschauer (1988, 2) found that "fully 1.0 percentage point of the total decline in [the average annual growth rate of] productivity of 1.2 percentage points can be attributed to the neglect of infrastructure."

As the production of goods and services has become increasingly dependent on the rapid and effective incorporation of technology, economists have begun to focus on the emerging and varied roles of the supporting *technology-based* infrastructure. Increasingly, industrialized nations are making substantial investments in this new infrastructure, and it is coming to be viewed as an essential element of a competitive economy. As a result, economists and policy analysts will have to make increased efforts to understand its roles in facilitating product and service innovations and in enhancing productivity growth.

M. Teubal et al. (eds.), Technological Infrastructure Policy, 59–86.

1. Types and Functions of Technological Infrastructure

A general definition of economic infrastructure is an element of an economic system that is commonly used by competing economic agents. Similarly, technological infrastructure is an element of an industry's technology that is jointly used by competing firms.[1] One category of this advanced infrastructure is *generic* technology. Generic technologies are the core product and process technologies from which specific commercial applications are developed through subsequent applied R&D by competing firms. Typically a generic technology embodies a laboratory-proved concept but not the subsequent market-specific products and processes that are eventually derived from it. Achieving a level of generic technical knowledge sufficiently reduces technical risk to allow positive investment decisions to be made with respect to subsequent applied R&D. The substantial technical risk, let alone commercial risk, frequently coupled with large economies of scale or scope in generic technology research leads to substantial market failure at the company level and thus underinvestment.

Thus, much generic technology has come to be viewed as a public or quasi-public good (i.e., as infrastructure) and is therefore developed jointly by firms that compete against each other later in the technology life cycle. The rationale is that sharing the early-phase R&D results (the generic technology) and having access to it early in the competitive life cycle is preferable to an "all-or-nothing" strategy in which each firm tries to independently develop the generic technology as a proprietary asset.

When the market failure and thus underinvestment is particularly severe, government participation in the cooperative research effort is frequently required. The policy rationale for subsidizing generic technology research is that (1) the process of producing information (i.e., the reduction in technical risk) leads to underinvestment decisions by competing firms at the applied research phase of R&D and beyond. Enough applied R&D investment by enough firms occurs to ensure a competitive market for commercial applications of the generic technology and thus efficient resource allocation.

A second category of technological infrastructure includes the various techniques, methods, and procedures that are necessary to implement the firm's product and process strategies. Methods such as total quality management (TQM) can be differentiated upon implementation within a firm, but they

[1]Tassey (1991, 9-11; 1992, Chap. 3) and Justman and Teubal (1993) have developed taxonomies.

must be traceable back to a set of generic underlying principles if customers are to accept claims of product quality. For example, demonstration to customers of compliance with ISO 9000, a set of international quality assurance standards, must be derived from a set of common principles even though implementation within the firm can be customized to fit internal production strategies. Thus, the ISO 9000 standards constitute a form of infrastructure.

A third type of technology-based infrastructure, and the focus of this chapter, is the set of "technical tools" for making the entire economic process more efficient, or in some cases, possible in the first place. Collectively these tools are called *infratechnologies*.[2] For most industries, all phases of R&D, production, and market development are supported by various infratechnologies.

2. Economic Roles of Infratechnologies

As indicated in Figure 1, infratechnologies are ubiquitous in the scope of impacts on the typical technology-based industry. This element of technological infrastructure becomes embodied in or supports generic technology and its applications. It also provides the technical basis for certain types of standards which directly affect process and quality control at the production stage, and is often essential for the efficiency of market transactions through risk reduction. Infratechnologies fall into four general categories:

- *scientific and engineering data* that are used for conducting R&D, controlling production, and consummating market transactions (the latter applications are especially important in continuous process technologies, such as chemical production and petroleum refining);
- *measurement and test methods* that are essential to conduct state-of-the-art R&D, monitor production, and execute market transactions (product acceptance testing);
- *production practices and techniques*, such as process models for understanding relationships among production parameters, thereby allowing more efficient design and control of production processes; and
- *interfaces* that permit the efficient physical and functional combinations of components into manufacturing and service systems.

[2]See Tassey (1982, 1991, 1992) and Link and Tassey (1987). This category of technological infrastructure was officially recognized for the first time in the United States in the 1994 Economic Report of the President.

Figure 1

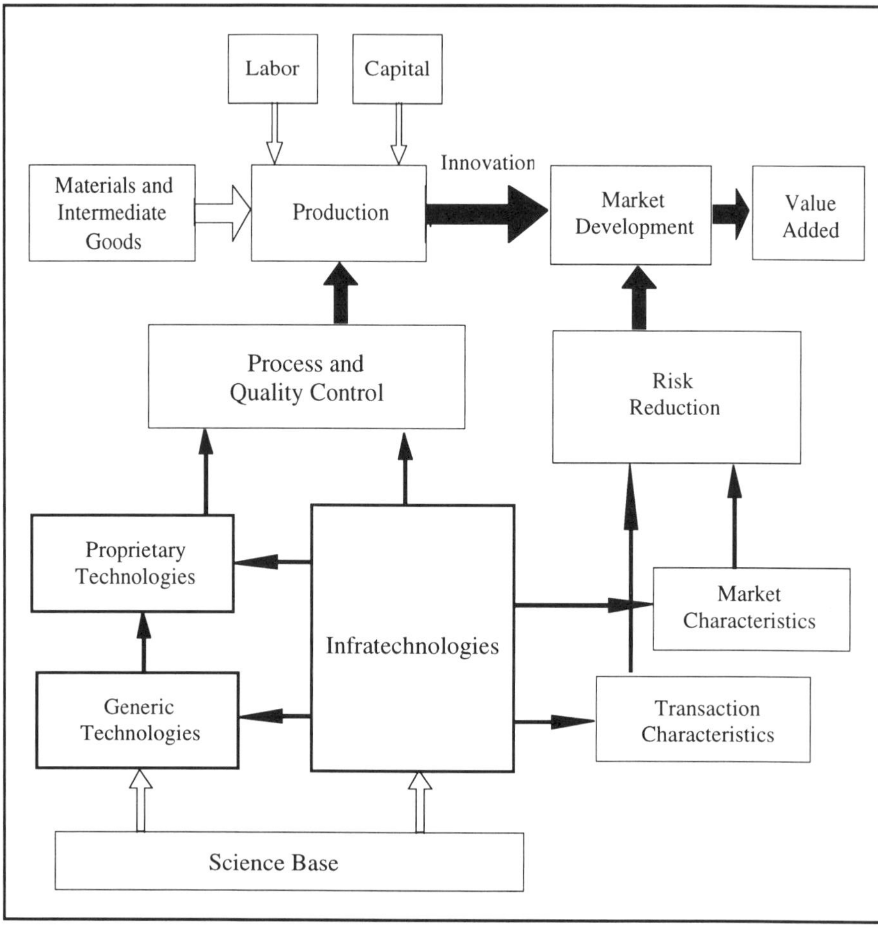

Source: Tassey (1992, Chap. 3).

As the complexity and pervasiveness of technology grow, infratechnologies are becoming ubiquitous in the modern economic system. As the arrows in Figure 1 indicate, measurement-related infratechnologies are embedded in each of the three major stages of technology-based economic activity, specifically: R&D, production, and market penetration.

2.1 R&D Stage

At the R&D stage, measurement science adds to the pool of basic knowledge that all scientists and engineers use. Continual improvements in the accuracy and precision of basic measurements standards are essential for the success of many product and process R&D efforts. For example, communication systems depend on highly accurate and reliable time and frequency standards.

More generally, demand on the R&D process and thus the tools used in this process are steadily increasing. Improvements in performance attributes, including manufacturing characteristics, must now be *designed in* during the R&D process with a much higher degree of differentiation than in the past. The trend in materials research is to *engineer* technologically-advanced materials, such as ceramics, polymers, and metal-matrix composites for increasingly differentiated applications. Design of these new custom materials is reaching the molecular or even the atomic level; that is, performance characteristics are being adjusted for specific uses by rearranging individual atoms and molecules. This level of design sophistication was hardly contemplated just a few decades ago when materials for most products were highly uniform in composition and alterations in composition were quite crude, consisting only of gross changes in the proportions of the components. Achieving such highly customized products requires highly differentiated materials, properties data, and extremely accurate test methods.

To engineer advanced materials at the molecular and atomic levels requires measurement accuracy with respect to the properties of these materials that is orders of magnitude beyond that of just a few years ago. In fact, totally new measurement techniques are needed. Without these new techniques, much state-of-the-art R&D would not be possible. For example, research in new high-temperature superconductors has been impeded by the lack of a widely accepted measurement method for determining the current-carrying capacity of the superconductors. In the absence of such a method researchers cannot readily compare and verify their findings. Firms will be reluctant to make R&D investments for commercial applications of the technology without a substantial reduction in performance uncertainty.

Such measurement technologies, standards, and associated materials-properties data make industrial research and development more productive by allowing industrial scientists and engineers to develop and apply new technological principles, and then to communicate the results to the corporate decision-making process and to suppliers of technologically advanced inputs.

Communication with suppliers is improving while interaction with competitors is increasing. The boundaries of the firm are dissolving to a significant extent as firms find it necessary to combine competitive assets at the R&D stage and beyond. Each of these areas of interaction requires standardized methods of communicating scientific and technical knowledge and then agreeing on the characterization of research results.

Enhancing the efficiency of R&D is essential to overall economic growth. In particular, economic studies have demonstrated the importance of R&D investment to long-term productivity growth and thus to growth in value added. More recently, international comparisons (Lichtenberg 1993) indicate that private R&D investment *within* a domestic economy is strongly correlated with both the growth rate and level of productivity in that economy. Such correlations imply that a nation benefits, at least initially, more from its own R&D than other nations do. Thus, having superior technological infrastructure to facilitate domestic R&D will attract both domestic and foreign capital to this critical stage of the econoomic process and benefit the domestic economy.

2.2 Production Stage

At the production stage, measurement methods, calibrations, standards, and data are crucial to quality assurance and process control. Modern production depends increasingly on precise measurement to increase productivity through a total systems approach. In the past, measurement for process control was mostly a static or "snap shot" activity, occurring mainly at the *end* of production. By contrast, modern processes require continuous measurement and the ability to respond instantly to measurement change *during* production.

Because products are increasingly based on advanced materials where defects even at the single-atom level can affect performance characteristics, their production requires measurement-intensive control. For example, accurate measurements of the thickness and electrical properties of the thin layers within integrated circuits are essential to controlling the quality of those circuits during manufacture. Accuracy good enough to distinguish the thickness of individual atomic monolayers will be required for the next generation of integrated circuits. The nation whose infrastructure provides such measurement capability will likely be the leader in semiconductor technology for this next generation.

Similarly, quality assurance was once mostly a matter of inspection at the end of the production line. Today the emphasis is on making it right the first time. Machines are being programmed to measure their own performance attributes, to measure the attributes of the products being processed, and to adjust the production process automatically to account for deviations from the prescribed design.

Moreover, automated production technologies are increasingly emphasizing *flexibility* in terms of product variation and systems-oriented control. Both attributes mean a greater need for sophisticated measurement and control algorithms. In addition, designing a flexible manufacturing system at reasonable cost requires interface standards to ensure purchasers access to multiple vendors, to promote competition among vendors, and to allow greater access to markets for these components by small firms.

Adoption of international interface standards which are compatible with domestic versions of the technology is essential for access by that country's firms to world markets. Interface standards are just as technically sophisticated as the other elements of a systems technology. This is evidenced by the fact that the National Institute of Standards and Technology (NIST) in the United States had to develop a state-of-the-art automated manufactuing research facility (AMRF) to provide industry with the technical basis for needed factory automation interface standards, as well as to provide a facility at which industry can test products for conformance with the standards.[3]

It is not just the role of measurement infratechnologies in quality assurance and process control that make them important at the production stage. They also constitute an important element of overall production cost. Measurement costs account for about 20 percent of the cost of producing optical fibers, about 25 percent of the cost of making silicon integrated circuits, and about 25 percent of the cost of making microwave integrated circuits. (Quick, Finan & Associates 1990). Thus, the efficiency of their use is important.

[3]Test facilities are a separate category of technology-based infrastructure, which is often essential for the diffusion and transfer of a substantial number of infratechnologies. The phrase "provide industry with the technical basis for... standards" is indicative of the voluntary standards process in the U.S. economy in which the infratechnology may come from a number of sources (such as NIST), but the standard itself is set through a voluntary consensus process undertaken by industry. NIST still sets the basic measurement standards (time, frequency, physical constants, etc.), but most of its laboratory research budget is now directed at developing the infratechnologies that serve as the basis for the *industry-led* standards process.

2.3 Market Penetration

Test methods, other types of standards, and data are also essential for market penetration. High levels of performance risk frequently accompany high technology products to market. Test methods, test structure, and industry standards help new technologies diffuse by reducing these risks.

For example, power transformers used by electrical utilities must have extraordinarily low electrical losses, because over the lifetime of a transformer those losses can exceed the purchase price. Advanced measurement methods are necessary to detect these losses, and severe penalties can be exacted by the buyer for the seller's failure to meet agreed loss specifications.

The objective of "equity in trade" has always been an important role for measurement. In the past this meant largely providing for an adequate system of weights and measures. Today, however, as more markets become technologically complex, market acceptance of new products requires technologically sophisticated acceptance testing equipment and procedures.

When test methods acceptable to both buyer and seller are not available, significant costs can be incurred as engineers and other technical staff attempt to resolve performance-measurement disputes. The higher transactions costs add to the effective price of the new product, so market entry is delayed by the disagreement itself, while the increased price due to inefficient acceptance testing further slows market penetration.

Such occurrences can be disastrous for firms that participate in international markets where product life cycles are increasingly shortened by intense competition. Being late in many of these markets can mean a permanent inferior market share. These measurement infratechnologies must be available at the critical points in a technology's life cycle.

3. Technology-Based Market Failures

Measurement-related infratechnologies operating at the levels of R&D, production, and marketing generate a wide range of substantial benefits to an economy. However, industry tends to underinvest in the research to develop these infratechnologies and in the resources to implement them. While the specific causes of underinvestment in infratechnologies differ from those that affect private-sector investment in generic technology or investment in technology generally, the problem can be attributed to the existence of market failure. Government subsidization of infratechnology development and dissemination is justified on the basis of correcting or compensating for these market failures.

Because infratechnology is one of several categories of technological infrastructure, the rationales for government assistance in providing it should be placed in the broader context of rationales for government intervention in technology-based markets. Policy analysis should identify and examine all market failures that are occurring in individual industries so that subsequent policy responses are integrated in a complementary and hence efficient manner. A portion of identified market failures will be the result of technology elements having characteristics of infratechnologies.

In general, a market failure is some malfunction of the private-sector resource allocation mechanism.[4] Market failures result in either too much or too few resources allocated to a particular area of economic activity, relative to the levels of investment required to maximize growth in output and the standard of living. Such malfunctions of the marketplace are caused by private-sector economic agents either not internalizing certain costs that they create or not capturing all of the benefits from their economic activity. As these *externalities* become significant, so do the distortions of optimal investment patterns.

When too few resources are directed to technological infrastructure, government policies to remedy the failure or even provide the technological infrastructure are appropriate. Market failures that result in an underallocation of resources to technological infrastructure may originate in the inherent nature of the technology itself, market structure, behavioral characteristics for both producers and users of the technological infrastructure, and problems of integration and coordination. More precisely, these include:

- **The inherent nature of the technology, as when there exist**
  - high levels of *technical risk*: the risk that the technology does not perform as intended;
  - high levels of *commercial risk*: the risk that the markets for applications of the technology are not as large as required for adequate return on investment;
  - large economies of scale or scope in R&D: technology development costs are high when R&D is capital-intensive (scale) or when the number of targeted applications is too limited (scope);

[4]For a summary of the economics literature on the concept of and possible policy responses to market failure, see *The New Palgrave* (Eatwell *et al.* 1987, 326).

- requirements for multidisciplinary R&D, especially at the systems-integration level or in the sense of combining several heretofore separate technologies (e.g., biosensors).[5]

- **The structure of the supply side of the market that must develop, produce, and deliver the technology, as when**
  - average firm size is too small relative to the capital intensity of R&D or production requirements;
  - the degree of horizontal and vertical integration may be inadequate for overall production-chain efficiency or for small-firm participation in specific markets that make up the production chain, thereby increasing technical and commercial risk associated with scale and scope economies in R&D, market timing, and system compatibility.[6]

- **Behavior on the supply side of the market, for instance when**
  - attitudes toward risk-taking are skewed;
  - learning economies are not realized, especially in industries where increasing levels and complexity of automation require major changes in organization of production, including massive use of information technologies;
  - multiple-market applications of a technology are broader than the existing strategic foci of individual firms, so that necessary R&D and market planning skills are not available in individual firms (economies of scope are not captured);
  - the market depends on easily diffused technologies, such as knowledge-based or "soft" technologies, which have lengthy R&D stages but relatively low production costs;

[5]For example, Japan's Human Frontiers Science program for research in biometrics is predicated on the promise of technology "fusion": study how the brain works and you can make a better computer; study how cells transfer energy and you can devise new energy conservation technologies. In one of the MITI projects, called "Ultra-Advanced Manufacturing Systems", the aim is to apply semiconductor manufacturing technologies such as ion beam implantation, CVD, and patterning with excimer lasers to the fabrication of physically large products, such as propellers, artificial organs, and pipes for use in clean rooms. Individual firms can hardly be expected to include such apparently unrelated applications in their market strategies, nor would they have the necessary technical and market capabilities. MITI views the low likelihood of such cross-applications as a major market failure.

[6]The term "production chain" refers to the vertical structure of industries and markets, for example: raw materials (silicon), components (semiconductors), equipment (computers), and services (data processing). The competitive position of each level in the chain depends on certain interactions with the levels above and below it.

  - "capturability" is difficult and market entry is easy at the production stage;[7]
  - coordinating the formalization of public technology goods (infratechnologies) as standards is impossible.

- **Behavior on the demand side of the market where purchase decisions with respect to technologies are made - as where**
  - attitudes toward risk-taking are skewed;
  - infrastructure for assessing product performance is not available;
  - the ability to integrate technology into existing production structures is inadequate.
- **Difficulty of integrating management, labor, and organization strategies with new types of technology, because**
  - many manufacturing firms do not have the multidisciplinary skills to integrate organizational, labor empowerment, and technology development strategies.

- **Difficulty of acquiring knowledge of foreign competition, due to**
  - the lack of benchmarking of foreign technology commercialization strategies.

Market failures must also be characterized by the point in the technology life cycle at which they typically occur, so that the *timing* of policy implementation can be determined. Market failures appear frequently in technology-based industries:

- during the early stages of a technology's development;
- during the sale of complex technology-based products and services whose attributes are difficult to measure, especially during the early phases of market penetration; and,
- when an element of an industrial technology becomes commonly used by a large number of firms as a standardized element of an infrastructure system, such as an information network.

[7]Two examples from NIST's Advanced Technology Program (ATP) portfolio are the rapid response manufacturing project at the National Center for Manufacturing Sciences and the advanced manufacturing of electrical products project at the South Carolina Research Authority. Both projects involve the extensive development of knowledge bases, product description models, and in the case of NCMS, process control models. Such research results are very difficult to hold proprietary and therefore need considerable cost sharing in recognition of the nature of research output (public technology goods).

Thus, large and diverse numbers of market failures can prevent a technology from realizing its economic potential at several different points in the technology's life cycle. However, the current practice in most industrialized nations is to rationalize government programs based on a limited subset of the above. In particular, risk reduction and intellectual-property protection have been the two most frequent policy thrusts in addressing market failures.

Even here oversimplification of the typical technology life cycle has resulted in confusion and misdirected policies. For example, risk reduction at the generic-technology phase of R&D frequently requires cost sharing through consortia, often with government subsidization, while risk reduction at the applied R&D phase typically requires a tax incentive. Failure to identify and articulate the correct R&D phase to be addressed by the policy response results in substantial policy inefficiency and adversely affects economic growth. In such situations the wrong mechanism is often employed; however, even if an appropriate policy mechanism is arbitrarily selected, it is likely to be criticized and therefore compromised.

Removing a market failure requires careful analysis of the nature of the technology development process, the nature of the production process, the existing or emerging industry structure, and the ways by which markets deliver the technology to users. Only through such comperehensive microeconomic analysis can growth policy be efficiently implemented over time.

Multiple market failures can occur at particular points in the economic process associated with the life cycle of a particular technology. The policy process often focuses on only one of these barriers, or when it does respond to multiple failures, it does not take into account the interactions among these response mechanisms. Moreover, the timing of the elimination of the market failure is often ignored in formulating a policy response.

Most financial support for the development of advanced technology has been directed at market failures that result from the high technical risk inherent in the early phases of the R&D process. This focus has led to the charge of government picking winners and losers. Part of the correct policy response is to point out that a large economy such as the United States must diversify its technology base to (1) achieve a horizonally diversified and hence more stable economic growth path, and (2) avoid the problems of insufficient vertical integration of specific production chains.

Even though the most frequently used market failure rationale is *high technical risk,* this claim by itself is not sufficfient. The proposed research area must also offer high potential economic benefits. Market failures associated with high technical risk occur when (1) industry's risk preferences are skewed to the low end, causing rejection of high-risk projects that also have high expected payoffs; (2) the high expected payoff is recognized but distant so that its present value is too low relative to technical risk to justify investment (i.e., the market failure is too high a discount rate); and (3) firms with a strategic interest in the technonology are too small to sufficiently absorb the risk. In all these cases of high technical risk government can mitigate the market failure by distributing risk over many projects or by lowering the discount rate.

Another category of market failure which causes significant underinvestment by the private sector has little to do with technical risk. Instead, it involves *high market risk.* A common cause of this type of market failure is the inability to earn a sufficient rate of return on the investment. Terms used to describe technology for which this category of market failure occurs are *capturability* and *nonproprietary.* However, another frequent cause is the inability to accurately and consistently measure the performance of complex, technology-based products. This inability adds significantly to transactions costs, that is, the costs of actually consummating a market transaction.

The capturability category of market failure, when it applies to proprietary products and processes, can be mitigated to a significant extent by appropriate changes to intellectual-property laws. For public-technology goods, however, other institutional responses are required. For example, the rationale for a national measurements and standards infrastructure is based on the existence of infratechnologies that commonly exhibit this economic imperfection; a large number of individuals and firms benefit from the same technology and must use it collectively. Thus, the aggregate benefit to the economy is high, but benefits are only realized through widespread and consistent use by market participants. This commonality and comprehensiveness of use of measurement-related infratechnologies and data give them a nonproprietary character which results in underinvestment by the private sector.

Moreover, the development of measurement infratechnologies frequently requires expensive equipment and highly skilled staff which generate limited benefits to a particular firm or even a single industry. The measurement-related research of a firm or industry therefore may have insufficient *scale* to be efficient. Furthermore, individual firms and industries tend to use a limited set of measurement methodologies, but the science and technology of measurement is highly interrelated and can be applied to the problems of many different industries. Thus, the measurement-related research of a particular firm or even an entire industry may also have insufficient *scope* to be efficient.

Another reason for inadequate investment in infratechnologies is the distributed nature of the economic impacts, which requires more resource-intensive policy analysis and strategic planning. As previously described, many infratechnologies are measurement related, but are required at different stages in the economic process from basic research through development and production to marketing. This presents a relatively complex policy modeling problem with respect to determining economic roles and impacts.

Finally, the *competitive neutrality* characteristic of many infratechnologies leads to underinvestment. Generic technologies are the basis for subsequent product and process applications. When a technological innovation occurs, it is a highly visible event and therefore attracts considerable attention. Thus, generic technologies are at the center of corporate strategic decision making and are very visible to policy makers as well. NIST's Advanced Technology Program (ATP), which co-funds generic technology research with industry, became instantly visible and grew rapidly in size. Infratechnologies, on the other hand, by virtue of their role as tools leveraging the efficiency of the various stages in the technology-based economic process, are ubiquitous and therefore collectively important but less visible.

Beyond problems with incentives to develop infratechnologies, market failures can occur in their implementation, say, as standards. The majority of industry standards in the United States are voluntary, that is, they are set by consensus among competing firms (and frequently their customers). Conflicting competitive strategies cause differences among firms with respect to their views on content and timing of a standard. Delays in promulgation and the existence of multiple standards for extended periods of time are two frequent consequences of these differences. Thus, the aggregate economic effect of an inefficient voluntary standards process can be substantial. (Hawkins and Mandell forthcoming; U.S. Congress 1992).

4. Mechanisms for Developing and Delivering Infratechnologies

Infratechnologies are supplied by a variety of sources: (1) individual firms, (2) industry consortria, (3) government laboratories, and (4) foreign sources.

Individual firms invest in infratechnologies as part of their overall competitive strategy, but often investments are limited and results qualitatively unacceptable. Consequently, individual firms systematically underinvest in infratechnologies. Much of the private investment that does occur takes place in the early part of a technology's life cycle when other sources of infratechnology have not yet instituted research programs or produced usable results. Such early-cycle investments are often stop-gap measures to support first-to-market competitive strategies. As competitors enter the market and the number of customers grow, the demand for more and better-quality infratechnologies overwhelm the ability of individual firms to supply them, or to get their own versions of these public-good technologies adopted as industry standards.

In response to the need to more rapidly and effectively supply this category of technological infrastructure, the consortium has recently gained prominence as a mechanism for developing and disseminating infratechnologies. Consortia are especially attractive when the infratechnology falls into the middle-ground category of *quasi-public* goods; in such cases the infratechnology has a sufficiently focused industry user population to raise the benefit per firm to a level that warrants the investment necessary to make participation in a consortium worthwhile.

NIST's early infratechnology research included two consortia, organized during the 1970s in the chemical engineering area, and a third consortium formed in the early 1980s. Within the past five years, however, NIST's use of this mechanism has proliferated. By 1994 twelve consortia were active with five more in the formation stage. The 12 currently ongoing consortia at NIST that pursue the development of infratechnologies and their implementation are listed in Table 1.[8]

Several factors explain the increased use of this mechanism. First, budget constraints in the 1980s prevented NIST from undertaking the development of many infratechnologies needed by industry, or from getting the research done in the time frame dictated by competitive conditions.

[8]NIST also develops infratechnologies under contract with trade associations. These associations are technically permanent consortia, but are not listed in the table.

Second, many infratechnologies can be appropriately classified as quasi-public goods, which means that their economic benefits accrue to a modest number of firms. On the one hand, the benefit to individual firms from many infratechnologies has grown as technology-based products have become more complex and demands by customers for higher quality and compatibility have increased, but the public-good content is high enough to create significant underinvestment within the individual firm. On the other hand, competitive pressures, such as reduced "windows" in time in which new products can be successfully introduced into the marketplace, plus the growing demand for open systems, have effectively raised the public-good content of categories of infratechnologies that previously had a relatively large private-good character. This movement from both ends of the public-good–private-good spectrum toward a quasi-public-good composition has led to increased use of cost sharing through consortia. Thus, firms from one or several industries are increasingly willing to share the costs of developing the infratechnology. Presumably, the contribution from each firm is rationalized by the private benefit ultimately derived.

Third, consortia are an excellent technology-transfer mechanism. NIST has found that many infratechnologies are developed and largely implemented before being officially adopted as an industry standard. By participating in the mechanism that develops the infratechnology, firms both influence the research and rapidly assimilate the research results.

Table 1 indicates the range of industries recently engaged in cooperative development of infratechnologies and the variety of infratechnologies being pursued. The financial participation by industry includes both cash and in-kind contributions (equipment, materials, labor). The majority of the government contribution comes from NIST in the form of staff time, facilities, and materials. Other federal agencies have made minor cash contributions to NIST to help support several of the consortia. The distribution of cost sharing for each consortium reflects a number of factors, including: number and average size of participating firms, size and time to realization of expected benefits to individual firms, and the value of in-kind contributions.

Table 1: Consortia Developing Infratechnologies

Consortium Name (start date)	No. of Partici-pating Firms	Description (type of infratechnology)	Financial Contribution ($000) Industry	Govern-ment
Flowmeter Installation Effects (1985)	23	Design and control of flowmeter peformance: characterization of fluid flows	$ 1,017	$ 600
Production of Powdered Metals (1987)	6	Control of production of advanced metals for jet engines: gas atomization control techniques	1,200	4,095
Microwave Integrated Circuits (1989)	5	Control of production: on-wafer testing methods and protocols/standards	413	3,668
Automated Analytical Laboratory Systems (1990)	14	Automation of analytical chemistry laboratories: communication protocols and improved data quality	1,725	1,000
Toxic Byproducts of Fluorinated Sulfur (1991)	5	Improved materials usage by electric utilities: safety and environmental procedures	1,620	1,545
Ceramic Machining (1992)	15	Machining of advanced structural ceramic materials: process control methods and materials characterization	1,050	1,396
Polymer Processing Measurement (1992)	5	Control of production of plastic and rubber products: in-line optical measurement techniques	200	1,320
Processing of Polymer Blends (1993)	6	Process models: techniques for mixing several polymers to achieve improved properties of toughness and strength	270	1,045
Casting of Aerospace Alloys (1993)	7	Process models, materials characterization, in-line sensing for real-time process control	1,920	5,095
Advanced Biosensors (1993)	7	Signal quality (elimination of background interference)	390	75
Ceramic Powders and Slurries (1993)	6	Process models and in-line process control for production of ceramic powders	138	1,130
BACnet Interoperability (1993)	12	Integration of multivendor building control systems: conformance testing for BACnet standards	N.A.	150

Source: NIST Program Office.

The majority of these consortia are formed using the Cooperative Research and Development Agreement (CRADA) mechanism, established under the 1986 Technology Transfer Act. Under a CRADA, a government agency may allocate resources internally in order to contribute to the project, but may not transfer funds to industry partners. Both government and industry, as well as university and other participants in CRADAs, contribute staff, facilities, and materials (and funds in the case of industry). The original members of a consortium bear the majority of the burden of negotiating the technical and non-technical aspects of the cooperative research project. Thereafter, additional firms typically join with a minimum of negotiation.

A comparison of the characteristics of participants in bilateral and consortium CRADAs involving NIST is shown in Table 2. These 506 CRADAs were implemented between 1988 and 1994. The first column lists the type or size class of the organization involved. The second column gives the number of bilateral CRADAs between NIST and some other organization. The third column lists CRADAs between NIST and another organization as part of a broader multi-organization, NIST-led consortium. In other words, this column shows that multiple CRADAs may be signed between NIST and different types and sizes of organizations who are members of a single consortium led by NIST.

5. Trends in Technological Infrastructure

A number of the infratechnology research projects, especially those focused on the development of process models, represent a trend toward integration of infratechnologies with the generic technologies that they support. A decade ago the generic technology and infratechnology elements of the typical industrial technology were largely separate. Infratechnologies were important for R&D and post-production acceptance testing. Especially in the case of the latter, they were a highly distinct activity and were often applied by separate organizational units within the firm, such as a metrology or analytical services laboratory.

However, over the intervening period trends toward automation, concurrent engineering, and total quality management have resulted in the integration of infratechnologies and the generic technologies that they leverage. A particularly important example of this trend is the integration of measurement functions into the production process in order to effect real-time process control. This approach enables higher yields and greater quality by correcting production errors as they occur, rather than waiting for the end of the production run to detect unacceptable products.

As a result of the integration of heretofore separate types of infrastructure, mechanisms formerly used primarily for development and dissemination of one type of infrastructure are beginning to be used interchangeably. That is, whereas government funding has been primarily used for advancing generic technologies and government laboratory research largely for developing infratechnologies, combinations of the two mechanisms are now more frequently employed. For example, NIST's Advanced Technology Program (ATP) can allocate up to 10 percent of its funds to the NIST laboratories for infratechnology research to support ATP-funded generic technology research projects. In Japan, tax incentives are used to promote cooperative generic technology research among several firms and both infratechnology and generic technology research projects between firms and government laboratories.

Table 2: CRADAs at NIST: Bilateral and NIST-Led Consortia

Type of Partner	Bilateral CRADAs	NIST-Led Consortia CRADAs
Small firm	160 (46.3%)	40 (24.9%)
Medium and large firm	135 (39.1%)	94 (58.4%)
University	14 (14.1%)	16 (9.9%)
Government	2 (0.6%)	9 (5.6%)
Other (trade associations, professional societies, external consortia)	34 (9.9%)	2 (1.2%)
Total CRADAs (percent of all CRADAs)	345 (68.0%)	161 (32.0%)

Source: NIST, Office of Technology Services.

6. NIST and the Benefits of Government Investment in Infratechnology

Government R&D is viewed by much of the economics literature as complementing private R&D, but the social rate of return has frequently been estimated to be significantly lower than the average for privately-funded R&D (Nadiri 1993). This finding is not surprising given that these studies focused on U.S. Government R&D, which historically has been largely mission-oriented (defense, space, etc.). Although spinoffs to the private economy from mission-oriented government R&D do occur, it is generally recognized that such R&D is not the most efficient way to complement private R&D for the pursuit of economic growth objectives. When market failures affecting private investment in R&D are directly targeted, the social rate of return to government R&D should be much higher. The case of NIST strongly supports this view.

Current strategies for global technology-based competition require that new products and processes be (1) high quality, (2) produced at low unit cost, (3) commercialized in short periods of time, and (4) compatible/interoperable with other products that comprise complex systems (e.g., telecommunications networks and automated factories). The model for implementing these strategies increasingly involves a complex set of technology-based infrastructures. In recognition of this trend, the U.S. Congress created the National Institute of Standards and Technology (NIST) in 1988 by combining the measurement-infrastructure role of the former National Bureau of Standards with a technology-funding role (the Advanced Technology Program) and a technology-transfer role (the Manufacturing Extension Partnership (MEP) Program). In providing these three categories of technological infrastructure, NIST supports all four of the above competitive factors.

NIST develops a wide range of infratechnologies for U.S.-based firms, including basic measurements, standards, production control methods, manufacturing process models, advanced measurement methods, validated technical databases, product performance tests, and quality assurance techniques. These tools are designed to increase industry's ability to rapidly and efficiently develop generic technologies from the underlying science base, develop market applications (products and processes) derived from these generic technologies, and, to rapidly penetrate the markets for those products and processes. Two examples of the scope and impacts of infratechnologies are given below.

6.1 Semiconductors

Semiconductor manufacturers participate in highly competitive markets. Success is determined by relative quality, price, and speed of response to changes in demand. All three of these factors are affected by infratechnologies. At the R&D stage, ability to understand and manipulate materials through precise knowledge of their properties is essential for the development of new generations of semiconducors, which have increasingly smaller physical dimensions but higher performance requirements. The semiconductor industry will not invest in the research necessary to characterize these materials because: (1) such research derives from a different science and technology base than generic semiconductor product technology, which is the industry's main area of expertise; (2) such material characterizations are expensive to develop and certify; and (3) such information cannot be directly embodied in the product, thereby lowering the potential to make it proprietary. Thus, the economies of both scale and scope in this research are typically not captured at the firm or even industry level. Similarly, as R&D proceeds, product characteristics such as electrical resistance and thermal conductivity must be accurately measured to determine if the R&D is, in fact, achieving its goals. Finally, these measurements must be made by standardized methods available to all, if they are to be widely accepted.

At the production level the automation of semiconductor manufacturing to increase both productivity and quality requires numerous types of infratechnology ranging from sensor technologies to data formatting and communications standards. Finally, the actual transactions in which many thousands of chips change hands require assurances to the buyer that performance specifications have been met. The uncertainty with respect to performance can lead to delays in market transactions and, more importantly, higher transactions costs. The result is slower diffusion of the technology in the marketplace. Both time delays and higher prices can allow foreign competitors a chance to gain market share. These transactions costs can be substantially reduced through standardized test methods, for example.

NIST's effort to provide the industry with the next generation of infratechnologies is the National Semiconductor Metrology Program. A significant aspect of this program is that targeted infratechnologies are specified in the semiconductor industry's "roadmap".[9] This is an industry-produced strategic plan for semiconductor technology development in the United States, including technology forecasting, barrier analysis, and requirements for infrastructure support.

6.2 Robotics

A successful robot depends on a number of infratechnologies. The generic technology is aimed at the basic tasks that a robot might perform (sensing, selecting a mode of operation, performing physical tasks, and turning the work piece over to another machine). For robots to perform these tasks effectively and integrate into the broader automated factory system, a number of infratechnologies are needed. Specifically, a task such as sensing (seeing or feeling) depends on independently developed sensor technology that must be adapted to and integrated with the other elements of the robot. However, a subactivity such as sensing is much less effective in terms of the robot's overall performance if the data generated cannot be efficiently transmitted to a control computer and the appropriate instructions sent back. Because the control computer and the sensors are made by different manufacturers, communications between the two will be inefficient and thereby greatly decrease the robot's overall productivity, regardless of how advanced are the generic technologies of the individual components.

Suppliers of robotics technology have little incentive to invest in communications protocols which are not part of their proprietary product technology and actually increase competition with other suppliers. Users of robotics would benefit from the increased competition, but typically lack the technical expertise and the ability to work effectively with competitors.

Such communications inefficiencies are solved by developing *interface standards*, based on various infratechnologies which facilitate data flows. When one considers that these interfaces occur between components of the robot, between the robot and other machines, between all machines and the factory-level (host) computer, and between manufacturing and R&D and

[9] Examples of infratechnologies specified for the National Semiconductor Metrology Program include material and exposure characteristics of photoresists (production), dimensional measurements of the masks that define the printed circuit pattern (production), methods for assessing electromigration of metal atoms used to create conductivity attributes of silicon (R&D), chip performance prediction models for CAD systems (R&D), and methods for verifying on-chip test circuits (market transaction).

marketing divisions of a firm (or several firms), the combined potential productivity impacts of the collection of underlying infratechnologies is easily seen to be substantial.

Another element of robotics where infratechnologies play an essential role is the controller. Controllers are basically on-board computers with accompanying software that perform several control functions for the robot. These control functions fall into two main categories: motion and logic. Motion controllers are responsible for the physical movements of the robot, while logic controllers effect decision-related actions such as changing a tool or turning on a lubrication subsystem. Controllers for robots and other machine tools have proprietary architectures and this market failure has not yet been dealt with, as it has been in the case of personal computers (PCs). The result is higher cost to the users of robotics technology and less than optimal configuration of the entire production system.

The industry is finally beginning to address the need for open systems architecture and is working with NIST to develop the necessary standards. As in the case of the PC, open architecture should significantly expand the availability of modular, easily integrated software and hardware from many vendors. The introduction of the required infratechnologies and associated standards should extend the life and enhance the performance of controllers by allowing replacement of updated modular components. It should also increase the flexibility of the design of the overall production systems when equipment from different vendors can be integrated and then replaced individually, thereby increasing the automated factory's productivity.

7. Estimates of Economic Impacts

The major categories of technological infrastructure (generic technologies, infratechnologies, and technical information) are provided through three basic institutional mechanisms: laboratory research, funding of research in industry, and technology transfer. As part of an effort to document the importance of providing technological infrastructure to industry, NIST has undertaken a series of economic impact studies to estimate its contributions to U.S. industrial competitiveness and also to provide insights into the generic roles of government in supporting competitive technology-based industries.

Table 3 summarizes quantitative estimates from these economic studies and compares the results to two sets of impact studies of industrial R&D

using the same methodology. The methodology used by NIST is also consistent with that in earlier studies of the social rate of return (SRR) from basic research.[10] The rates of return from NIST research are high compared with those found in the NSF studies of cross-sections of private investments in technology and other public technology investments. These results are not surprising, given that NIST targets its research at specific industry-wide infrastructure problems and builds a technology-transfer strategy into each project.

Economic analyses conducted to date have found that the SRR for NIST infratechnology research is about twice the average SRR for private-sector innovations. Assuming the conventional negatively-sloped marginal efficiency of investment curve, the implication is that investment in technological infrastructure of the type provided to industry by NIST is too low relative to private-sector investment in technology. Expansion into related areas, presumably with somewhat lower SRRs, would reduce the average SRR for the project area as a whole until the rate of return approached the SRRs for other types of investment - at least for other types of technology investment.

Table 3: Social Rates of Return to Investments in R&D

Impact Studies by Technology Area	Social Rate of Return (median)
Academic Research (76 firms)	28%
Private-sector innovations (38 innovations)	78%
NIST infratechnology research (11 projects)	167%

Source: See footnote 10.

[10]The social rate of return used in these evaluations is defined as the discount rate that reduces to zero the net present value of the flow of economic benefits generated by the investment over a specified period of time (Ruegg and Marshall 1990). The estimates of the SRR from academic research are from Mansfield (1991a, 1991b), estimates of the SRR from industrial innovations are from Mansfield (1977) and Tewksbury *et al*. (1980), and the NIST estimates are from various contractor/consultant reports.

Some other studies of the impacts of government laboratories have been attempted using bibliometric techniques. While such approaches may provide some information on the usefulness of government research to other researchers and hence on the productivity of research, or when patents are tracked, on the *potential* for economic impact, they do not provide information on the relative economic impacts actually achieved.

An example of these impacts can be seen in NIST's development and transfer of measurement infratechnologies to the U.S. optical fiber industry, serving as the basis for over 20 standards. Users of optical fibers - primarily communications systems companies - have demanded increasingly higher levels of performance. Performance variables, such as attenuation (signal loss), backscatter, and core diameter are difficult to measure accurately. The result in the past has been disagreements between optical fiber manufacturers and their customers (communications systems firms). These disagreements cause delays in the diffusion of optical fiber technology. For example, the rate of signal loss is extremely important to communications systems firms because it determines the number of amplifiers that must be placed in a telecommunications network. Each amplifier adds to the overall cost of the system and the total number required, therefore greatly affecting the private-sector investment decision.

In the case of optical fibers, the measurement of product performance was so important to executing market transactions that the industry and NIST focused almost entirely on developing standards that target market transaction costs and hence the market penetration stage of economic activity. A 1991 study of the economic impacts of NIST's infratechnology research in support of the U.S. optical fiber industry estimated a social rate of return of over 400 percent (the highest to date of any study of NIST impacts on U.S. industry). Moreover, during the study an official of a leading optical fiber manufacturer stated that not only had the NIST research produced the documented transactions-cost savings, but that the reductions in performance uncertainty had resulted in an optical fiber market that is 25 percent larger than would otherwise be the case. (Link 1991a).

In summary, the high social rates of return estimated for technology investments leads one to conclude that more resources should be allocated to investment in all categories of technology research, including the public technology elements.[11] Also, the estimated high macroeconomic rates of return to investment in technology, while providing a general rationale for increased investment in R&D, do not provide the ability to allocate funds *among* technologies. Such resource allocation is a *microeconomic* process, requiring analysis by both industry and government at the technology and industry

[11]The rate of return to investment generally in the U.S. economy is in the range of 8-14%, depending on the particular measures of profits and investment chosen. This estimate is an average as opposed to a marginal rate of return for capital, but it gives an approximate reference point for comparison purposes.

levels. In other words, strategic planning is required - using economic impact studies such as those cited in the above table as one input.

8. Conclusions

Infratechnologies increase the efficiency with which technology-based economic activity is conducted. They act collectively as a leveraging agent on all three stages of the technology-based economic process: R&D, production, and market development.

The multifaceted economic role of infratechnologies increases the economic, and hence, policy importance of this type of infrastructure. However, industry underinvests in them because (1) their efficient development frequently requires the capture of significant economies of scope (in R&D) and (2) the necessity of their widespread use (frequently as industry standards) prevents proprietary ownership by individual corporations. Moreover, measurement, test, and calibration methods must frequently be traceable to a technically credible and competitively neutral entity, usually a government laboratory or institute.

Governments usually underestimate the importance of infratechnologies. Several industrialized nations have recently considered reducing or even eliminating existing support in this area. However, as the typical product life cycle shortens and as demands for product flexibility and greater productivity and quality increase, governments will need to supply more rather than less of this category of technology-based infrastructure.

With respect to future research, the ongoing economic impact studies at NIST show high rates of return to investment in this category of technological infrastructure, but direct economic impact data do not provide much information on how infratechnologies interact with the proprietary technology strategies of industrial firms or with other elements of technological infrastructure within different corporate organizational or industry structures. Modeling the interactions among different categories of infrastructure and market-specific technology development strategies under different organizational structures is needed.

References

Aschauer, D.A. 1988. Rx for productivity: Build infrastructure. *Chicago Fed Letter* (September).

—. D.A. 1989. Is public expenditure productive? *Journal of Monetary Economics* 23: 177-200.

Cordes, J. and N. Vonortos. 1994. *Factors Affecting Demand for NIST Calibration Services.* Gaithersburg, MD: The National Institute of Standards and Technology.

Eatwell, J., M. Milgate, and P. Newman. 1987. *The New Palgrave: A Dictionary of Economics*. London: Macmillan Press.

Justman, M. and M. Teubal. 1993. *Technological Infrastructure Policy (TIP): Creating Capabilities and Building Markets.* Jerusalem: The Jerusalem Institute for Israel Studies.

Hawkins, R. and R. Mansell, eds. Forthcoming. *Standards, Innovation, Competitiveness: the Politics and Economics of Standards in Natural and Technical Environments.* London.

Lichtenberg, F. 1993. R&D investment and international productivity differences. In *Economic Growth in the World Economy*, edited by H. Siebert. Tubingen: J.C.B. Mohr.

Link, A.N. and G. Tassey. 1987. *Strategies for Technology-Based Competition: Meeting the New Global Challenge.* Lexington, MA: Lexington Books.

Mansfield, E. 1977. Social and private rates of return from industrial innovation. *Quarterly Journal of Economics* 91: 221-240.

—. 1991a. The social rate of return from academic research. *American Economic Review* 81.

—. 1991b. Academic research and industrial innovation. *Research Policy* 20: 1-12.

Nadiri, M.I. 1993. Innovations and technological spillovers. NBER Working Paper No. 4423, National Bureau of Economic Research. New York.

Ruegg, R. and H. Marshall. 1990. *Building Economics: Theory and Practice.* New York: Van Norstrand Reinhold.

Quick, F. and Associates. 1990. *U.S. Investment Strategies for Quality Assurance.* Gaithersburg, MD: Program Office, National Institute of Standards and Technology.

Tassey, G. 1982. Infratechnologies and the role of government. *Technological Forecasting and Social Change* 21: 163-180.

—. 1991. The functions of technology infrastructure in a competitive economy. *Research Policy* 20: 345-361.

—. 1992. *Technology Infrastructure and Competition Position.* Norwell, MA: Kluwer Academic Publishers.

Tewksbury, J., M. Crandall, and W. Clare. 1980. Measuring the societal benefits of innovation. *Science* 209: 658-662.

U.S. Congress, Office of Technology Assessment. 1992. *Global Standards: Building Blocks for the Future*, TCT-512. Washington, D.C.: U.S. Government Printing Office.

Information Distribution and the Growth of Economically Valuable Knowledge: A Rationale for Technological Infrastructure Policies

Paul A. David and Dominique Foray*

Science and technology policies in the West in recent years have been directed toward fostering the generation of new knowledge as the basis for commercially exploitable innovations, rather than toward improving the distribution of existing scientific and engineering knowledge and increasing the accessibility of the latest additions to the knowledge stock. This particular policy thrust has been maintained for too long, and there now is a pressing need to restore some balance; in other words, to undertake measures that would raise not only the marginal social rate of return on future R&D investments, but also would increase the social payoffs from past R&D expenditures.

This chapter demonstrates that a key determinant of the economic performance of an innovation system is its capacity to effectively and efficiently distribute knowledge; we call this the *knowledge distribution power* of the system. Among the factors that determine the distributive power of the system are: the nature of intermediation between knowledge holders and users of knowledge, the reward structures and institutions that support the creation and transfer of knowledge, and the openness of the information system. Openness is a particularly important characteristic of knowledge distribution power because it is vital for the efficient use of costly research resources in creating reliable knowledge. Open access that distributes knowledge widely and rapidly:

- facilitates independent replication of findings;
- promotes swift generalization of results; and
- avoids excessive duplication of research.

* This paper draws upon the Report "Accessing and expanding the science and technology knowledge base" (DSTI/STP/TIP(94)4) prepared by Paul A. David and D. Foray in the context of the OECD project on national innovation systems. An expanded version of this chapter is published as David and Foray 1995a.

M. Teubal et al. (eds.), Technological Infrastructure Policy, 87–116.

Utilization of existing knowledge stocks, both in production applications and as the basis for further research, requires greater attention in order to counterbalance conventional policies that give incentives to innovators by enabling them to appropriate economic rents on new knowledge. *Technological infrastructure policy* (TIP) should, therefore, focus on knowledge distribution power so as to provide potential innovators with timely and easy access to relevant knowledge bases.

In the following section we describe the economics of knowledge distribution, identifying the important relationship between the stock of knowledge and the temporal flow of innovation. This is not a purely definitional relationship as in most endogenous growth models, but rather one that is shaped by the institutions that mediate between the stock of knowledge and the flow of innovations.[1] We show that the shaping of that relationship is an *organizational and institutional problem*, best understood in the context of the *optimal utilization of a nonrival good*.[2] Subsequent sections focus on a variety of factors that govern the ability of knowledge to be distributed in the system, and on particular policy issues pertinent to improving knowledge distribution. The final section provides a synthesis of TIP options that are designed to increase the knowledge distribution power of an innovation system.

1. The Economics of Knowledge Distribution: Interactions and Positive Externalities

Knowledge usually is described by economists as a *nonrival good*, that is, a good which is infinitely expansible without being diminished in quality, so that it can be possessed and used jointly by as many as care to do so.[3] An economic good that can be concurrently possessed and utilized by all cannot be treated as if its transfer was a zero-sum proposition. The corollary to this property is that once knowledge is generated, there is no value added from

[1]In most endogenous growth models that deal with the accumulation of knowledge, the flow of innovation is the time derivative of the stock of knowledge (Romer 1994; d'Autume and Michel 1994).

[2]The notion of *optimum utilization of knowledge* was first employed by Machlup (1983).

[3]On historical precursors of the modern appreciation of the infinite expansibility of ideas, see David 1993. Callon (1994) offers a discussion of the notion of nonrival goods from a sociological point of view.

subsequent discoveries; the first discoverer is the sole contributor to social surplus (Dasgupta 1987).[4]

The process of knowledge generation in science and technology is cumulative and integrative. Knowledge cannot only be considered an output; it is also the main input of any process of knowledge generation. Cumulative forms of knowledge are those in which today's advances lay the foundation for succeeding rounds of progress.[5] A synthetic knowledge is generated through convergence or collateral integration between previously independent pieces of knowledge. Thus, for many industries advances take place in a generational process; new products or processes are not radical departures from existing modes, but rather build on and extend the knowledge and technology used in the production of the products or processes they supersede. This property gives rise to successive improvements defining a trajectory.

Due to these properties the process of knowledge generation produces positive learning externalities. The generation of a new piece of knowledge increases the probability of creating useful new products, processes, and ideas arising from novel and unanticipated combinations. As Machlup (1984) put it, "the more is invented the easier it becomes to invent still more."

These positive learning externalities are based on various knowledge interactions. Following David and Foray (1994) let us compare production of knowledge to the discovery of new mineral deposits. First, some knowledge is like surveying; it generates maps that raise the return to further investment in exploration and exploitation (David, Mowery and Steinmueller 1992).[6] Second, it has been observed that mineral production in an area leads to localization of exploration, so that at least for some time, mineral reserves

[4]Alternatively, one may consider the property of *indivisibility in use* by looking at research from a process point of view rather than from a result perspective: scientific research tasks do not exhibit the property of indivisibility in the sense of needing to be performed only once. Their reproduction and replication in many laboratories is a necessary part of the generation of reliable knowledge (Foray 1992).

[5]Scotchmer (1991) and Merges and Nelson (1994) accept the same definition of *cumulativeness*, a concept which can be traced back at least as far as Arrow (1962). In economics literature it is commonly noted that information as an output of research, also serves a major role as an input into further research.

[6]The extent of spillovers from certain kinds of knowledge may be dependent also on the structure of the phenomena to which the knowledge pertains. For example, in David, Mowery and Steinmueller (1992), the applicability of "homotopy", "metanymy", and the absence of "lumpyness" (i.e., indivisibility) suggest that there will be substantial spillover effects from knowledge acquisition. But these are not universal characteristics of every branch of science and technology.

increase in the territory where exploitation is underway (David and Wright 1992). When discovery increases the probability that others will undertake exploration in the neighborhood, producing knowledge is likely to generate positive externalities for the explorers and each agent has an interest in diffusing the product of his discovery so as to profit from the results of others. At the very least, information about where others have failed to make a discovery will be valuable in guiding one's own search. The third aspect of positive externalities deals with the migration of young investigators into new fields – colonizing new areas with tools and concepts developed elsewhere. As in the case of exploitation of minerals, this "frontier expansion" requires falling yields in colonized areas to generate breakthroughs into new territory.

There has been a recent resurgence of the view that in the realm of knowledge production, positive externalities dominate. This is reflected in Romer's (1993) emphasis upon not only the nonrival character of ideas available for application in conventional production, and the consequent increasing returns to investment in R&D, but also in the notion that new knowledge promotes increasing product specialization and essentially unbounded growth of productivity and welfare improvement from that source. Scotchmer's (1991) emphasis on the cumulative nature of inventive activity is part of the same reorientation of economic analysis away from insistence on diminishing margin returns in all directions of the convexified world. David (1993) draws attention to the impediments that intellectual property rights protection poses for innovation, based on the notion that a major application of new knowledge is in the further production of knowledge; he points out that this has greatest force in situations in which the returns to increased "round-aboutness" of knowledge production are not diminishing at the margin, but rather increasing.[7]

Thus, we may conclude that there is a particular causal relationship between the stock of existing scientific and technological knowledge and the flow of innovation at a given period of time. This relation requires qualification; to exploit the opportunity of innovation provided by the stock of knowledge it is necessary to provide low-cost access to existing and new knowledge.

[7]Knowledge interactions can also produce a *bounded learning* phenomenon, reducing the magnitude of the positive learning externalities. What can be learned at any one point in time is strictly bounded. Scientific and technological advances are interdependent and the fact that a crucial knowledge process may be missing may diminish the returns to further advances in other areas. This consideration has been spelled out by David (1975) as an argument for the localization of technical progress in a core around the existing technique; one can't advance except in a balanced way, which means that a constraint on advance in one element of a system will block all advance, rapidly diminishing the R&D effort.

1.1 Access to Knowledge

An efficient system of distribution and access to knowledge is a *sine qua non* condition for increasing the amount of innovative opportunities. Wider distribution of knowledge and timely, inexpensive access to new findings reduce wasteful duplication of research efforts; putting information into the hands of a more diverse population of researchers also tends to increase the probability of useful new products and processes arising from novel and unanticipated combinations. Thus, *knowledge distribution is the crucial issue of the relation between the stock of knowledge and the flow of innovation.*

The conditions for the efficient distribution and utilization of knowledge cannot be expected to arise automatically from the interplay of market forces. Various obstacles impede the distribution power of a system of innovation:

- First, building up information structures generates access costs. These costs increase as stocks of knowledge evolve and become more specialized, requiring mastery of new technologies, notational conventions, and conceptual apparatuses. Beyond the contextual setup costs, there are additional costs of managing the ever-growing data streams arising from the exponential expansion of the volume of codified knowledge. Thus, the actual search for information carried out by any particular research project is most often quite localized, carried out within a limited portion of the total space that could, in principle, be explored.
- Second, a great deal of knowledge is tacit, inseparable from the agents and organizations which developed it. The transfer and acquisition of tacit knowledge constitute costly operations that require the active participation of the knowledge holders.
- Third, the creation of intellectual property rights – designed in order to allow agents to capture rents from their innovative efforts – create access problems that impede distribution.
- Fourth, institutional incompatibilities among organizations impede the flow of knowledge transfers as the result of distinctive rules of disclosure, institutional goals, and reward structures. The compatibility problem is especially severe when knowledge distribution calls for institutional "border crossings".

The activity of diffusing economically relevant knowledge is not itself a natural one. Rather, it is socially constructed through the creation of appropriate institutions and conventions, such as open science and intellectual property rights (David 1994). Improving what David and Foray (1994) termed the *distribution power* of the system of innovation often has been portrayed as a desirable objective that has to be sacrificed in order to provide stronger market incentives for private investment in organized R&D, since copyright, patent, and trade-secrecy laws create obstacles to access and restrict the commercial utilization of knowledge. But those same impediments also stand in the way of using existing knowledge for innovation itself. These market failures result in knowledge being used in only a tiny part of the potential space for its exploitation, limiting the realization of the potential external benefits which arise from the nonrival nature of knowledge.

1.2 Institutional Intermediation

Policymakers seeking to speed up the generation of new products and processes might well consider modifying the workings of institutions that deal with intellectual property in ways that speed disclosure and reduce the marginal costs of access to new technological information for purposes of research. Efforts could also be made to manage intellectual property as a vehicle for cooperation and investment coordination to expand productive capacities, rather than for the capture of monopoly rents as is currently the case. There may well be a need for specialized "bridging" institutions and technology-transfer incentives, designed expressly for the purpose of supporting swift and low-cost access to existing knowledge stocks in particular fields; the agricultural experiment and extension stations that were introduced to serve the technological needs of the U.S. farm industry are a classic example. But specific reforms of legal and other institutions cannot be undertaken without considering how they will interact with diverse elements of the institutional infrastructure. Because sweeping reforms tend to impose heavy costs in terms of uncertainties about formal and informal property rights and incentive structures, incremental changes are more likely to be favored. The process of change in the institutional infrastructure tends to be evolutionary and path-dependent (David 1992). Thus, even before considering the possibilities of fundamental innovation in the institutional sphere, it is important to explore the possibilities of making further use of the capabilities of existing institutional infrastructures.

2. Knowledge, Institutions, and Knowledge Distribution Power

What governs a system's capability of distributing scientific and technological knowledge? The key factors are those concerning the social organization and reward structures of the institutions supporting the generation of such knowledge. Rather than emphasizing epistemological distinctions between *technology* and *science*, we find it more illuminating to consider the implications that the modes of organizing research activities have for the costs of transferring the knowledge that is produced. Following the analytical framework developed by Dasgupta and David (1994) and further elaborated in David and Foray (1995a), this particular conception of economically valuable knowledge focuses on the characteristics of knowledge affecting production and acquisition, such as the degree of explicit codification (contrasted with tacitness), the degree of publicness (contrasted with privateness), and the degree of disclosure (contrasted with secrecy). These three dimensions define a space within which types of knowledge and the agencies creating them can be located (see Figure 1).

It is important to stress that these characteristics are not inherent in the knowledge itself but are rather the product of social organizations and of the attendant reward structures of those institutions supporting the production of knowledge.[8] The true nature of new knowledge does not stem from any intrinsic differences between knowledge that is scientific rather than technological, nor between basic and applied scientific knowledge. The critical factors governing the distribution and the utilization of new findings are those regarding the rule structures and behavioral norms about information

[8]Thus, as Dasgupta and David (1994) suggest, what gets brought into focus (and codified) and what remains in the background (as tacit knowledge) is to be explained endogenously, by considering the structure(s) of pecuniary and non-pecuniary rewards and costs facing the agents involved. Similarly, if it is true that "scientific research is 'theory dependant' while engineering research often operates beyond the bounds illuminated by prevailing theory and the accumulation of understanding takes place in localized experiments of a trial and error kind" (Metcalfe 1992), then the degree of theoretical dependance is no more than a social norm, a product of the competition between scientific communities. It is thus possible to explain the changes in the norms of academic engineering training in the present century in regard to dependence on formal mathematical modeling. There are, as with technical standards, degrees of incompatibility (of freedom) in the accumulation of scientific knowledge (Loasby 1986).

disclosure that dominate in the particular social organizations within which the new knowledge is found or improved.[9]

Figure 1: The Knowledge-Product Space

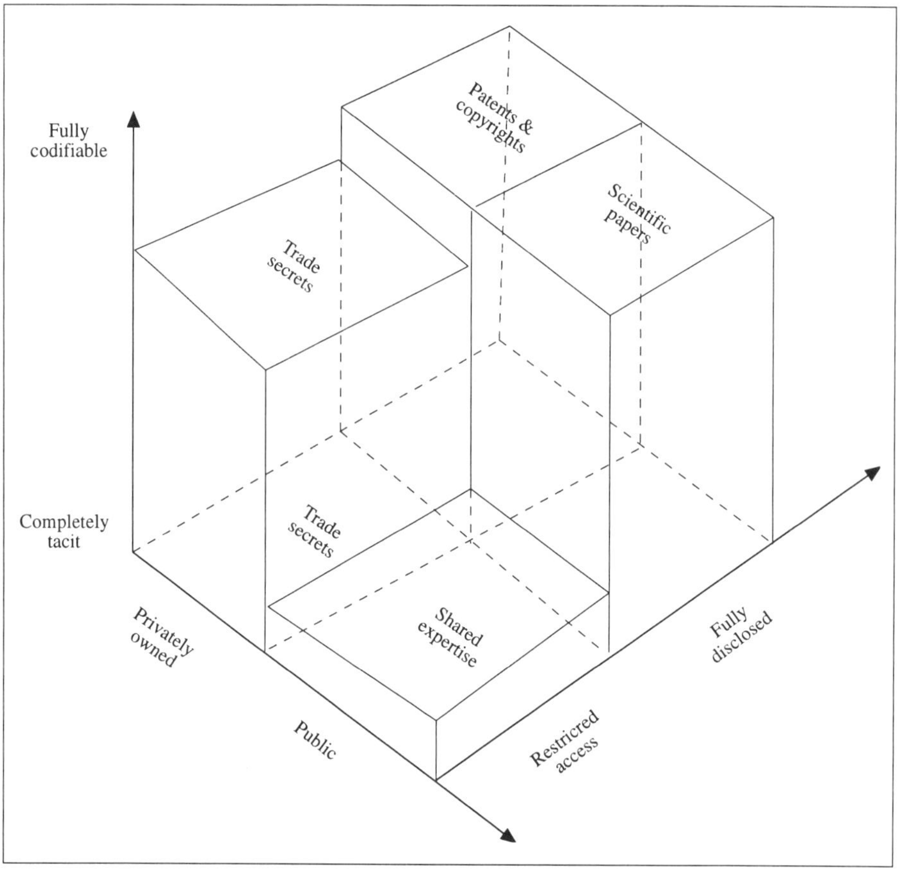

2.1 Codification and Tacitness

Codification of knowledge is a step in the process of reduction and conversion which renders the transmission, verification, storage, and reproduction of information especially easy. Codified information is typically organized and expressed in compact and standardized format in order to facilitate and reduce

[9]However, while these characteristics are endogenous and not epistemologically grounded, David, Mowery and Steinmueller (1992) have pointed out that epistemological considerations do bear on the extent of positive externalities (see footnote 5).

the cost of these operations. In contrast, tacit knowledge cannot be dissociated from the work practices of research and production units. Tacit knowledge is acquired experientially and transferred by demonstration, rather than being reduced – either immediately or eventually – to explicit and codified methods and procedures. Tacit knowledge therefore cannot readily be expressed outside the production context in which it is generated (David 1993; David and Foray 1995a).

Codification and tacitness may be treated as locations along a continuum whose endpoints are represented by *complete systematization* on one side and by *procedural knowledge* on the other. Complete systematization of cognitive content uses a vocabulary of defined terms such that the interpretation of the text is unambiguous. Unsystematized procedural knowledge exists as expertise that particular individuals possess. On the demand side, insofar as codified and tacit knowledge are substitutable inputs (at the margin) both in the production of further knowledge and in practical implementations, the relative proportions in which they are used are likely to reflect their relative costs and availability to the users. Similarly, on the supply side, differences in the extent to which knowledge generated by researchers in various fields gets codified for packaging as information, rather than retained in a tacit form, will reflect both the reward structures within which researchers are working and the costs of codification.

The standardization of language and expressions under different forms strongly reduces the costs of codification. Preexistent standards of reference (numerical, symbolic, pictorial, geometrical languages, and taxonomies of any kinds), performance standards, and vocabularies of precisely defined and commonly understood terms all contribute greatly to reducing the time and effort required to produce unambiguously codified messages. Protocol and format standards for the communication of information also play a strategic role in codification. When available in a timely fashion they greatly increase the incentives and the opportunities to move toward codified forms of knowledge. Hence, variations in the relative importance of codified and tacit knowledge in the work of different research communities has no necessary connection with the "hardness" or "softness" of their respective disciplines.

The digital revolution has continued and intensified the move toward codification (Cowan and Foray 1995; Ergas 1994; Arora and Gambardella 1994). The revolution in information technology has substantially influenced the way in which knowledge is codified by such advances as the substitution of graphic representation for natural language and the development of expert systems. The development of new algorithms, one the greatest achievements of modern mathematics, has greatly enhanced capabilities to formally represent or codify knowledge.[10]

2.2 Ownership Status

The ownership status of knowledge should also be treated as a continuous variable rather than as a discrete state of legally public or private property. Different intellectual property rights systems provide knowledge holders with bundles of rights that approach the concept of perfect property to varying extents. Patent rights, for example, are temporally delimited and may be granted subject to requirements of compulsory licensing at "reasonable" fees. Copyrights, likewise, exist for only a fixed period and do not protect against use of the material by independent originators – a trespass against which patents do offer recourse. Even within the patent system there are big differences among the national systems regarding scope, length, and disclosure implications (Foray 1995a; Ordover 1991).

New systems of property rights are emerging that warrant protection for innovators while maintaining zero-marginal-cost access for research use. A recent example of such an institutional arrangement is the system of prepaid, lump-sum access fees in biotechnology. Its successful introduction attests to the possibility of according protection to an inventor, while at the same time maintaining zero-marginal-cost access to knowledge-products required for research. This model maintains the free circulation of genetic resources through a system of dependency licenses in the case of dependent innovations (Joly and Hermitte 1993). Another example of an intermediate form involves trade-secrecy law, which some legal theorists maintain does not create property rights. They argue that in as much as a secret is something that cannot be disclosed in public without destroying, it thus cannot be described sufficiently to enable identification of its nature in a determination of ownership at any point in time. Under this interpretation, knowledge can be protected as a trade secret because the law protects the right to enter contractual relationships of confidentiality. Following such a reading, we would not characterize trade secrets as *private* knowledge.

[10]These new algorithms share the characteristic that while the analytical complexity of the problems being solved increases exponentially in the number of variables, the computational requirements of the algorithm increase by some linear function of the problem's size (Ergas 1994).

2.3 The Disclosure–Secrecy Dimension

The extent of disclosure is a continuous variable bounded by full disclosure at one limit and total secrecy at the other. The degree of disclosure required is not uniform across intellectual property rights regimes, and even with a given regime, different kinds of text may be protected by varying completeness of disclosure. Computer software, for example, may be copyrighted without revealing the source code, and in some instances even the full body of object code does not have to be disclosed. Standards of disclosure may be defined not only by the statute laws and intellectual-property agencies such as patent offices, but also by the policies of the journals in which scientific papers are published. For example, professional journals may or may not insist upon disclosure of the exact coordinates of complex proteins whose molecular structure is being reported, or require disclosure of the computational algorithm used in analyzing experimental observations.

2.4 Factors Determining the Knowledge Distribution Power

As the preceding sections illustrate, the characterization of knowledge is a complex matter that is defined by the detailed stipulations of the intellectual property rights regime, and by the norms governing knowledge disclosure among members of different social organizations engaged in research activities.

There is, however, no deterministic characterization of a knowledge-product with respect to the institution that supports its generation. Institutions and norms of behavior have a certain degree of suppleness, which means that knowledge generated in a particular institution can be expressed in a broad range of forms depending upon the incentive structures and institutional compromises which characterize the institutional context. For example, knowledge-products created in a university environment may take the form of scientific papers, patents, or shared expertise, while access may be restricted or delayed. Thus, a great deal of detailed institutional information is required in order to undertake a comparative study of the character of knowledge-products that different research communities create. For this reason, analysis of the distributional capabilities of a system requires a detailed description of the elementary social mechanisms that support coordination among agents engaged in research and innovative activities.

The distribution power of a system can be assessed by reference to a certain configuration of norms, institutions, and incentive structures. A system with high distribution power would be characterized by the predominance of the norm of disclosure (versus secrecy and access restriction), strong incentives toward codification (versus tacitness in maintaining the knowledge stock), and an intellectual property rights systems that enhances disclosure and coordination functions (such as systems that maintain free access to new findings for research use). Such a core of norms and institutions can reinforce the collective perception that it is pro-innovative, reducing obstacles to information dissemination due to properties and secrecy, and replacing them with devices for coordinating the actions of innovative agents. In such a system the distribution of knowledge is established as an institutional goal: universities act as open nodes in global information networks; firms diffuse their new practices and findings very quickly through a variety of market and nonmarket mechanisms; and researchers from different organizations and locations are drawn easily into collaborative institutional activities (standardization committees, cooperative research networks) that enable them to solve coordination problems posed by collective actions.[11]

The norms of openness and free access are very fragile, however, in the case of knowledge. Policy measures that make the knowledge stock more socially useful will in many instances conflict with those measures designed to make it more economically valuable to private parties. This is especially likely in the short run. There is, therefore, a persistent, market-driven pressure toward reducing the distribution power of the innovation system by altering the norms, incentive mechanisms, and property rights systems in directions that encourage noncooperative modes of pursuing and transmitting scientific and technological knowledge.

A distribution-oriented system is one whose institutions, incentive mechanisms, and coordination arrangements have the following three proximate objectives:

- limiting the extension of intellectual property rights that erode the public-good nature of knowledge and therefore impede its distribution, into fields and sectors that were initially organized under noncommercial norms of openness and cooperation – examples include university research and sectors of emerging science-based technologies;
- designing systems of property rights in ways that do not impede the pooling of knowledge, since intellectual-property systems have disclosure

[11]See David and Foray (1995b) for an analytical exploration of the question how a system can get locked into a dominant convention of openness, disclosure, and knowledge distribution.

features that can be developed in ways supporting coordination of the actions of innovative agents; and

- improving the efficiency of information search and evaluation in view of the incipient increased costs of storing, retrieving, and using knowledge caused by the exponential expansion of the science and technology knowledge-base.

3. Coping with the Extension of Property

One crucial aspect of the extension of the system of private property into the realm of scientific and technological knowledge is the penetration of market-driven rivalry into domains of knowledge production where rivalry among researchers or among organizations was previously restrained under the noncommercial norms of openness and cooperation. Thus, an essential goal of TIP is to counterbalance this tendency by providing incentive mechanisms and creating institutions that allow the system to maintain its distribution power. There are two domains where this function is an issue of great importance: in university research and in sectors of emerging science-based technologies.

3.1 The Penetration of Market-Driven Rivalry into University Research

A number of studies have examined the relationship between university and industrial research. On the side of cooperation, three important roles for universities emerge:

- university research produces knowledge that is potentially useful for industry, including scientific results, engineering technologies, prototypes, scientific instrumentation, software, and infratechnologies;
- the university sector establishes, supports, and maintains information networks for the circulation of knowledge that firms connect to and exploit;
- universities produce human capital utilized by industry, through the training of researchers, by enabling the establishment of scientific reputations, and by providing retraining in areas where knowledge-obsolescence is rapid.

Reciprocally, university efforts to advance commercial technology often end up stimulating basic research because of the bottlenecks and opportunities for basic science that are inevitably encountered. Thus, short-run reallocations of resources from basic toward applied research need not undermine basic science in the long run (Rosenberg and Nelson 1993).

Conversely, David, Mowery, and Steinmueller (1994) have identified sources of inherent conflict between university and industry research institutions. These involve the incompatibility of the reward systems which underlie the organization of research each domain. The social organization of university research is based upon *open science*, a term that expresses the idea of free pursuit and open disclosure of knowledge. This is accompanied by a system of reputational reward and resource allocation based on validated claims to priority in discovery or invention. By contrast, achieving commercial gain from knowledge in the industrial sector requires rewards and resource allocation based on the appropriation and control of knowledge for competitive advantage. This inherent tension in cultures means that:

> ...university participation in industrial research involves a series of restrictions aimed at balancing, regulating, and containing these conflicts, including guarantees of (eventual) disclosure of research after predetermined delays, negotiation of intellectual property rights deriving from research outcomes, and scrutiny of potential personal conflicts of interest in faculty research activities (14).

University involvement in commercial applications of research often influences and even distorts the traditional norms of university-based research, with potential adverse implications for knowledge distribution power. Empirical studies have identified a variety of manifestations of this problem, including:

- the expansion of university patenting and research contract activities and its implication for the organization of research (Henderson, Jaffe, and Tratjenberg 1994);
- the increasing industrial influence and control on the research agenda of university laboratories through the expansion of university–industry research centers (Cohen, Florida, and Goe 1994);
- the adoption of the practice of restricting access to so-called intermediate results, that is, research-related information that cannot be published in journals, such as experimental materials and innovative instruments, locally produced software, and data sets (Hilgartner and Brandt-Rauf 1994); and
- the increasing tendency of scientific publications to not fully disclose relevant findings that would permit replication by other scientists, as evidenced by the decreasing proportion of published molecular structures whose coordinates are disclosed, and by the reduced disclosure of

algorithms used in mass spectrographic analysis (Dasgupta and David 1994).

These assorted examples are particularly worrying since universities are logical candidates for enhancement rather than curtailment of their open-node function in global scientific and technological information networks. As observed by David, Mowery and Steinmueller (1994), the interaction between university and industry is likely to prove most dysfunctional when universities internalize the norms of highly proprietary research. This can occur either because the research is closely linked to the strategic aims of the collaborating firm, or when universities adopt such practices for their own financial interests. Through their role of preserving, interpreting, and transmitting knowledge, universities have always been a strong element in the institutional infrastructure designed to improve the distribution power of the systems of innovation. Furthermore, the traditions of cooperative, disinterested scholarship that the academies support, along with the institutional norms of open-science communities to which university regulations have been adapted, make these institutions particularly well suited to the tasks of integrating internationally distributed sources of knowledge and training people how to locate and use it. However, the tendency for universities to be increasingly involved in industrial collaborations and commercial applications of research, leads us to be concerned about their ability to continue to fulfill these essential roles.

3.2 Spontaneous Transformation of the Appropriability Regime[12]

Increasing fragmentation of the knowledge stock as the result of market-induced restrictions on access may also arise within individual sectors where rivalry previously was restrained under *information-pooling* and openness. Some recent cases attest to the possibility of a spontaneous diffusion of a norm of secrecy and access restriction in such systems.

The current transformation in the software industry from a comparatively weak intellectual property rights regime to a much more restrictive regime is one example. Originally, protection of intellectual property rights in the software industry was accomplished by resort to copyright law; this institution prohibits users of software from making unauthorized copies, preventing the appropriation and resale of the intellectual property. But the existence of a copyright does not prevent other programmers from using algorithms or

[12]Substantial parts of this section draw upon Kahin (1990) and Garfinkel, Stallman, and Kapor (1991).

techniques contained in the program in their own work. A single software technique can be implemented in different ways to do totally different jobs. Thus, the institutional system of the software industry was diffusion-oriented, encouraging information dissemination and facilitating coordination among the innovative projects. Even though trade secrecy was used extensively for program source code in the software industry, technical confidentiality was not considered to be a significant problem. Under this regime, it was normal for computer scientists in both the commercial and academic sectors to publish their discoveries, with universities playing a critical role by developing software and distributing it without charge. The system was thus perceived as being rich in positive externalities, in which norms of cooperative behavior and the formal institution of copyright provided self-reinforcing mechanisms to lock the system into to this dominant convention of openness and disclosure.

Within the last few years, however, software developers have been surprised to learn that hundreds, even thousands, of patents have been awarded for programming processes ranging from machine-instruction sequences to user-interface features. Many of the patents cover processes that seem conventional or obvious, and developers now fear that any of the thousands of individual processes in their programs may be subject to patent-infringement claims. There are two notable features of this transition phase in the intellectual property rights regime of the industry: first, it is a process which exhibits the main characteristics of a spontaneous mode of diffusion; and second, the process is conducive to a dramatic disruption of the cooperative research networks that sustained the dissemination of information and the coordination of activities among innovators.

The spontaneity of the process is caused by chains of reprisal which encourage and reinforce the new form of behavior, turning it into a norm. (David and Foray (1995b) suggest that this process can be formally analyzed as a process of additive interdependent Markov chains.) Thus, although very few programmers and entrepreneurs believe that patents are necessary for improving the social organization of R&D in the industry, there is a proliferation of patents. This paradox is due to the fact that most patenting by companies is done in order to have something to trade or as a defense against other patent-infringement suits. As a result, a new game, with new rules and new norms of behavior, is emerging spontaneously and rapidly.

Second, this new game leads to a radical disruption of the networks of cooperation. The expansion of the use of software patents could mean the end of software developed at universities and distributed without charge.

Patents could also mean an end to public-domain software which has played an important part in making computers affordable to public schools. Software patents would pose a special danger to small companies which often form the basic structure of software development but cannot afford the cost of patent searches for litigation. And finally, software patents introduce a major timing problem in an industry where product cycles are very short; application times for software patents take an average of 32 months.

A crucial question arising from this story is whether the industry is evolving toward a cross-licensing system where firms secure narrow patents to trade for the rights to other patents; such a system might be capable of sustaining the existing cooperative networks. However, given the characteristics of the product – modern software packages may contain thousands of separately patentable processes, each of which adds to the risk of infringing on patents that are already in the pipeline – it is hard to imagine that the patent system will be able to become a positive coordinating factor in this industry.[13]

4. A New Role for Intellectual Property Rights

Protection of intellectual property rights need not be an obstacle to the distribution of knowledge. Registration of intellectual property requires disclosure, and procedures can be designed to induce rapid disclosure and wide dissemination of innovations. Furthermore, protection may be granted in ways that encourage patentees and copyright-holders to make their innovations available for use by others at reasonable cost. This latter goal can be approached by restricting both the breadth or scope of the patent grant, and the height or degree of novelty required of the typical patentable innovation. Narrow patents, and patents involving low degrees of novelty tend to favor re-use and recombination of existing knowledge stocks because they promote pooling and cross-licensing of innovations (Ordover 1991). Other means are available too, such as introducing *sui generis* legislation adapting property rights specifically to meet the needs of sectors (e.g., software) where the recombination model of innovation is particularly important, and extending the principles of compulsory patent and copyright licensing at "reasonable cost" (see the example of the system of prepaid, lump-sum access fees discussed below). The dilemma of trying to design intellectual property rights protection

[13]On the opportunity of creating a new kind of intellectual property for computer programs, see Samuelson (1985) and Schmidt (1985).

so that it enhances the distribution power of knowledge is that such measures tend to diminish the private value of individual increments to the privately owned knowledge base, thus reducing incentives to their creation, while at the same time they raise the social value of those contributions.

At the other extreme of the disclosure spectrum is trade secrecy, which directly inhibits knowledge distribution; examples include the various forms of access restriction found in private firms and in scientific institutions engaged in military research. Strong trade-secrecy laws reinforce the attractiveness of holding back R&D results until filing broad patent applications on largely novel systems. Aside from a strategy of directly weakening trade-secrecy protection, two policy other approaches can be utilized to enhance knowledge distribution. One approach is to make the forms of intellectual property rights protection that require detailed disclosure more attractive than secrecy; this can be accomplished by providing assistance and expertise in patenting procedures, and by facilitating the policing and enforcement of patent infringement. The second approach is to promote collective forms of pre-competitive R&D such as high-tech consortia and industrial collective research, as well as interfirm cooperation in development of anticipatory technical standards.

The dynamics of using intellectual property rights as a means of promoting innovation via cooperative knowledge sharing is characterized by positive feedbacks and multiple equilibria. The properties of the system can be considered in terms of the linkages illustrated in Figure 2a. As the number of patents rises, the quantity of information available increases. As a result, the economic returns on innovations will rise in response to the increased probability that the innovator will be able to profit from new knowledge adapted to his specific project. The number of innovators, and thus the number of patents filed, will subsequently rise. The strength of the system lies in the fact that not only is the number of patents a function of the number of innovations, but the number of innovations would also seem to be a function of the number of patents. The dynamics of the system is illustrated in Figure 2b; the two curves representing the supply of innovations and patents both slope upward. The system is therefore characterized by the possibility of multiple equilibria. It may be trapped into a low-level equilibrium if one of the positive feedback effects is weak or not assured. This is what happens when an increase in the number of patents does not give rise to an increase in the amount of knowledge available, as for example, when the access costs of patents and inventions are too high. It can also occur when the increase in the number of innovations are not accompanied by a similar increase in the number of patents, as is likely if patenting costs are high or when the existence of trade-secrecy laws makes these a more advantageous means of appropriating knowledge.

Alternatively, under conditions of strong feedback effects the economy will be situated within a virtuous circle of generation and diffusion of new knowledge as indicated by point B in Figure 2b. For example, in Japan the link between the number of patents and the amount of knowledge available is reinforced by early-disclosure requirements and rapid on-line access to patent-office data bases, while the link between innovation and patent is governed by the provisions relating to the low costs of patents, their lower exclusionary value, and the absence of legislation on trade secrets (Ordover 1991; Foray 1995a).

5. The Information Search Problem

In a system characterized by high distribution power the proportion of public and private knowledge that is disclosed relative to restricted-access knowledge should be high. Nonetheless, capabilities to retrieve, evaluate, use, and maintain knowledge may still constrain its efficient use. According to Richardson (1960) and Simon (1982), the principal problem concerns the storage of information and its accessibility. How should organizations be designed for an information-rich world? As Justman and Teubal suggest elsewhere in this volume, in the case of SMEs the "capabilities for identifying, selecting, and absorbing novel production technologies, equipment, and raw materials are becoming increasingly important given the exponential growth in the options available."

Simon (1982) provides an excellent illustration of the difficulty of retrieving specialized technological information. He examined the lag between the level of sophistication of statistical techniques applied by data users and the level of sophistication among experts in statistics. He finds that the sophisticated programs are not retrieved even when they would be appropriate, either because users are not aware of potential advances, or because access to the existing knowledge through appropriate inquiry procedures has not been institutionalized. Among other things, users may not be able to attract or properly compensate experts whose professional interest is usually oriented toward developing new techniques rather than working on routine applications of existing procedures.

Figure 2a: Patenting, Innovation, and the Growth of Knowledge

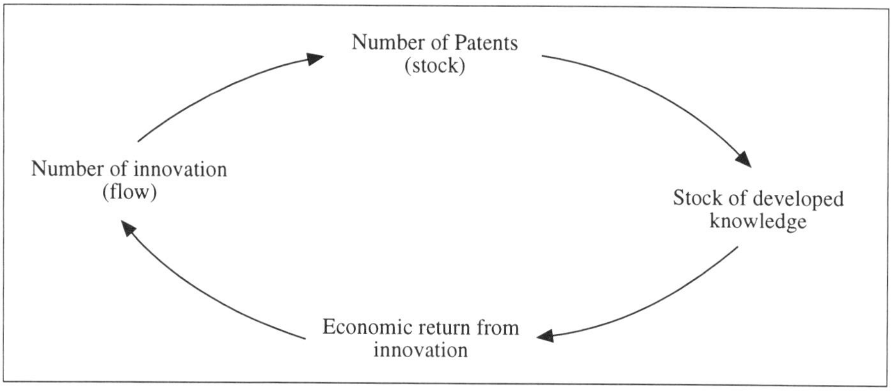

Figure 2b: Positive Feedback and Multiple Equilibria in the Supply of Patents and Innovations

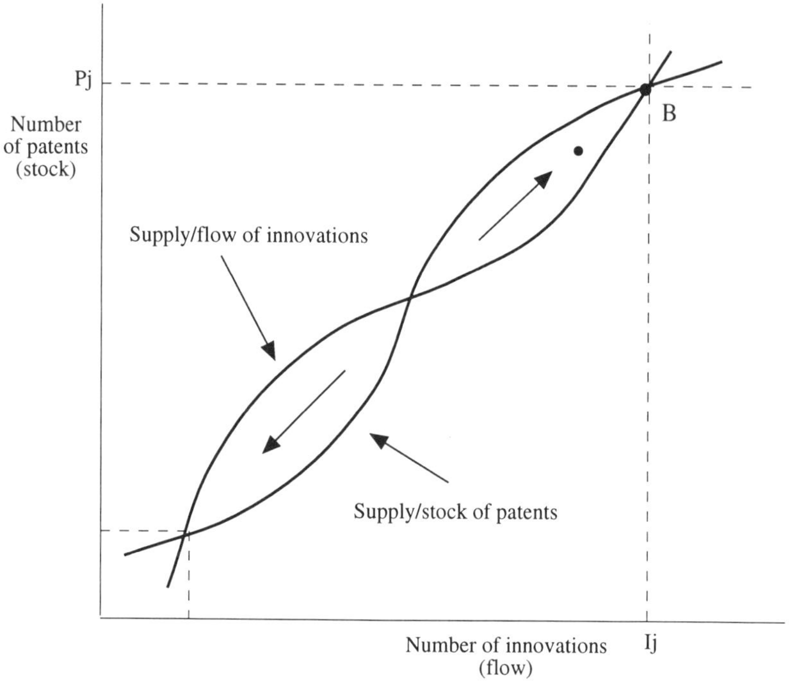

This particular example suggests a more general lesson: the search for pertinent information which is adapted to specific projects is most often localized and rarely expands into the entire potential space in which the sought-after information may be located. But in an environment where useful knowledge may be generated in domains that are both professionally and geographically distant from the individual firm, the optimal allocation of new knowledge among innovative firms requires a broadening of the knowledge-distribution process. The challenge is twofold:

- increase the amount of technological information available to each agent, and at the same time
- "buffer the agents from the overrich environment of information in which they swim" (Simon 1982, 44).

The apparent contradiction in these two constructions of the problem is more imagined than real, inasmuch as it should be possible to improve simultaneously both the richness of the information environment and the capacity of economic agents to deal with it.

The first objective follows Richardson's emphasis on enhancing the availability of information for coordinating market action (1960). A variety of policies could accomplish this goal, including:

- modifying the intellectual property registration system to improve information disclosure features;
- extending information infrastructures to build a universally accessible digital library; and
- more effectively using the standards-setting process as a forum for the exchange of technical information both within industry and between users and suppliers.[14]

The second challenge arises from Simon's observation that the scarce resource has become attention rather than information. Economizing on the resources that agents must expend to sort through and select information can be accomplished by:

- reducing redundancy produced by the *lawful* property of facts;[15] and
- developing information-processing systems as devices for supporting attention-conserving functions.

Simon argues that we need store only the fraction of information needed to predict the rest: "With each important advance in scientific theory, we can reduce the volume of explicitly stored knowledge without losing any

[14]The system of industrial standardization, in other words, functions as a means of placing ongoing pressures on firms to upgrade their products, while providing them with the technical information required to do so (Ergas 1987).

[15]Facts are considered to be *lawful* if certain of them can be predicted from certain others; redundancy which is produced by the *lawful* property of the facts can be eliminated without loss.

information whatsoever" (1982, 45). Reducing redundancy is efficient when retrieval costs exceed the costs of recreating information from new experiments or deriving it from theory. Codification is one important method of reducing redundancy that promotes the efficient retrieval of scientific and technical knowledge. Alternatively, information-processing systems such as the *consolidation of general discriminating capabilities* (Teubal and Zuscovitch 1994) through the creation of reference standards and artificial agents will also increase the productivity of the search processes for technological information (Steinmueller 1992). Furthermore, the institutionalization of access to the existing stock of knowledge must be developed in the form of technology centers (for basic technological infrastructure) and transfer science (for advanced technological infrastructure), involving, in particular, new incentive mechanisms for providing expertise (as recommended in the Simon example on statistical techniques).

6. TIP Options to Increase Knowledge Distribution Power

Technological infrastructure policy can be used to increase the distribution power of knowledge by moving the innovation system toward the *disclosure* region of the knowledge-product space (Figure 1). We have identified ten lines of action that could be pursued in this regard.

1. *Promote the protection of intellectual property rights*: It is clear from our discussion that intellectual property rights emerge as a crucial TIP issue since the cumulative expansion of the codified knowledge base has increased the importance of market transactions as a channel for knowledge distribution. Thus, TIP should promote intellectual property rights protection that includes:
   - widespread licensing for commercial exploitation;
   - fixed access charges for research use;
   - intellectual property registration-system enhancements to improve information-disclosure features; and
   - international harmonization of patents and copyrights in order to benefit informational aspects.

   It should be noted, however, that support for effective protection of intellectual property rights does not imply the desirability of providing a high level of protection.
2. *Encourage the development of modern information infrastructures such as uniform protocol and format standards*: Modern information

infrastructures facilitate the efficient communication of information and play a strategic role in the expansion of the codified knowledge-base. This objective must be explicitly considered by policymakers in regulatory areas such as network access, standards and protocols for communication, privacy, security, and intellectual property rights. International cooperation on electronic infrastructures must be encouraged. Also, issues concerning international data flows for science and technology should be separated from those concerning the transmission of commercial data; these include: the use of communication satellites, the operation of electronic networks, and conventions on transborder data transmission.

3. *Support increased training in the utilization of electronic information systems*: Both assuring the durability of and maintaining accessibility to knowledge stocks will require expanding investment in educational facilities to train people in the variety of skills necessary to establish, maintain, and utilize the digital libraries in which knowledge is increasingly stored. The realization of the potential benefits from the knowledge stock depends as much on human capital for exploitation as on physical capital for storage, and information infrastructure for transmission. The capabilities required of technological-information users for searching, evaluating, selecting, manipulating, and interpreting potentially useful knowledge are becoming increasingly complex, requiring more specialized training. At the same time these skills will be required of larger portions of the labor force. Further, skills for identification and retrieval of useful knowledge that is not currently utilized will require familiarity with older systems and "languages". All of these requirements contribute to the growing importance of formal and continuous training in information management, replacing the mode of learning through demonstration and personal experience that was appropriate to a time when relevant knowledge was largely tacit.
4. *Facilitate exchange of technical information through reference standards institutions*: Besides simplifying communications among engineers and scientists, reference standards serve to increase interoperability of systems designed in different countries by reducing nonstrategic sources of variety in design. Standards institutions can actively promote knowledge distribution because they often serve as forums for exchange of technical information and vehicles for collective R&D in the design of future systems (Foray 1995b; Tassey 1995 and this volume).

5. *Help firms to construct new forms of coordination*: This is an especially important option in a system in which the norm of openness has been altered (Weder and Grubel 1993). Romer (1994) considers this aspect of technology policy as a process of creation and selection of institutions that allow firms to coordinate changes in their environment. The goal is to provide institutional frameworks capable of generating industry-specific public goods. Examples of the type of collective action that can be achieved though such frameworks include: development of particular technical skills and competencies, sponsoring distinctive fields of basic research, design of specialized equipment, establishment of technical standards, and even self-organizing investment boards.[16]
6. *Promote university–industry–government R&D collaborations*: An important component of this recommendation is the preservation of the role of universities as nodes in open knowledge networks. As potentially powerful as this tripartite form of collaboration is, its success will turn on questions concerning the allocation of property rights. The determination of whether cooperatively produced knowledge will be covered by university or private-sector norms regarding intellectual property rights will affect participation incentives and ultimately define the conditions under which results will be diffused. This question will be especially difficult where knowledge-production is intrinsically cumulative and the university contribution is part of a generic-research program, leading to potential conflicts regarding disclosure and proprietary rights.
7. *Support educational programs in the transfer sciences*: Transfer sciences and engineering technologies are devoted, by definition, to seeking interconnectivity between industrial and academic research while preserving the dominant social rules of the open science world. Thus, they can play a key role in the information search problem through incentive mechanisms that connect innovative agents to particular bodies of knowledge. Transfer sciences facilitate industrial transferability of university research through training oriented toward the utilization of established knowledge and research methods to solve practical problems

[16]Elsewhere in this volume, Grindley, Mowery and Silverman consider the evolution one such coordinating institution, SEMATECH. The shift from horizontal research cooperation to vertical cooperation between its members is an example of Romer's depiction of firms creating and adapting institutions to coordinate changes in their environment. SEMATECH is becoming an institutional tool that allows firms to coordinate their actions in the process of generating industry-specific public goods.

arising from industrial needs. This approach to industrial problem-solving as a joint product of university research differs from university research undertaken expressly to solve industry's problems. Opportunities exist within many universities to change the balance of resource allocation and prestige in favor of the transfer sciences.

8. *Improve the resource allocation of mission-oriented technology programs*: Large technological programs represent more than half of public R&D resources in some countries, yet there rarely exists any systematic organization of how alternative technological concepts are developed, tested, and chosen. Selection of best-design technologies is frequently fragmented and fortuitous, often based on historical circumstance or on the strategic capacities of decisive agents. This weakness can be ameliorated by building coordinating structures that could implement mechanisms and procedures for exchanging and distributing information. They can also establish centralized assessment procedures to ensure that design selection is the product of systematic evaluation of all possible variants. Additionally, the distribution power of mission-oriented programs can be enhanced by expanding the number of participants involved, opening new channels for technology diffusion, total quality management implementation, and spinoffs (Foray 1994).[17]
9. *Encourage international cooperation for megascience*: Large projects create common practices and personal contacts among researchers. These can lead to the emergence of efficient formal and informal networks for the distribution of scientific knowledge. Using modern electronic communications and collaboration technologies, these networks are easily maintained, promote swift diffusion of information, and encourage potentially fruitful interactions among diverse and scattered experts.
10. *Facilitate investment in intangible knowledge products*: Through the development of new financial instruments, institutions, and accounting standards, investing in the generation and distribution of knowledge can be made more attractive to the private and even eleemosynary sectors. Careful consideration also needs to be given to how the legal and financial regulatory environments affect the formation and financing of collaborative R&D enterprises.

[17]An alternative approach would be to switch from mission-oriented to diffusion-oriented programs. However, this is likely to be costly and perhaps inefficient with respect to the information structures of the country considered (Foray and Llerena 1995).

7. Conclusion

More than two decades ago Richardson (1960) presciently observed that informational constraints represented a critical barrier to the coordination of investment decisions among firms, and consequently played an important role in determining both private and social rates of return on capital formation. At the time, the emerging information-theoretic approach to the study of innovation and technological change, which would come to be identified with the pioneering work of Arrow (1962), stressed the peculiarities of information as a commodity, foremost among them the fact that information could be transmitted at virtually zero marginal cost. The widespread acceptance of this premise among academic economists had a two-fold influence upon the evolution of science and technology policy. On the one hand it reinforced the perception of the automaticity of spillovers, and the associated problems of appropriability of returns and insufficient incentives for private investment in the creation of knowledge. The conventional set of remedial policy prescriptions included public funding of R&D and the strengthening of intellectual property rights protection. On the other hand, the assumption that information moved at negligible cost also led to the view that it was inherently difficult to control. From this it followed that attempts to constrain existing knowledge could not be significant impediments to allocative efficiency, contrary to Richardson's claims.

We have advanced slowly to the point of appreciating Richardson's insights in these matters. The costs of information acquisition and processing, and the bounded rationality behaviors to which those costs lead are now widely recognized, due to the work of Simon (1982). It is increasingly understood that although the secular trend toward greater codification of knowledge continues, there remain important elements of tacitness surrounding both the generation of new knowledge and the utilization of familiar procedures. Rather than being completely slippery and free-flowing, some data remains "sticky", to use von Hippel's (1992) characterization of the difficulties of transferring engineering knowledge from the sites at which it is generated. As a consequence, neither geography nor culture can be disregarded when economists consider the determinants of the behavior of economic agents and corresponding resource allocation decisions relating to the generation, exchange, and utilization of knowledge.

This gradual analytical reorientation has brought us to a reexamination of the economics of the distribution of knowledge which serves as a useful corrective to the long-standing preoccupation with the production of new knowledge. As we have shown in this chapter, this change in perspective has significant implications for science and technology policy, and for many of the fundamental institutional and organizational structures of modern societies. The design of policies to relieve information constraints in the creation, distribution, and application of knowledge turns out to be a central challenge not only for business enterprises, but for public decision makers striving to enhance the performance of national systems of innovation.

References

Arora, A. and A. Gambardella. 1994. The changing technology of technological change. *Research Policy* 23: 523-532. Special issue in honor of N. Rosenberg.

Arrow, K. 1962. Economic welfare and the allocation of resources to invention. in *The Rate and Direction of Inventive Activity: Economic and Social Factors*. Princeton: NBER.

——. 1969. Classificatory notes on the production and transmission of technological knowledge. *American Economic Review*.

d'Autume, A. and P. Michel. 1994. Politique économique, recherche-développement et croissance. *Journées de l'AFSE*. Aix en Provence.

Callon, M. 1994. Is science a public good? Fifth Mullins Lecture. *Science, Technology and Human Value* 19 (4).

Cohen W.R., R. Florida, and W.R. Coe. 1994. University–industry research centres in the United States. Carnegie Mellon University. Typescript.

Cowan, R. and D. Foray. 1995. The changing economics of technological learning. IIASA WP. Laxenburg.

Dasgupta, P. 1987. The economic theory of technology policy: an introduction. In *Economic Policy and Technological Performance*, edited by Dasgupta and Stoneman. Cambridge University Press.

Dasgupta, P. and P.A. David. 1994. Towards a new economics of science. *Research Policy* 23: 487-521.

David, P.A. 1993. Knowledge, property and the system dynamics of technological change. World Bank Annual Conference on Development Economics. Washington D.C.

——. 1994. Why are institutions the "carriers of history"? Path-dependence and the evolution of conventions, organizations and institutions. *Structural Change and Economic Dynamics* 5 (2): 205-220. Stanford Institute for Theoretical Economics.

David, P.A. and D. Foray. 1994. Dynamics of technology diffusion through local network structures. In *Evolutionary Economics and Chaos Theory: New Developments in Technology Studies*, edited by Leydesdorff. Pinter.

—. 1995a. Accessing and expanding the science and technology knowledge-base. *STI Review*. OECD. Forthcoming.

—. 1995b. *Institutions, Incentives and the Nature of Externalities in the Process of Knowledge Generation*. Presented to the Conference in Honour of George B. Richardson, 5-7 January 1995. St. John's College, Oxford.

David, P.A., D. Mowery, and W.E. Steinmueller. 1992. Analysing the economic payoffs from basic research. *Economics of Innovation and New Technology* 2 (4): 73-90.

—. 1994. *University-industry research collaborations: Managing missions in conflict.* Presented to the CEPR-AAAS Conference on University Goals, Institutional Mechanisms and the 'Industrial Transferability' of Research, 18-20 March, 1994. Center for Economic Policy Research, Stanford University.

David, P.A. and G. Wright. 1991. Resource abundance and American economic leadership. CEPR Discussion Paper, no. 267. Stanford University.

Ergas, H. 1987. Does technology policy matter? In *Technology and Global Industry: Companies and Nations in the World Economy*, edited by B. Guile and H. Brooks. Washington, D.C.: National Academy Press.

Ergas, H. 1994. The "new face" of technological change and some of its consequences. Mimeo.

Foray, D. 1992. General introduction. In *Technology and the Wealth of Nations*, edited by Foray and Freeman. Pinter.

—. 1994. University-industry research relations: institutional blocks, mission-oriented policy and the transferability of knowledge. CEPR-AAAS Conference on University Goals, Institutional Mechanisms and the 'Industrial Transferability' of Research, 18-20 March, 1994. Center for Economic Policy Research, Stanford University.

—. 1995a. The economics of intellectual property rights and the systems of innovation: The inevitable diversity. In *Technical Change and the World Economy*, edited by Hagedorn. London: Edward Elgar.

—. 1995b. Coalitions and committees: How users get involved in information technology standardization. In *Standards, Innovation and Competitiveness*, edited by Hawkins, Mansell, and Skea. London: Edward Elgar.

Foray, D. and P. Llerena. 1995. Information structure and coordination in technology policy. *Evolutionary Economics.* Forthcoming.

Garfinkel, S., R. Stallman, and M. Kapor. 1991. Why patents are bad for software. *Issues in Science and Technology* (Fall).

Henderson, R., A. Jaffe, and M. Tratjenberg. 1994. Numbers up, Quality down: Trends in university patenting 1965-1992. CEPR-AAAS Conference on University Goals, Institutional Mechanisms and the 'Industrial Transferability' of Research, 18-20 March, 1994. Center for Economic Policy Research, Stanford University.

Hilgartner, S. and S. Brandt-Rauf. 1994. Controlling data and resources: access strategies in molecular genetics. CEPR-AAAS Conference on University Goals, Institutional Mechanisms and the 'Industrial Transferability' of Research, 18-20 March, 1994. Center for Economic Policy Research, Stanford University.

Hortsmann, I., G. MacDonald, and A. Sliwinski. 1985. Patents as information transfer mechanisms: to patent or (maybe) not to patent. *Journal of Political Economy* (October).

Joly, P.B. and M.A. Hermitte. 1993. Plant biotechnology and patents in Europe: An economic analysis of alternative models of intellectual property rights. *Biotechnology Review* 1 (November).

Kahin, B. 1990. The software patent crisis. *Technology Review* (April).

Kitch, E. 1977. The nature and function of the patent system. *Journal of Law and Economics* 20.

Loasby. B. 1986. Organization, competition, and the growth of knowledge, In *Economics as a Process*, edited by Langlois. Cambridge University Press.

Machlup, F. 1983. What do we mean by the optimum utilization of knowledge. In *The Optimum Utilization of Knowledge*, edited by Boulding and Senesh. Westview Press.

——. 1984. *The Economics of Information and Human Capital.* Princeton University Press.

Merges, R. and R. Nelson. 1994. On limiting or encouraging rivalry in technical progress: The effect of patent scope decisions. *Journal of Economic Behavior and Organizations* 25.

Metcalfe, S. 1992. Competition and collaboration in the innovation process. Discussion Paper no. 79, University of Manchester.

Ordover, J. 1991. A patent system for both diffusion and exclusion. *Journal of Economic Perspectives* 5 (1).

Richardson, G.B. 1960. *Information and Investment.* Clarendon Press.

Romer, P. 1993. The economics of new ideas and new goods. Published as a supplement to *World Bank Economic Review*, edited by Summers and Shah. Washington DC: IBRD/World Bank.

—. 1994. Implementing a national technology policy strategy with self-organizing industry investment boards. *Brookings Papers, Microeconomics* 2.

Rosenberg, N. 1992. Scientific research and university instrumentation. *Research Policy* 21 (4): 381-390.

Rosenberg, N. and R. Nelson. 1993. American universities and technical advance in industry. CEPR publication, no. 342. Stanford University.

Sabel, C. 1992. Studied trust: Building new forms of cooperation in a volatile economy. In *Technology and the Wealth of Nations*, edited by Foray and Freeman. Pinter.

Samuelson, P. 1985. Creating a new kind of intellectual property: Applying the lessons of the chip law to computer programs. *Minnesota Law Review* 70: 471.

Schmid, A. 1985. Intellectual property rights in bio-technology and computer technology. *Journal of Institutional and Theoretical Economics* 141 (1).

Scotchmer, S. 1991. Standing on the shoulders of giants: Cumulative research and the patent law. *Journal of Economic Perspectives* 5 (1): 29-41.

Simon, H. 1982. The economics of information processing. Part VI of *Behavioral Economics and Business Organization*. The MIT Press.

Steinmueller, E. 1992. The economics of production and distribution of user-specific information via digital networks. In *The Economics of Information Networks*, edited by Antonelli. North-Holland.

Tassey, G. 1995. The roles of standards as technology infrastructure. In *Standards, Innovation and Competitiveness*, edited by Hawkins, Mansell, and Skea. Edward Elgar.

Teubal, M. and E. Zuscovitch. 1994. Demand revealing and knowledge differentiation through network evolution. In *Patterns of a Network Economy*, edited by B. Johanson, C. Karlsson, and L. Westin. Springer Verlag.

von Hipple, E. 1992. Sticky information and the locus of problem soving: Implications for innovation. *Management Science* 40.

Weder, R. and H. Grubel. 1993. The new growth theory and Cosean economics: Institutions to capture externalities. *Weltwirtschaftliches Archiv*.

Technological Infrastructure in Information Technology Industries

W. Edward Steinmueller

Both traditional and modern uses of the term infrastructure are related to "synergies", what economists call positive externalities, that are incompletely appropriated by the suppliers of goods and services within an economic system. The traditional idea of infrastructure was derived from the observation that the private gains from the construction and extension of transportation and communication networks, while very large, were also accompanied by additional large social gains. Thus, society as a whole had an interest in promoting these networks because they created new opportunities for economic choice and growth through spillovers that were involuntary in the sense that they could neither be avoided nor entirely captured by the creators of transportation and communication networks.[1] Over the past century, publicly regulated and promoted investments in these types of infrastructure have been so large, and the resulting spread of competing transportation and communications modalities have become so pervasive, that they have come to be taken as a defining characteristic of industrialized nations. The size and pervasiveness of these traditional infrastructures is now so great that they no longer command the social attention and priority that previously supported their public subsidization and regulation.

During the past half century a new type of infrastructure has come to be recognized as important within industrialized nations.[2] This new infrastructure is based on knowledge that spans the boundaries of individual enterprises and that originates either from explicit efforts to create spillovers through the public disclosure of knowledge or, implicitly in involuntary knowledge

[1]The involuntary character of spillover benefits has, of course, also been the result of public regulation of the monopoly profits of transportation and telecommunication industries. See McGraw (1984) for an outstanding history of the development of such regulation.

[2]The watershed contribution in justifying the extension of public support for basic science was "Science: The Endless Frontier" (Bush 1945). More recently, Tassey (1992) has articulated a vision of how technological infrastructures are characteristic of modern "system" technologies. Both Bush and Tassey are concerned with distinguishing clear boundaries between public and private roles.

M. Teubal et al. (eds.), Technological Infrastructure Policy, 117–139.

spillovers from the commercial activities of enterprises. As with physical infrastructures, the infrastructure arising from knowledge creates private as well as social gains. Using knowledge requires *capabilities*, business enterprise skills, and experience in adapting and absorbing knowledge, which are used for the development of new or the improvement of existing products and production processes.[3]

The form of knowledge is also relevant to this discussion. Knowledge may be codified as in the case of operating procedures, patents, or blueprints; or it may remain more tacit and intrinsic as in problem-solving experience or accumulated skill that distinguish the operating enterprise from the abstract construct most often encountered in economic analysis.[4] Although both codified and tacit knowledge have the potential for creating positive spillovers across enterprises, the mechanisms by which these different types of knowledge are disseminated and received differ. Because codified knowledge may be widely and quickly disseminated, property right mechanisms are necessary for preserving the creator's economic interest in such knowledge. These property rights form the basis of market transactions in the exchange of such knowledge and preserve the incentives for its creation and dissemination. Tacit knowledge, because it is more difficult to disseminate, may be more readily linked to other forms of business relationship such as joint ventures and technology exchange agreements. Both types of knowledge, however, signal competitors that commercial opportunities exist and provide knowledge that may be useful for developing substitutes for available knowledge. Moreover, when the technological foundations of an industry facilitate a variety of technological solutions, the advance of knowledge is likely to be broadened and diversified through processes of adaptation, imitation, and reproduction outside the boundaries determined by intellectual property rights or contractual relationships.

The major theme of this chapter is that advances in information technology have created knowledge spillovers that have opened more opportunities for competitive entry than they have closed. Technological infrastructure in several of the information technology industries involves substantial contributions

[3]There has been considerable discussion in recent years of the idea of "core" competences, (Teece 1987; Prahalad and Hamel 1990), which are capabilities that are difficult to transfer or reproduce and that therefore serve as stable anchoring positions for investment in business enterprise.

[4]Winter (1987) offers a very useful introduction to the distinctions between and consequences of tacit and codified knowledge.

from the involuntary spillovers of knowledge in the process of both creating and using such technologies. This method is illustrated with examples drawn from the computer and integrated circuit industries where significant involuntary spillovers of knowledge have influenced technological progress and industry structure. The implications of these examples for our understanding of the influence of infrastructure on horizonal market structure and competition policy are explored from a general viewpoint. The influences of infrastructure on vertical market structure are also examined with a further example highlighting the role of interface-compatibility standards. The chapter concludes by identifying some unresolved issues and qualifications that suggest new directions for research and policy.

1. Involuntary Knowledge Spillovers and The Theory of Technological Infrastructure

From the viewpoint of a society, knowledge creates positive externalities within national economies or, under conditions of liberal global trade and scientific exchange, within the worldwide economy.[5] For our purposes there are two key features of knowledge that make it part of the technological infrastructure of an industry. First, the externalities it creates are incompletely captured or appropriated by individual enterprises or even networks of firms, so that it may be said that they are one dimension of an economy's social capabilities.[6] When knowledge is incompletely appropriated it is possible to distinguish between private and social gains. When knowledge is effectively

[5]Many of the proponents of capabilities or core competences are particularly concerned about the consequences of erosion of these capabilities within particular domestic economies, a signal of which is often purported to be the use of foreign-produced inputs (Cohen and Zysman 1987). It is nonethelesss true that the pattern of international trade among advanced industrial economies is increasingly based upon intrasectoral exchange, a pattern that exists only under conditions of intrasectoral specialization among trading partners, in other words, the development of complementary capabilities for similar economic activities. However, the assumption that this pattern of trade in particular, or liberal world trade in general, will continue indefinitely, may be a basis for concern.

[6]The term "social capabilities" has been employed by Abramovitz (1989) in explaining national differences in long-term economic growth performance. His much broader view of social capabilities encompasses institutional, cultural, and educational factors that contribute to a nation's ability to catch up or overtake other nations in economic growth. Here, we will consider only the problems of gaining access to technology within the broader issue of social capabilities.

owned and only transferred through market transactions, there is little to distinguish it from other specialized assets that, within a competitive framework, should earn returns commensurate with their marginal revenue product.[7] Our concern, however, is with knowledge that makes a contribution to social capabilities above the returns earned by its creators or owners.

The second feature of knowledge infrastructure is that it persists over time. While knowledge in general may be constructed and dismantled rapidly, knowledge that becomes part of technological infrastructure needs to persist long enough that it can be recognized and exploited by enterprises and other organizations not directly involved in its creation. In other words, confining attention to knowledge that is persistent focuses attention on those attributes of knowledge that take time for individuals and organizations to acquire. Moreover, investments in knowledge acquisition provide an incentive to find new applications for this knowledge after it has been acquired. Persistence is thus linked to the cumulative and incremental character of technological progress.

In the information technology industry several types of knowledge play an important infrastructural role because they share the characteristic of being both persistent and incompletely appropriable. Information technology industries – the collection of enterprises offering products and services based on electronic methods of acquiring, storing, processing, and distributing information – have benefitted from advances in integrated circuit, optoelectronic, and magnetic technologies. In addition, each of these *enabling* technologies has important roots in materials science.[8] Both the enabling

[7]Nonetheless, the commercial availability of specialized inputs is an important dimension of social capabilities, even if the returns from these inputs are privately captured.

[8]In Tassey's framework, materials science, including solid state physics, is the most "generic" technology underlying information technology (Tassey 1992, 98-99). Tassey also uses the term "infratechnology." Tassey defines infratechnologies as "practices and techniques, basic data, measurement methods, and measurement-related concepts which increase the productivity or efficiency of each phase of the R&D, production, and the market development stages of economic activity." Tassey asserts that infratechnologies are "generally competitively neutral, in large part because of their typical widespread and uniform use." If, however, such technologies are a sufficient technological basis for entry, they cannot be competitively "neutral" but will enhance competition. Unfortunately, within Tassey's framework, technologies are either "proprietary" (i.e., subject to ownership) or "infrastructural" (i.e., "generally" competitively neutral and widely available). But involuntary spillovers of knowledge that enable entry are both available and competition enhancing. Thus, while such spillovers fit Tassey's epistemological definition, they fail to conform to this behavioral hypothesis of competitive neutrality.

technologies and materials science have the features of persistence and incomplete appropriability that are necessary for the emergence of a technological infrastructure. Thus, one approach to the problem of information technology infrastructure is to examine how these technologies have developed.[9] The common technological foundations of information technologies have contributed to a blurring of boundaries among different types of information and thus among different segments of the industry. Streams of data representing voice, still and moving images, and text – once only associated with large and remote computers – have become linked through common information storage, processing, and transmitting technologies. Increasingly, data flow over a global network of telecommunications links and through a forest of computers, large and small, near and remote. This process of convergence in the *use* of the technology has created an important knowledge pool that is distinct – yet related – to the knowledge used in the production of information-technology products.

The significance of technological convergence suggests two complementary approaches to the analysis of technological infrastructure. One approach examines the *generation of the technologies* themselves, that is, the knowledge of materials and enabling technologies that support technological convergence among the information technology industries.[10] The other approach focuses on the *use of these technologies*, and the creation of common pools of knowledge regarding their application. This chapter concentrates almost exclusively on the latter approach since it is the use of new technology where the experience of the information technology industries offers a unique contribution to our understanding technological infrastructure. Moreover, it is the use of these technologies that provides many of the inducement mechanisms for generating innovations. This choice of focus leads, however, to an important qualification. The enormous advances in materials and facilitating technologies, combined with their breadth of applicability, are both key elements in the evolution of knowledge demand and utilization in the information technology industry. Caution is required in

[9]Moreover, many industry observers would concur that the boundaries of the information technology industry should be further enlarged to include enabling technologies produced by segments of the scientific instrument and capital goods industries that permit the fabrication, testing, and quality control of these materials, as well as complementary technologies such as electronic displays and fibre optics that are also closely linked to new materials.

[10]Pursuing this first approach would likely lead to a greater emphasis on what will later be termed the "traditional" approach to understanding knowledge-based infrastructure in which publicly-funded scientific advance plays a central role.

applying lessons from the experience of the information technology industries to industries where technologies are more heterogenous and subject to slower and less uniform rates of advance.

The idea that knowledge is infrastructural is largely a contemporary construct reflecting the growing importance and complexity of knowledge used in the production of goods and services. Earlier ideas, such as the "industrial district" of Alfred Marshall or even the division and specialization of labor depicted by Adam Smith, captured important elements of knowledge as infrastructure. Two modern trends in industrialized economies differentiate the modern use of the term technological infrastructure from earlier externality-generating developments such as industrial districts. The first trend is the growing internationalization of the *public-good* character of knowledge, its availability throughout the world to an ever larger number of organizations.[11] It has become increasingly likely that wherever research developments are achieved, they are monitored and imitated somewhere else in the world. This trend applies not only to research knowledge created as the result of public funding, where the social value of knowledge is now well established, but also to knowledge developed in the private sector. Commercially generated knowledge creates voluntary and involuntary spillovers that also have public-good features.[12] These spillovers from commercial research activities are available globally, subject to the problems of adaptive and absorptive capacities as well as intellectual-property restrictions.

The second trend is the increasing importance of capabilities for imitating, adapting, reproducing, or independently recreating knowledge first discovered or generated elsewhere. This trend is directly related to the growing commercial use of scientific and technical knowledge as an input in product and production process innovation.[13] Because a vast amount of knowledge has become

[11]The other characteristic of public goods, nonrivalrous consumption (the possibility, within limits, of increasing consumption without affecting the abilities of others to consume) is an inherent characteristic of knowledge. Knowledge is a "purer" public good in this respect than textbook examples like clean air or water that are subject to congestion externalities.

[12]The economic analysis of involuntary spillovers has received considerable attention recently. See van Meijl (1994) for a review of several contributions as well as a model of the impact of such spillovers on the R&D investment choices of firms. Rosenberg (1990) offers a number of reasons why firms perform basic research with their own funds. Several of these, including participation of company researchers in scientific communities that require public disclosure of research results, encourage *voluntary* knowledge spillovers.

[13]Kuznets (1966) argued that the development of "science-based industry" was an epochal change, distinguishing our modern era from the past and unrivaled since the development of agricultural cultivation in its impact of economic growth.

available, the relative importance of capabilities for utilizing and accessing this knowledge has increased so much that it may now be said that capabilities to access knowledge are essential to its infrastructural character (Cohen and Levinthal 1990; Barabaschi 1992).

The growing significance of capabilities for accessing knowledge suggests a reorientation of the analysis of technical change toward the study of barriers and enabling mechanisms for such access (David and Foray 1994). The same reorientation is implied for the study of infrastructure. Thus, despite the continued importance of infrastructural investments in generating new knowledge as a public good, more attention should be devoted to barriers and enabling mechanisms for knowledge utilization. This chapter examines how technological trajectories and technical interface methods contribute to the creation of a technological infrastructure in the information technology industry. Other mechanisms, however, in the areas of intellectual property and methods for financing technological progress deserve equal standing in a complete picture of the technological infrastructure of the information technology industries.[14]

The infrastructure created by the public-good character of knowledge has an influence on industrial structure. In particular, the extent of asymmetries in the accumulation of knowledge and capabilities seems to be one determinant of patterns of competitive entry and the ability of incumbents to retain a technological lead over new entrants. In several segments of the information technology industry – computer subsystems, personal computers, and semiconductors – substantial competitive impetus comes from new entrants. Competitive entry highlights the nature of involuntary spillovers from commercial activities that create a technological infrastructure without explicit coordination mechanisms. Although the process of accumulating capabilities through involuntary spillovers followed by entry, in the absence of explicit coordination, is one means of stimulating technological progress and competitive fitness, it is not necessarily the only method. Other industry segments such as telecommunication switches, large computer systems, and massive software projects may require knowledge accumulation *within* a single large enterprise. Moreover, it is clear from international comparisons that different industry structures may generate comparable innovative

[14]For example, see Mowery and Steinmueller (1994) on the effect of the 1955 consent decree that led to the release of AT&T's intellectual property in semiconductor technology. The ability to finance new enterprises has a major impact on whether existing enterprises can extend their control over new knowledge without challenge from competitive entrants.

capabilities.[15] As a result of these qualifications, this chapter does not attempt to create a normative theory of which industry structure is best suited for exploiting the infrastructure that arises from involuntary spillovers of knowledge. Instead, it has the more limited purposes of demonstrating how such infrastructures come into existence and tracing some of their impacts.

2. Development of Technological Infrastructure in The Computer Industry

Whatever page one selects from the early history of digital computers one finds a solitary researcher or a team, in a university setting, with government support, attempting to translate the abstract idea of a stored program digital computer into working electronic circuitry.[16] This history is, however, subject to some important qualifications. While one may trace the path of development of the digital computer from the university research laboratory to the commercial products of Burroughs and IBM, the latter company had, during the 1920s, developed commercial applications for automated data processing using the electromechanical technology of tabulating card punches, readers, sorters, collators, and accounting engines. Among the legacies of this earlier history was the use of tabulating cards as input and output media. For IBM, however, the earlier experience provided a more important advantage – a unique insight into how data processing equipment could be utilized in the business environment. The integration of early data-processing methods with the use of digital computers highlights an underappreciated characteristic of technological change: the initial implementation of new technologies have little relation to their ultimate use (Rosenberg 1976; Ceruzzi 1986).

The computer as developed with government support, was essentially a scientific instrument serving as a substitute for human "computers" in the calculation of reference tables for scientific and military application. Transforming this scientific instrument into an engine for the everyday processing of data streams originating from the operation of a business enterprise is now seen as an obvious derivative application of computer

[15]See Steinmueller (1988) for an argument that this is the case with the U.S. and Japanese semiconductor industries.

[16]See Flamm (1987; 1988) for an economic history, and Ritchie (1986) and Williams (1985) for more general histories of these developments.

technology (Ceruzzi 1986).[17] In practice, however, this application required a decade of experimentation before it could be systematized in the development of the IBM/360 and the competing mainframe computers of IBM's rivals (Pugh, Johnson, and Palmer 1991).

IBM had learned from its previous experience that commercial applications of data processing were largely focused on *data reduction* rather than *data generation*. Much of the use of computers as scientific instruments was devoted to the solution of discrete approximations of continuous mathematical functions that generated an output stream of data larger than the input stream. By contrast, in commercial applications the incoming stream of data is larger than the output stream. The punch card, which had played a secondary role to paper tapes or direct "hard wiring" techniques for programming early computers, became the basis for IBM's commercial computers because it was well suited to the creation of large-scale input data streams and thus had a persistent influence in the development of commercial digital computers.[18]

The data-reduction role of computers suggested a particular organizational model for their use. A commercial enterprise often generates data through a geographically extensive network of operations. Data-processing activities required centralization to properly utilize the capacity of the relatively expensive digital computer, but centralization was also the consequence of past experience in the commercial operation of data reduction in data-processing applications in banking, insurance, and other commercial fields. In these applications, media such as punch cards, magnetic tape, and optical readable paper forms were physicaly transported to central locations for data processing. This operating procedure generated important bottlenecks whose alleviation would offer technological improvements as data processing was further automated using the digital computer. One method of relieving the bottleneck of physical input streams was the decentralization of data input through the use of *remote job entry* stations connected through telecommunications links. In other words, the disadvantages of centralized data collection provided an incentive for the development of new computer hardware that reduced costs,

[17]Ceruzzi quotes one of the pioneers of the computer industry, Howard Aiken, as claiming "...if it should ever turn out that the basic logics of a machine designed for the numerical solution of differential equations coincide with the logics of a machine intended to make bills for a department store, I would regard this as the most amazing coincidence that I have ever encountered" (Aiken 1956).

[18]For three decades most of the data that flowed into large computers were packaged in columns of 80 characters, the length of a punch card; even today many data display terminals are designed for 80 characters in width.

speeded up transfer, and increased the reliability of the data stream. The existence of a constraint served to define a standard of performance for existing technologies by which alternatives could be assessed.[19]

The central roles of the punch card as input medium and data reduction as operating procedure manifest important features of an infrastructure. The punch card provided a known standard for encoding data while the problems of centralized data collection and data reduction provided a comprehensible target for innovative efforts as well as a variety of defined interfaces or links among subsystems. Moreover, the incorporation of these technologies into organizational routines meant that their use was persistent and growing over time as more commercial enterprises adopted these methods.

The variety of technological solutions created within this framework created a foundation for further technological improvements. Remote-job-entry stations would become interactive data terminals where data could be entered directly under the control of time-shared central computers, and the physical transport of data-entry media was often replaced by increasingly sophisticated telecommunications links. These improvements were, to an important extent, motivated by the bottlenecks of data transfer and served to define a trajectory for improvements; but these were not the only technology improvement mechanisms that operated in the commercial development of digital computers. Others that were important included: the development of minicomputers for real time control of industrial equipment, which alleviated the bottlenecks of input channels to mainframe computers; the development of faster and higher capacity mass storage devices such as the magnetic disk which relieved the constraint of fast and reliable temporary storage for program instructions and data; and the continued improvement of magnetic tape storage devices which solved the limitations of the punch card itself by offering inexpensive and faster mass storage. Each of these subsystems were connected to digital computers through interface specifications defined by computer manufacturers, but they nonetheless became points of entry for competitive products.

[19]See Rosenberg (1976) for a discussion of the relationships between technological bottlenecks and trajectories of technological improvement.

3. Technological Infrastructure in Integrated Circuits and Digitization

The invention of the integrated circuit (IC) in 1961 opened another important trajectory in the information technology industries. The IC was initially a "niche" device whose early applications, such as the guidance control units for intercontinental ballistic missiles, justified a relatively high cost to deliver reliability and miniaturization performance. During the 1960s, this cost fell dramatically, so that by the end of the decade a new company, Intel, was formed with the aim of developing a solid-state computer memory that would be price competitive with magnetic core memory subsystems for mainframe computers. Like the data-reduction trajectory described above, the trajectory of IC improvement served to focus attention on the opportunities for new applications. The bottleneck in this case was the cost and time required to develop electronic systems.

One of the most influential developments in reducing system development time was the conversion of electronic signals to streams of binary data or *digitization*. While digitization was already a well-established principle in the design of electronic systems, it gained enormous impetus from the cost reduction of logic elements in integrated circuits. A leading authority of the mid-1970s summarized this development by the maxim that it "invalidates the traditional goals of logic design," most of which were related to minimizing the number of circuit elements (Blakeslee 1975, 15). The number of such elements had to be minimized because the costs of the electronic system were directly proportional to the number of electronic components required to perform a single operation.[20] With transistor technology a handful of components were required for every elementary application; with the first ICs dozens of operations could be performed with a single component, and soon thereafter thousands of such operations could be performed by a single IC.

The particular set of skills associated with digitization have become a central feature of electronic-systems design across a very broad spectrum of applications. Where electronic-systems designers previously approached each design problem as having unique requirements that required an integrated design, digitization allowed the designer to compartmentalize design problems into stages or subsystems that communicate with one another using a relatively

[20]See Braun and Macdonald (1982, 99) for an excellent discussion of this problem whose severity set definite technological limits to transistor-based electronic systems.

limited set of interface methods. The set of design skills, and the menu of available IC devices for implementing designs, have accumulated steadily during the past two decades so that the designs of almost all electronic systems, regardless of level of complexity, are now implemented using such design methods.[21] These techniques are likely to persist indefinitely due to the convergence of information transmission and storage methods noted in the introduction to this chapter. Digitization has had a marked effect in creating a common focus for the product and process innovations of the integrated-circuit industry, and has been particularly important in achieving user acceptance of new products. The technical compatibility of new products is greatly facilitated by the evolution of a common approach, digitization, to the construction of electronic systems. This approach has created both substitution and complementarity relationships among ICs regardless of their manufacturer.

Digitization has thus created the sort of knowledge spillover from commercial activities that we have identified as infrastructural in being both persistent and creating widespread involuntary knowledge spillovers. A new entrant may examine the range of available products and the possibilities for new applications and devise a product innovation strategy that fills a new application need or establishes a new market niche, often without any need to be concerned about claims of intellectual property since the variety of feasible techniques is very large.[22] The process of entry in the IC industry has continued, at a varying pace, for over two decades despite the strong position of existing producers, and despite recurrent predictions that additional new entry will be foreclosed by entry barriers from accumulated experience, the growing costs of state-of-the-art facilities, or the profusion of available product solutions.

4. Implications of Information Technology for Technological Infrastructure

The examples from the technological history of computers and integrated circuits have important implications for how we view the economics of

[21]This development was summarized by one industry pundit, Nicholas Negroponte, with the aphorism: "There will be many more MIPS in the nation's appliances than in its computers." MIPS is million-instructions-per-second, a standard measure of computer power (Brand 1983, 5).

[22]The technologies for manufacture of ICs are, however, better protected and entry increasingly requires a strategy of innovation to gain access to established knowledge (Mowery and Steinmueller 1994).

infrastructure. After its initiating role, public investments in the development of digital computer technology were displaced by evolution along a trajectory of improvement defined by "focusing devices" of data reduction underlying the commercial application of digital computers and the interactions among the subsystems comprising the digital computer. Throughout these developments, public funding continued to play important roles in the creation and standardization of programming languages, the development of data communications technology, and the improvement of digital computers through the use of transistors and integrated circuits. Nonetheless, the rate and direction of innovation in the computer industry was powerfully influenced by the logic of data reduction and subsystem integration which were fundamental characteristics of commercial data-processing applications and by the evolution of the digital computer as a *system* with accompanying bottlenecks and complementarities.

Similarly, the early funding of IC research from the sales of high-performance devices to the U.S. Department of Defense set in motion the development of the IC as a practical commercial device and provided the U.S. with important first-mover advantages in this technology. Nonetheless, the subsequent history of the IC industry has demonstrated that this first-mover advantage was insufficient to preclude entry, either by strong foreign competition or by a succession of domestic firms in the U.S. (Steinmueller 1988; Okimoto, Sugano, and Weinstein 1984; Braun and Macdonald 1982). The inability to retain a first-mover advantage was the outcome of several developments, including the limited number of interface methods for linking ICs together to make electronic systems, and the related growing significance of digitization as a design principle in the creation of electronic systems. Digitization has permitted users (electronic-systems designers) to rapidly understand and accept new product innovations and has provided a framework for product innovation.

From an economic perspective, the knowledge generated from attempts to manage data streams in order to achieve data reduction and to advance digitization as a principle of electronic-systems design were very imperfectly appropriable. The requirements for standardizing data flows and creating compatible subsystems were the points of entry for competing products in both the computer and IC industries. Knowledge of the performance of the existing techniques was available to anyone who had experience with actual commercial applications. In short, the broad diffusion of technology brings with it a widespread knowledge of the limits of existing solutions and the corresponding opportunities for improvement. It is therefore not surprising that this process should encourage the entry of enterprises offering improvements within the framework of existing methods.

The inability of incumbents to dominate these new developments is largely explained by the dynamics of product improvement. New innovations based upon incremental improvement often appear in implementations that are only tangentially related to the eventual applications. For larger enterprises, the landscape of technological change is often defined by opportunities for significant improvements in costs, sales, and profits. In practice, however, few innovative products are immediately able to deliver very significant increments to revenues or profits. Innovations are therefore continuously born at the periphery of economic significance where the largest firms are less likely to employ their advantages of scale and experience. Experience with these innovations, in turn, creates localized knowledge that becomes a source of first-mover advantage when such products become successful in the market. This mechanism reverses the competitive advantage of incumbent and entrant, significantly increasing the uncertainty that the incumbent will catch up with the entrant. The entrant's risks, though still large, are substantially reduced by the existence of accepted solutions and their related defined trajectories of improvement. In the computer industry, these possibilities for entry were enormously facilitated by the modularity of subsystems and the definition of interface methods by which these subsystems could be integrated into larger systems.

Interface methods for achieving technical compatibility are enormously important in the information technology industries as a means of flexibly combining diverse collections of components and subsystems into information processing systems. Interface methods create spillovers which are imperfectly appropriated by their designers and which become entry opportunities for other firms. The entrant benefits from the existence of interface methods since they define a point of market entry and a target for innovative efforts. The range of entrants' innovative opportunities is, however, likely to be constrained by the fact that specific technological solutions, including interface methods, are accompanied by tacit particularities that only emerge as barriers when entrants attempt variations. For example, the creator of an interface method may or may not choose to anticipate possible improvements in the performance of devices that may be connected to such an interface. One approach to regularizing interface methods is to develop a voluntary standard, a publicly available specification of the interface method, that attempts to completely define the technical requirements necessary for compatibility. In the framework of this chapter, the creation of a voluntary standard for interfaces transforms the knowledge spillover from involuntary to voluntary. The desirability of this conversion for the interface designer depends upon whether such a voluntary standard will increase their revenues in products embodying or purchased as complements to the standard (David and Steinmueller 1994).

A voluntary standard may not be chosen in an effort to profit directly from the control of the interface method. The *strategic* use of interface methods has a mixed history. During the 1970s, IBM was accused of manipulating interface methods to hinder competitors without major harm to its own sales of computers. During the 1980s several companies including Texas Instruments sought control of the technical interface between software and personal computers with disastrous consequences to their computer products. More recently, computer game manufacturers such as Nintendo and Sega have sought to control interface methods for their products without apparent harm to their commercial success. It is difficult to conclude from this experience what choice a business enterprise should make with regard to control of interfaces. Similarly, the public welfare implications are uncertain. Control of interface compatibility is a means of appropriating quasi-rents that might otherwise be eroded by competitive opportunities, but these quasi-rents may finance innovative improvements that deliver higher levels of public welfare in the future.[23]

5. Implications of Technological Infrastructure for Industry Structure

The externalities created by involuntary knowledge spillovers that are a central feature of important segments of the information technology industry have

[23]The term quasi-rent is used here to designate the existence of economic profits in excess of competitive returns that exist by virtue of an "articifial" barrier to competitive entry. In the case of interface compatibility standards it is assumed that market incumbents may obstruct potential entrants without doing themselves serious damage in competition with other incumbent producers, i.e., their choice does not lead to a marked deterioration of the technical quality of product available to purchasers. In addition, it may be the case that an incumbent has accumulated market power relative to rivals through mechanisms other than the technical qualities of its product such as superior sales and marketing efforts. In this case, the use of interface stations to generate quasi-rents is an extension of the market power originating from other sources.

implications for industry structure beyond the specific issues of interface methods. With respect to horizontal structure, involuntary spillovers must be examined in relation to the economies of scale that are typical in the information technology industries. In examining vertical structure, the existence of spillovers complicates the competitive process by creating the potential for vertically related firms to vie for profits from the control of markets or technological leadership. The existence of a common infrastructure reduces the determinacy of vertical structure, bringing *component* suppliers into competition with *system* producers because the effect of the common infrastructure is to blur distinctions between components and systems.

Cost reduction through accumulated production experience (learning by doing) is a common feature in manufacturing industries generally, and in many of the information technology industries specifically. While learning often provides advantages to incumbents who can exploit their previous investments in establishing sales and distribution networks to rapidly increase cumulative output, learning will also offer cost advantages to any firm that is able to rapidly increase its share of a market relative to competitors. Static economies of scale occur when increases in current output result in less-than-proportionate increases in total cost so that the average cost of each unit of output is declining. In information technology industries, both dynamic and static economies of scale are often present. Learning-related dynamic advantages arise from the reduction of the costs of making continued progress through the incremental process described above as well as the accumulation of knowledge of how components can be refined and systems more efficiently integrated. Static scale economies are most often the direct result of spreading fixed product development costs over larger output.[24]

Unexploited economies of scale may suggest a public-welfare motive for limiting the extent of competition. After all, a single firm can achieve higher levels of cumulative or current output and therefore lower costs, so why not limit competition to assure the greatest possible cost reduction? There are two answers to this question. The first, and simplest, applies to the case of those parts of the information technology industries where product differentiation plays a modest role in competition. Under these conditions

[24]While the spreading of fixed costs is, properly speaking, a static economy of scale, it is often only realized through links to dynamic processes such as positive feedbacks that occur through the adoption process. One common feedback mechanism is that although current period demand is limited, demand over a longer term shifts outward as many users adopt the product and the externalities from the accumulation of specific skills and complementary products encourage further adoption.

there is a direct tradeoff between the advantages of learning or other dynamic scale economies, and the creation of monopoly power and pricing. The impact of involuntary spillovers is to limit the realization of such scale economies and it may therefore seem to be in the public interest to limit the extent of such spillovers. On the other hand, if product differentiation is substantial, as is most common in the information technology industries, it is unclear which product should receive the benefits of scale – choosing one may lead to an inferior outcome for consumers or disadvantage the industry in international competition. This suggests a negative answer to the question; limiting competition will reduce the variety of products and is therefore likely to provide a solution that is suboptimal from a social welfare viewpoint. The horizontal market structure implications of involuntary knowledge spillovers therefore depends upon the value of product variety – the greater the value associated with variety in terms of customer demand or competitive fitness, the more likely it will be that horizontal structure will be unconcentrated.

Involuntary knowledge spillovers have had a major impact on the vertical structure of the personal computer industry which has experienced a shifting vertical locus of innovative activity. IBM initially sought to establish a commonly recognized standard for the use of microprocessors in small computer systems by adopting the Intel 8088 microprocessor which was also "second sourced" from AMD and eventually other companies. The innovative lead passed upstream to Intel and Microsoft in the next generation of microprocessors based on Intel's 80386 which relied on the interaction between microprocessor and software technology while IBM's contribution to system integration was effectively bypassed by many imitators. IBM's efforts to retake its dominant position using a new system architecture based on the Micro Channel Bus for connecting peripheral devices such as display monitors and communication ports, failed. At present, the future of Intel's position is subject to uncertainty as Microsoft may choose to develop operating system software that allows other IC companies to devise competing microprocessors and as other IC companies reduce the duration of Intel's first-mover advantage. Similarly, although Microsoft's position is reinforced by substantial scale advantages, it faces growing competition from other software companies that have benefitted from the involuntary spillovers that have been generated in the broad diffusion of personal computer systems.

The competition among vertical levels for innovative and commercial leadership has clearly been of enormous benefit to the users of personal computers who have seen steady reductions in price per unit of performance for systems available in the market. The example of the personal computer is perhaps the most dramatic in illustrating the role of involuntary spillovers in encouraging competition and entry in "upstream" industries, but it has been reproduced, on a small scale, in many related information technology markets such as mass-storage devices, printers, and display technology. This example suggests the potential value of encouraging competition despite the apparent gains available from scale advantage.

This discussion of the implications of infrastructure for market structure provides a basis for understanding the competitive intensity of information technology industries despite the interests of individual firms in building and sustaining market power. The most important features of the information technology industries are the continuing rapid incremental advance of enabling technologies (e.g., integrated circuit, magnetic storage, and optical), and the potential for entry that can be simultaneously imitative (creating products that can be recognized as similar to those already serving useful purposes) and progressive (creating products that improve upon the performance or capabilities of existing products). In response to these features of the enabling technologies, almost all firms in the information technology industry have pursued a similar two-part strategy. The first part of the strategy involves either being among the leaders in introducing new products or being a "progressive" imitator; both outcomes produce temporary quasi-rents that encourage further entry and competition. The second part of the strategy employs compatibility with existing systems encouraging the development of complementary products that enlarge the market and prolong the life of the innovation before its competitive advantage is dissipated through imitation. The second part of the strategy contributes to the persistence of an innovation and thus its attractiveness for competitive imitation; ideally, this will be reinforced by positive externalities originating in the creation of complements and the self-reinforcing widespread adoption of the innovation.[25] This two-part strategy is a direct response to the involuntary spillovers that characterize the technological infrastructure of important sectors of the information technology industry. Without these spillovers, both industry structure and

[25]The positive externalities generated by widespread adoption include the creation of human capital that is specific to a particular innovation, the reduction of uncertainty of potential adopters (who can observe that many others have chosen the solution and thus infer that it must be workable), and, for hardware, the likely reduction of input costs as suppliers compete for the larger market. All of these externalities contribute to the persistence of a technology and thus to its infrastructural features when the creator is unable to capture all of the returns from this process.

competitive strategy would be different in the information technology industries.

Throughout this discussion the part of technological infrastructure based upon abstract knowledge has served only as background to a dynamic process that is largely determined by the innovative activities of commercial enterprises and the involuntary spillovers of knowledge that occur in the course of these processes. This emphasis has been intentional. While it is possible to identify very specific contributions of publicly funded research to the improvement of integrated circuits, software, and even personal computer systems, much of the technological infrastructure in the information technology industry has emerged from commercial development activities.

6. Public Policy Implications

This chapter has stressed the role of involuntary spillovers between enterprises in the information technology industry as a fundamental source of technological infrastructure. Such involuntary spillovers may occur at a very practical level, as in the case where data reduction served as an impetus to incremental technological change, a scenario which is closely related to the economic view of the relationship between learning and technological progress. Particularities have played a large role in our description of the technological infrastructure of the information technology industry. Nonetheless, it is such particularities that establish the practical foundations for commercial progress in the industry.

Several public policy conclusions follow from accepting that *involuntary spillovers* are an important characteristic in the technological and competitive performance of the information technology industry. It seems straightforward to conclude that public policies encouraging such spillovers, or at least preventing their reduction, are desirable. There are, however, two problems with this conclusion. First, interventions that seek to preserve such spillovers are equivalent to policies that would prevent the extension of private property rights in the development of knowledge, despite the role of such property rights in providing an incentive to create knowledge. Here, there is no easy solution. Existing systems of intellectual property rights were shaped by the dominant technologies of earlier eras and it is likely that reforms that expanded such rights in the area of information technology would stimulate research undertakings. Moreover, it will be unclear *ex ante* whether any particular reform will substantially interfere with the involuntary spillover characterizing the information technology industry. The variety and frequency of both incremental and major innovations in this industry suggest that protection through the extension of intellectual property rights is likely to be short-lived. Nonetheless, a deeper recognition of the role that involuntary spillovers have played in technological progress in this industry suggests a cautious approach in extending such intellectual property protection.

Second, the ubiquity of increasing returns to scale in the information technology industries argues for limiting spillovers and entry in order to tap unexploited scale efficiencies. Here, the answer is more straightforward. The existence of unexploited returns in the information technology industries is a principal incentive for innovation. It functions best, however, when it remains as a potential rather than actual reward. The ability to leverage the gains found in a dominant position into the control of technological advance and the suppression of variety are the costs that offset the potential gains from tapping the unexploited economies of scale in the information technology industry. While the day may come when fundamental technological limits suggest a rationalization and simplification of the many redundant production and innovation efforts underway in this industry, it has not yet arrived. So long as the rate of cost reductions and performance improvements continue in this industry, it would be enormously hazardous to intervene in ways that might discourage the very mechanisms responsible for this progress.

Caution is also in order in applying the experience of the information technology industry described here to other industries. The common infrastructure developed through the pervasive use of computer systems with independently functioning subsystems, the role of digitization in enabling market coordination of diverse IC producers, and the impact of common approaches to the design of personal computers may be relatively rare and fortuitous outcomes. Moreover, the fact that these applications have been enabled by a relatively few materials-related technologies having enormous potential for performance improvement and breadth of applicability are likely to distinguish the experience in information technology from other industries. Nonetheless, this discussion provides a useful starting point for examining the infrastructure of other industries where systems are constructed through the use of components and subsystems united by common interface methods and sharing underlying technologies. Such industries include a broad range of capital equipment industries in materials, transportation equipment, and aerospace.

By using the information technology industry experience described here as a guide, it is possible to examine the extent and consequences of involuntary spillovers in these other industries, as well as the current limits to such spillovers. These limits are one basis for technology policy. Policies that would encourage variety and discourage integration must be weighed against the value of scale economies and design integration. The complementarities between the public-good feature of publicly-funded knowledge creation and the involuntary knowledge spillovers from commercial activities suggest that there are no *neutral* technology policies with regard to technological infrastructure.

References

Abramovitz, M. 1989. *Thinking About Growth and Other Essays on Economic Growth and Welfare*. Cambridge: Cambridge University Press.

Aiken, H. 1956. The future of automatic computing machines. In *Elektronische Rechenmaschinen und Informationsverarbeitung*, proceedings of a symposium published in *Nachrichtentechnische Fachberichte* (4): 32-34. Braunschweig: Vieweg.

Barabaschi, S. 1992. Managing the growth of technical information. In *Technology and the Wealth of Nations*, edited by N. Rosenberg, R. Landau, and D.C. Mowery. Stanford: Stanford University Press.

Blakeslee, T.R. 1975. *Digital Design with Standard MSI and LSI*. New York: John Wiley and Sons.

Brand, S. 1983. *The Media Lab: Inventing the Future at MIT*. New York: Viking.

Braun, E. and S. Macdonald. 1982. *Revolution in Miniature: The History and Impact of Semiconductor Electronics*. Cambridge: Cambridge University Press.

Bush, V. 1945. *Science: The Endless Frontier*. United States Office of Research and Development. Washington, D.C.: U.S. National Science Foundation (reprint 1960).

Ceruzzi, P. 1986. An unforseen revolution: Computers and expectations, 1935-1985. In *Imagining Tomorrow*, edited by Joseph Corn. Cambridge: MIT Press.

Cohen, W.M. and D.A. Levinthal. 1990. Absorptive capacity: A new perspective on learning and innovation. *Administrative Sciences Quarterly* 35: 128-152.

Cohen, S. and J. Zysman. 1987. *Manufacturing Matters.* New York: Basic Books.

David, P.A. 1975. *Technical Choice, Innovation and Economic Growth.* Cambridge: Cambridge University Press.

David, P.A. and D. Foray. 1994. Accessing and expanding the science and technology knowledge-base: A conceptual framework for comparing national profiles in systems of learning and innovation. Working Group on Innovation and Technology Policy, DSTI/STP/TIP (94) 4, Paris: OECD.

David, P.A. and W.E. Steinmueller. 1994. Economics of compatibility standards and competition in telecommunications networks. *Information Economics and Policy* 6: 217-241.

Flamm, K. 1987. *Targeting the Computer: Government Support and International Competition.* Washington, D.C.: The Brookings Institution.

——. 1988. *Creating the Computer: Government, Industry, and High Technology.* Washington, D.C.: The Brookings Institution.

Kuznets, S. 1966. *Modern Economic Growth.* New Haven: Yale University Press.

McGraw, T.K. 1984. *Prophets of Regulation.* Cambridge: Belknap Press of Harvard University Press.

van Meijl, H. 1994. The influence of spillovers, product differentiation, and entry on technological change. MERIT Research Memorandum, 94-026. Maastricht, The Netherlands.

Mowery, D.C. and W.E. Steinmueller. 1994. Prospects for entry by developing countries into the global integrated circuit industry: Lessons from the United States, Japan, and the NIEs, 1955-1990. In *Science and Technology Policy in Interdependent Economies*, edited by D.C. Mowery. Boston: Kluwer Academic Publishers.

Nelson, R.R. and S. Winter. 1982. *An Evolutionary Theory of Economic Change.* Cambridge: Harvard University Press.

Okimoto, D.I., T. Sugano, and F.B. Weinstein, eds. 1984. *Competitive Edge: The Semiconductor Industry in the U.S. and Japan.* Stanford: Stanford University Press.

Prahalad, C.K. and G. Hamel. 1990. The core competence of the corporation. *Harvard Business Review* 90 (May-June): 79-91.

Pugh, E.W, L.R. Johnson, and J.H. Palmer. 1991. *IBM's 360 and Early 370 Systems*. Cambridge: MIT Press.

Ritchie, D. 1986. *The Computer Pioneers*. New York: Simon and Schuster.

Rosenberg, N. 1976. *Perspectives on Technology*. Cambridge: Cambridge University Press.

—. 1990. Why do firms do basic research (with their own money)? *Research Policy* 19 (2): 165-174.

Steinmueller, W.E. 1988. Industry structure and government policies in the U.S. and Japanese integrated-circuit industries. In *Government Policy Towards Industry in the United States and Japan*, edited by John B. Shoven. Cambridge: Cambridge University Press.

Tassey, G. 1992. *Technology Infrastructure and Competitive Position*. Norwell, Massachusetts: Kluwer Academic Publishing.

Teece, D.J. 1987. Profiting from technological innovation: Implications for integration, collaboration, licensing and public policy. In *The Competitive Challenge: Strategies for Industrial Innovation and Renewal*, edited by D. J. Teece. Cambridge: Ballinger.

Williams, M.R. 1985. *A History of Computing Technology*. Englewood Cliffs, New Jersey: Prentice-Hall, Inc.

Winter, S.G. 1987. Knowledge and competence as strategic assets. In *The Competitive Challenge: Strategies for Industrial Innovation and Renewal*, edited by D.J. Teece. Cambridge: Ballinger.

Government Technology Procurement as an Instrument of Technology Policy

Charles Edquist*

Government procurement practices are known to have significant economic consequences beyond the proximate objective of acquiring commodities and services. For example, when governmental authorities place orders for products and systems that have yet to be developed, the resulting technological development work may generate benefits that spread beyond the contracting firms or their industry. That portion of government procurement that involves substantial technological change has the potential to be an important governmental policy instrument for both innovation and industrial strategy. This *government technology procurement*[1] has great potential to contribute to technological infrastructure.

This chapter develops the concept of government technology procurement and shows how government technology procurement has historically been a very powerful policy instrument in Sweden in two respects. First, it has played an innovation-policy role, triggering technical change and thus contributing to the satisfaction of social needs and solving societal problems. Second, technology procurement has played an industrial policy role through the development of new products and systems which have strengthened the competitiveness of Swedish industry.[2] Traditional government support for research and subsidies for technology development operate on the supply side, seemingly distant and somewhat disconnected from the satisfaction of immediate societal needs. By contrast, government technology procurement works from the demand side and results in specific products. This characteristic makes government technology procurement politically attractive and widely

* I am grateful for comments on an earlier version of this paper from Mark Elam, Patrik Hidefjall, Staffan Jacobsson, Maureen McKelvey, Lennart Sturesson, Elisabeth Sundin, and Morris Teubal.

[1]Other terms used to describe the same phenomenon are: *user-governed technical change*, *technology-developing government procurement*, and *innovative procurement*. The defining element is that the supplier is obliged to undertake technological development work in order to satisfy the requirements of the buyer.

[2]Examples include ASEA's technique for transmission of high-voltage direct current; Ericsson's development of the AXE electronic telephone exchange system; and ASEA/ABB's development of the X2000 rapid train. These are discussed more fully below.

M. Teubal et al. (eds.), Technological Infrastructure Policy, 141–170.

accepted by the general public, by a large segment of industry, by trade-unions, and by most Swedish political parties.

Primarily, technology procurement is undertaken to satisfy a specific social or economic need through technical change. As a result, the technology base of the contracting firms is enhanced. The *protodemand* articulated by the public procurer greatly contributes to the generation of technological capability at the firm level. In the Swedish case, this market building substantially contributed to the consolidation of large multinationals like ABB and Ericsson. Government technology procurement has stimulated user-led capability creation.

The competence to specify functional characteristics of products and systems is a crucial element in technology procurement and can be considered to be a part of the technological infrastructure (TI) in a country. So can the technological capability which is generated within a specific firm as a result of the procurement process, if it becomes available to other firms as well. This technology diffusion may occur for a variety of reasons: spill-overs to other firms or to technology centers; collaboration in R&D consortia; or movement of employees between firms or between firms and public organizations.

Provided that government technology procurement is practiced in areas where there are genuine socioeconomic needs, and therefore an authentic and sustainable demand follows upon the protodemand, this competence can greatly enhance the competitiveness of national firms in export markets. Technology procurement can, in other words, be an important part of technological infrastructure policy (TIP) which stimulates user-led capability creation. Implementing government technology procurement so that it contributes to TI is not uncomplicated. Among the problems that need to be solved are:

- How is the necessary buyer competence developed?
- How can technology procurement best be organized?
- Does technology procurement imply that technology develops in different directions, or only that the process of technical change is speeded up?
- How are relevant societal needs identified and translated into functional requirements of new products and systems?

Solutions to these problems all require that the procurer has a "vision of the future" (defining needs) and can then translate this into functional characteristics (implementing the policy). Research projects that include natural scientists and engineers, as well as social scientists, can be created with such aims. Such projects could be a way to both systematically articulate needs –

what is socially desirable, technically possible, and economically feasible – and develop implementation mechanisms. This requires organizational innovations such as the creation of public organizations in which procurer competence is generated and recreated. In this way government technology procurement involves generating infrastructures which, in turn, create capabilities at the firm level and support competitiveness. Government technological procurement may also be an important element for building a technological infrastructure for specific strategic industries, although it must normally be combined with other instruments.

This chapter begins with a brief discussion of various kinds of technology policy instruments followed by the presentation of a number of historical examples of government technology procurement in Sweden. We then turn to the future of government technology procurement, including possible examples of procurement projects. Finally, practical issues which affect the implementation of technology procurement are addressed, including suggestions for dealing with problems that are likely to arise in the future.

1. Innovation Theories and Government Technology Procurement

Theories of the innovation process are often classified as being *linear* or *systems* oriented. Further, technology policies themselves can address the *supply side* or the *demand side*. A detailed treatment of the relationship between government technology procurement and theories of technological innovation and diffusion would ideally require extensive systematic treatment. In this section, we only attempt to place government technology procurement in its theoretical and policy context before turning to the Swedish case.

According to the so-called linear view of the innovation process, basic research leads to applied research and then to diffusion, including production and sale of products embodying the new technology. The resulting improvement in dynamic efficiency is thought to lead to economic growth and welfare. We now know that this model is overly simplistic and even an incorrect depiction of the innovation process, especially in regards to the causal factors which lie behind innovation and technical change. Despite this, the linear view is – often implicitly – common among research policy makers. This leads to an overemphasis of factors which operate on the *supply-push* side – at the expense of *demand-pull* side factors – of the process of technical change.

Technological change is far more complex than a sequential innovation process wherein different stages follow in a well-behaved and organized fashion. Instead, technical innovation and diffusion is an elaborate and often obscure process relying on interaction and feedback between different stages and with a great number of actors involved. Innovations can be seen as the result of dynamic, cumulative, and path-dependent processes. Both institutional and market forces are important in the systems where these processes occur. In this *innovation systems* perspective, the interactions between users and producers of new technologies are often considered central (Klein and Rosenberg 1986; Lundvall 1985; Cohen and Levinthal 1990; von Hippel 1988; Teubal, Yinnon and Zuscovitch 1991). These interactions are not limited to the mechanism of price signals as in classical market models, but also include exchange of qualitative information between a variety of agents, including users, producers, and nonfirm organizations which serve facilitating or catalytic roles in technology development.[3]

Effective design and implementation of innovation-policy instruments is critically dependent upon a correct understanding of the nature and characteristics of the innovation process. A linear view leads to innovation-policy instruments that emphasize the supply side of technology development and diffusion. Conversely, theories which are characterized by interaction and a systems-view tend to feature prescriptive instruments that work from the demand side to a larger extent. If the distinction between supply- and demand-side policy instruments is combined with a distinction between *technology development* and *technology diffusion*, we can classify technology policy instruments by the resulting four combinations shown in Figure 1 (Edquist 1993a, 55-58; Edquist 1994a).

(1) **R & D Policy**: Directing research and development activities is the most common technology policy instrument on the supply-push side of technology development. Typically this has taken the form of government funding for basic research at universities and public research institutes, or as subsidies to private-sector initiatives.

(2) **Accessing Technology**: This refers to expanding the spectrum of technologies – whether created locally or abroad – which is accessible to users. One method used in Sweden is the system of technical attachés abroad. They are located in important technologically advanced countries

[3]The *system of innovation* approach has developed during the last several years. Some contributions are: McKelvey 1991, 1994; Carlsson and Stankiewicz 1991; Lundvall 1992; Nelson 1993; Edquist and Lundvall 1993; and Edquist 1993b.

and report home to Sweden on the existence and use of new technologies, thus increasing the technological choices accessible by Swedish actors.

Figure 1: Technology Policy Instruments

	Technology Policy Instruments Operating on:	
	The Supply Side	**The Demand Side**
Technology Development	(1) **R & D Policy:** (Government Funding of Research)	(4) **Commanding Technology:** (Government Technology Procurement)
Technology Diffusion	(2) **Accessing Technology:** (Technical Attaches)	(3) **Implementing Technology:** (Government Procurement of Goods and Services)

(3) **Implementing Technology**: The determination of which technologies will be implemented is critically dependant on the market demand for goods and services. Government procurement is an innovation-policy instrument on the demand side which significantly influences the choice of technology and thus its diffusion. Examples of these influences include specification of energy sources that determine which (existing) energy conversion technologies are used, as for instance, when streetcars or subways are chosen instead of diesel buses. Government can also influence private demand indirectly through laws, taxes, and subsidies. A ban on direct-acting electric heating or higher taxes on certain energy forms obviously influence the choice between different energy conversion technologies.

(4) **Commanding Technology**: When the requirements of clients necessitate the development of new technologies, the technologies can be said to have been indirectly *commanded.* This demand-pull incentive originates from both private and public sector activities; but it is a powerful innovation-policy instrument when used by government to specifically influence technology development. Unlike *simple* government procurement which affects the choice between existing technology alternatives, *government technology procurement* also influences the development of new technologies. When undertaken systematically,

government technology procurement enables governmental authorities to use their own future demand as an instrument of TIP.

Despite the powerfulness of the demand-side instruments, technology or innovation policy is too often considered to consist only of R&D-policy, that is, public research funding.[4] This means that the supply-push perspective dominates the demand-pull oriented view of technological change and technology policy,[5] whereas it is the very interaction between demand-pull and supply-push forces which is central in innovation processes. Technology procurement is an important force on the demand side, precisely because it involves interaction between users and producers. One of its principal advantages lies in the fact that it establishes a direct connection between innovation policy and usable outcomes. Government technology procurement not only has great potential, but as we shall see in the next section, historically in Sweden it has been extremely effective.

2. Examples of Government Technology Procurement

This section presents four cases of government technology procurement in Sweden in order to illustrate the nature of the process. The cases described here deal with civilian government technology procurement, although military technology procurement is also very common in the areas of weapon systems, military aircraft, etc. The examples that follow stand out as relatively successful, although this not always the case. Government technology procurement projects may be unsuccessful for either technical or market reasons.[6] Nonetheless, the effectiveness of government technology procurement as a component of technology policy should be judged by the overall contributions that it makes to technological infrastructure.

[4]This can be traced back, at least in part, to the theory of market failures and the linear view of technical change. See Nelson 1959, Arrow 1962; and Edquist 1994a, section 3.1.

[5]If this dominance were broken it would probably become easier to integrate social and environmental elements and objectives into technology policy.

[6]Two Swedish examples of unsuccessful government technology procurement are: the TUDIS project, in which a personal computer for school use (COMPIS) was ordered; and the Tjorven project, involving a specially designed car for postmen. Outside of Sweden, one of the most famous failures of government technology procurement has been the Concord, the civilian supersonic aircraft. After the development of the plane, initiated by the French and British governments, market demand proved to be too low for successful commercial production of the aircraft.

Two of the cases involved ASEA,[7] an industrial equipment manufacturer whose main products include equipment for power generation and transmission, transportation, and other industrial applications. Traditionally, a limited number of customers in Sweden purchased the major part of ASEA's output, and many of these customers developed close relations with the producer, helping to increase the effectiveness of existing products and to develop new ones. ASEA's technological history is, to a large extent, the result of very successful interaction between the company and several technologically oriented clients requiring diverse electrical equipment. Initially, the majority of these clients were private Swedish companies – electrical power producers, steel works, paper producers – and later also included foreign companies. Two government clients, Vattenfall and Statens Järnvägar, became key partners in ASEA's development of new technologies, and their cases illustrate the potential gains from government technology procurement (Glete undated; Fridlund 1993).

2.1 Vattenfall and Electrical Power Transmission

During the period 1930-1950 ASEA and Vattenfall collaborated very successfully. At the time ASEA produced mainly for the domestic market and was not exposed to especially strong competition from abroad. ASEA's technological advances during this period are, to a great extent, explained by the demands of their Swedish customers who were increasingly requiring technologically improved equipment; Vattenfall, the largest producer of electric power in Sweden, was one such customer. The company had as its task the exploitation of Sweden's substantial waterpower resources, located mainly in the north. Since most electricity consumers were in central and southern Sweden, this created a need for safe and effective systems of power transmission over very long distances. In order to solve this problem Vattenfall chose to collaborate with ASEA, the largest domestic producer of electrical equipment, in the area of breakers for high-tension current. In comparing ASEA's product with those of foreign competitors, Vattenfall discovered that ASEA was not leading. Vattenfall therefore invited ASEA to take part in a research project in order to catch up (Glete undated, 17; Fridlund 1993).

In 1940 ASEA started a development program for high-tension direct current. Although considered a possible alternative to alternating current for electrical transmission over long distances, ASEA had been unable to

[7]Through a merger with Brown-Bowery of Switzerland ASEA was transformed into ABB in 1989.

successfully develop the technology for practical use. Vattenfall became interested in the technology, seeing a possible solution to its long-distance electrical transmission problem, and the two entered into a long-term collaboration wherein Vattenfall provided the test resources and ASEA the material. The project was successful and in 1956 the first system in the world for high-tension direct current was inaugurated between the mainland and the island of Gotland. (Ironically, the system was finished too late to be used for transmission between northern and southern Sweden.)

Following World War II Vattenfall and ASEA also developed a system for high-tension alternating current at 400,000 volts, improving on the existing highest capacity system (285,000 volts). In 1953 the thousand-kilometer-long 400,000 volt line between Harsprånget and Hallsberg was put into operation. Its capacity was three-times larger than a line of 200,000 volts, with 40% lower transmission costs.

The collaboration between ASEA and Vattenfall is a classic case of government technology procurement. The development of the new technology was in response to specific long-term demands from a government authority to solve specific problems. The collaboration between the supplier and the customer had shown itself beneficial for both partners. ASEA's research and development program was adapted to Vattenfall's future demands and this client remained a faithful buyer of the new systems and products. The collaboration also contributed greatly to the generation of technological capability within ASEA and helped the firm to establish a very competitive position in export markets. These achievements gave ASEA a leading position in transmission technology which it could then exploit in international markets. During the following three decades ASEA was the world's leading producer of equipment for high-tension direct current (Glete undated; Fridlund 1993).[8]

2.2 Statens Järnvägar and Rapid Trains

Once electrification was completed in Sweden, domestic demand decreased for ASEA's power generating and transmission equipment; ASEA sought to compensate by expanding into new export areas, including the market for transportation equipment. Until the middle of the 20th century, domestic sales of rail equipment to Statens Järnvägar (SJ) – the Swedish state railway

[8]Collaboration between ASEA and Vattenfall in the field of nuclear power was not as successful. The Swedish government supported a government-owned company (AB Atomenergi) and thus the ASEA – Vattenfall collaboration did not receive any government support in this field.

company – had consisted of "old-fashioned" equipment that had been specially designed for Swedish conditions. ASEA was concerned about not being able to offer locomotives that were as advanced as those available in other countries. Following a period of research and development, ASEA succeeded in interesting SJ in ordering a light and fast locomotive. Although the design was successful, no serial production for SJ was in sight. However, the development of new diode technology made it possible for ASEA's new locomotive to be used on railway lines with different electric systems. During the 1960s ASEA received large orders for these products from Eastern Europe (Glete undated, 20-21).

During the 1960s ASEA worked on thyristors – a technology area in which ASEA was a pioneer – and offered to design and produce a thyristor locomotive for SJ. SJ ordered a prototype and also a series of finished locomotives. ASEA-SJ thus became world leaders in thyristor locomotives, which have been exported to a large number of countries. Again, the protodemand from SJ gave rise to a capability which resulted in increased international competitiveness (Glete undated, 20).

In 1978 SJ began to investigate the technical and economic possibilities for using rapid trains on existing rail lines in Sweden. Very-high-speed train technology – as for example in Japan (Sinkansen) and in France (TGV) – requires building new, straight track. Because of Sweden's low population density, the cost of massive track replacement would be too high. If Sweden were to have any rapid trains at all they would have to run on the existing tracks lines with their curves, slopes, and inclines. The technical problem was to obtain rolling stock that suited the existing rail lines while allowing higher speeds. New technology was needed.

SJ's economic calculations not only included the costs of rolling stock, but also costs for infrastructure, workshops, education, personnel, and maintenance. Other factors that were considered included quality, repair costs, and delays. This approach focused on the total or *life-cycle cost* (LCC), meaning that the notion of quality was brought into the picture. Taking into account projected increased traveling and income, SJ's calculations showed that the project would be profitable, both privately and socially (Andersson 1982, 46-47).

SJ solved part of the technical problem by developing the ATC (automatic train control) system and by supplying its lines with heavier, fully welded rails without joints. In order to "eliminate or at least reduce the centrifugal forces effect on the passengers we must slope the carriage bodies in relation to the frame with a certain angle depending on the 'over-speed' in the curve" (Andersson 1982, 45-6). However, other requirements also had to be fulfilled in order for the train to conform to the Swedish tracks (e.g., reduced axle loads and softer bogies with radially adjustable axles). A train that met these specifications did not exist; it had to be developed.

Gradually, SJ formulated specifications for a rapid train, including exacting requirements regarding size and performance characteristics. SJ provided ASEA with the following types of information:

- drawings and specifications for interiors, cross-sections, and all equipment to be included;
- performance characteristics on all technology;
- time tables and circulation plans (i.e., the production that the train should manage);
- characteristics of the railway lines including curves, track quality, and acceptable noise levels; and
- both interior and exterior design.

The intent was to provide the supplier with a clear picture of what the train was expected to accomplish and how the trip would be experienced by the passengers. The demand specification also included information on LCC in the form of a model with around sixty equations that described the cost relationships for purchase, maintenance, and operation. ASEA's top management was, however, initially not particularly interested in calculations based on LCC (Andersson 1982, 47).

Despite the detailed articulation of functional requirements, the demand specification still did not contain any definition of the technical design. It was incumbent on the ASEA to design every detail in accordance with functional requirements. The result of this technology procurement process is the X2000, Sweden's high-speed train designed and produced by ASEA/ABB, which has been operating on Swedish railroads for several years. The X2000 has become a very successful competitor to air travel in Sweden, although it has not yet been sold to any other countries.

2.3 Televerket and Telephone Exchanges

Collaboration between the telecommunications equipment producer, Ericsson, and Televerket, the Swedish national telephone operator, started at the beginning of the 20th century. In the early 1920s L.M. Ericsson developed an automatic exchange system which proved to be internationally competitive.

This development was built in large part on experience that Televerket had acquired through extensive experimentation. From this time, Televerket and Ericsson collaborated, rather informally, in the development of telephone systems (Glete undated).

In the 1960s the two firms began joint development of electronic systems. The collaboration was formalized in 1971 by the founding of a joint development company, ELLEMTEL, to develop the electronic and flexible AXE-system. The AXE-system was first installed in Saudi Arabia in 1978 and is now used in nearly 100 countries; it developed into the most important base for Ericsson's current international competitive position. The system is so central to Ericsson's activities that the company would probably not have existed as an independent company without it (Vedin 1993). Once again, the demands of the procuring firm led to the generation of a crucial technological capability within a Swedish firm, a competence that was later used in export markets.

2.4. NUTEK: Technology Procurement with an Energy-Saving Aim

In 1991 the Swedish Parliament decided to implement a five-year program to promote more effective energy use. In 1993 the program was extended to seven years. The Unit for More Effective Energy Use, which is a part of the Swedish National Board for Industrial and Technical Development (NUTEK), is responsible for the program. The program includes technology procurement, demonstration projects, and efforts to bring new technology to the market.

An example of this activity was the procurement of a new refrigerator in the early 1990s with two specific requirements. The first was to reduce the use of freon because of its damage to the earth's ozone layer. The second was to substantially reduce energy use in the operation of refrigerators in comparison with earlier designs. A bidding contest was announced with a commitment to order at least 500 refrigerators from the company that could satisfy the procurement requirements. A satisfactory design was presented by Electrolux within a relatively short time. This refrigerator example illustrates clearly that innovation policy through technology procurement can have environmental (or other) objectives as well as economic ones.

In this case it was not a question of using the government's own demand as it was in the earlier examples. Instead support was given to the buyers, builders, and administrators of apartment buildings, so that they could influence the suppliers. It was a matter of organizing and directing customer demand.

The individual buyers' interest in better products was stimulated and combined to get the suppliers more interested. This process eliminated a part of the risks that are associated with technology development. The state agency contributed to the formulation of the functional demands and to the administration of the procurement. The product-development potential of supplying companies was, in this way, mobilized.

NUTEK has also used this type of technology procurement for other household appliances and equipment: refrigerators and freezers, washers and dryers, windows, ventilation equipment, control equipment for fluorescent tubes, and automatic-shut-off computer monitors. The primary aim of the procurement in these cases has been energy conservation.

An important difference from the earlier examples is that NUTEK – which initiated and organized the technology procurement – was not the end user of the products that were developed. This was not a case of a skilled buyer, like SJ or Televerket, which contacted a few potential suppliers in order to procure a complex product or system; instead, NUTEK's role was that of a coordinator and technical resource to facilitate the procurement of simpler mass-market products that met socially desirable specifications. This places special demands on the organization of the procurement process which will be discussed further in subsection 4.3.

In both types of procurement the technological capability is generated in one or a few firms and may remain locked in there. This begs the question of transforming firm-based capabilities into technological infrastructure (TI). The answer requires the creation of an organization responsible for making the technology capability available to other actors and associations.

3. The Policy Roles of Government Technology Procurement

The case studies in the previous section illustrate the two important but distinct roles that technology procurement has played in Sweden. First, government technology procurement helped to develop physical infrastructure systems, influencing technical development so that socioeconomic problems could be solved and societal needs could be met. This function we call the *innovation policy role* of government technology procurement. Of course, not all societal problems have technical solutions, and thus the effectiveness of government technology procurement to solve those problems is limited by political, economic, and social constraints as well. Government technology procurement has also been instrumental in the growth of major Swedish companies, contributing to their international competitiveness through the formulation of user demands that led to product development. This aspect we call the *industrial policy role* of government technology procurement.

Although these two roles of government technology procurement are very closely connected, it will be useful to distinguish between them as we explore how government technology procurement can be employed as an important element of technological infrastructure policy. The fact that technology procurement historically worked well as an innovation and industrial policy instrument does not necessarily mean that it will in the future. This section examines the types of problems that can be addressed by government technology procurement in the future, discussing both its potential uses in innovation policy and industrial policy.[9]

3.1 Technology Procurement as an Innovation Policy Instrument

Government technology procurement is distinct from ordinary market demand in three ways: the nature of the needs that are being satisfied, the direction of technological change, and the pace of technological change. When used as an innovation policy instrument, technology procurement is an effort to directly meet specific societal needs that are unlikely to be met by the market. Further, government technology procurement can stimulate the development of alternative technological paths that will expand the range of technical solutions that become available in the future. And finally, government technology procurement should shorten the rate at which new technologies are created and implemented both by early identification of specific needs and by assuming a share of the supplier's risk.

The point of departure in the application of government technology procurement must be the satisfaction of *genuine* social needs; the products and systems that are developed – and the technical change that enables their provision – as the result of government technology procurement must be targeted to solve specific problems. Although the specific targets of government technology procurement should be narrowly defined, the potential set of

[9]Annex I briefly lists a number of areas where technology procurement can be used in the future in the Swedish context. In all cases the procurement would contribute to solving societal problems while at the same time strengthen both the technological infrastructure and technological capabilities at the firm level. These examples are dealt with more extensively in Edquist (1994b).

objectives is extensive. Historically, the principal goal of civilian technology and innovation policy has been increased economic growth, but this need not always be the case. Innovation policy can certainly be used to solve environmental and social problems, as for example, promoting less dangerous and more stimulating work life or a better environment.[10] Thus, to be effective innovation policy requires clearly and distinctly formulated objectives; otherwise, the work will lack consistency and fall victim to a variety of pressures.[11]

Government technology procurement should ultimately provide something the public "wants", but which it is unable to articulate effectively through the market. What distinguishes the needs being fulfilled is that they require *intermediation* and *interpretation* by a political process, and that the solutions must then be specified and "translated" into functional requirements. In this process the future users ought to play an important role. In this respect government technology procurement supplements ordinary private demand. It differs in that the perspective can be more long term; qualitative information can more easily be exchanged between buyer and seller; and the supplier's risks can be reduced by guaranteeing markets for the resulting products or services. The buyer simply creates incentives for the supplier to invest in the development of new systems and products. As we saw in Figure 1, these incentives come from the demand side.[12]

Given that government technology procurement results in some specific technological outcome – say for example, the development of energy efficient refrigerators – questions arise whether the policy has influenced the direction of technical change (would the innovation have come about as the result of market forces) or whether it has influenced the speed of technical development (how much sooner was the innovation introduced than it otherwise would

[10]That innovation policy can have noneconomic objectives is evidenced by the long and substantial history of defense-related government intervention in the process of technical change. Such policies have by no means been governed by goals of economic growth or increased competitiveness, but rather by national security, military strength, and independence objectives (Edquist 1993a, 19).

[11]Clearly articulated objectives are also necessary for effective evaluation in order to decide if the policy has been successful or unsuccessful (Edquist 1993a; 19).

[12]This does not exclude the possibility that the initiative to technology procurement in certain cases can come from the potential supplier. Neither does it preclude the use of supply-side policies – as for example, financial support for the technology development itself – in coordination with the demand-side use of government technology procurement. The distinguishing characteristic here is that the expected result of the procurement must nevertheless be the satisfaction of a specific need or the solution of a well-defined societal problem.

have been). Successful government technology procurement influences firm behavior in a nonmarginal way by creating incentives for firms to try different ideas. If the direction of technological change is substantially affected, then in an evolutionary theoretical perspective, technology procurement leads to increased technological variety and the development of additional *technology paths* that would not have otherwise been available. If a part of this increased technological diversity survives the internal firm and market selection, then government technology procurement has effectively added new technological choices to those already existent. In turn, these new paths can give rise to further innovations that would not have been possible otherwise. In this way, government technology procurement may accomplish the realization of technological change that far exceeds the specific shift in technology induced by the policy. On the other hand, even if government technology procurement only accelerates the development and diffusion of new technologies, its impact may be felt far into the future through its effect on the rate of economic growth.

The extent to which government technology procurement in practice can influence the direction and pace of innovation and technical change is difficult to determine with precision, requiring detailed empirical studies that are beyond the scope of this chapter. Nevertheless, casual observation of one of the most famous examples of government technology procurement indicates its potential impact on the direction of technical change.

Space technology would never have developed without government orders. Enormous efforts to "put a man on the moon", among other things, led to the realization of a new technological trajectory. This in turn led to new innovations that are used outside of space activities, for instance, certain types of integrated circuits and Gore-Tex clothing. In principle, there is nothing that precludes such grandiose efforts from being employed to solve environmental problems among others. This type of innovation policy would influence the character of innovations and the direction of technical change.

3.2 Technology Procurement an Industrial Policy Instrument

We next turn to the question of how government technology procurement can promote industrial competitiveness by generating technological capabilities within firms. Previously we saw examples of technology procurement strengthening the competitiveness of Swedish companies: ABB's high-tension current-transmission system and Ericsson's AXE-system. A general form of the argument was expressed in a report published by the Swedish Academy of Engineering Sciences in 1982:

> Technology procurement which is correctly implemented means an indirect support of the development capacity of industry, often more to the purpose than other forms of public support to the development of industry. The industrial capability generation which technology procurement implies, can have large industrial policy significance as a basis for export production, etc. (IVA 1982, 6)

The phrase "indirect support" implies that strengthening the competitiveness of Swedish industry has not been the primary objective of technology procurement. The primary purpose of government technology procurement has been *innovation policy* rather than *industrial policy*; regardless, the implications for industrial structure and competitiveness can be substantial. The crucial element of government technology procurement that makes it an important tool of industrial policy is that it exposes the industry to demands to do something new. If firms can fulfill these demands they can gain a competitive advantage in international markets.

It is crucial that the procurement involve a substantial element of competition among potential suppliers. The requirement for competition is usually met where the procurement takes place through open – and even international – bidding. If orders are given to a single company with whom the supplier has a long-term relationship, there is a risk that the procuring firm becomes "locked" into a collaborative arrangement which leads to the development of inferior products and systems. In such a case, technology procurement may fail in both its innovation policy and industrial policy goals. Neither should procurement be restricted to host-country firms; rather, for the sake of efficiency, bids must be solicited from abroad as well. Since successful industrial policy necessitates that the products that are developed meet international standards in order to be competitive in other countries, this element of international competition creates a necessary and effective pressure on potential domestic suppliers.

When the submission of bids is genuinely open and international, orders will sometimes go to foreign companies. If neither the development of the technology nor the resulting production take place domestically, it may be argued that the procurement failed in its national industrial policy function, although it may still be successful in its innovation policy role. Historically though, government procurement orders tend to go to domestic firms, even when bidding is international; this has most often been the case in Sweden. This is natural on account of the importance of geographic, linguistic, and cultural proximity as well as due to the existence of networks of long-standing relationships between suppliers and users. One mechanism that tends to favor domestic firms is the fact that firms active in Sweden are most likely well informed about plans in the earliest stages of a procurement process. There is, therefore, reason to believe that despite internationalization of bidding, a large part of government technology procurement will continue to be filled by domestic firms.

Once procurement becomes "national" (Swedish) in this sense, its indirect industrial-policy function operates through several mechanisms. First, government technology procurement promotes a structural transformation of national (Swedish) industry as the newly developed systems and products lead to new economic activities. Second, these newly developed products are often (almost by definition) characterized by a high R&D intensity. Hence government technology procurement can change the R&D-intensity of industrial production. High-technology sectors are usually characterized by higher productivity, higher productivity growth, faster market growth, and even higher salaries, in comparison with industry as a whole (Edquist 1993a, 44-45). This is a very important objective in Sweden since the R&D intensity of Swedish industry is considerably below the OECD average and decreasing (Edquist and McKelvey 1994).

Thirdly, technology procurement leads to increased demands for a highly skilled and technically trained labor force. In this respect, technology procurement accordingly promotes structural change towards an economy which is characterized by high productivity, high qualification demands, and high salaries.[13]

Therefore while at the same time that technology procurement presents interesting possibilities for solving specific social or environmental problems, it also creates opportunities for changing the structure of industrial production with respect to product attributes, technological intensity, and employment characteristics. This process of structural transformation creates important technical capabilities at the firm level which serve as an important basis for international competitiveness. As part of an technological infrastructure policy,

[13]It can also be shown that technology procurement often leads to more jobs, thanks to investments in new activities and new production. This means new employment if the new product does not functionally replace an old one which demanded more labor for its production and – in the case of an investment good – if the investment good does not save more jobs when used than were needed for its production. The ability of product innovations to create jobs distinguish them from process innovations. The very complex relations between different kinds of innovations, productivity, and employment is investigated in Edquist (1993c).

government technology procurement thus contributes to the increased competitiveness of Swedish firms.

4. Technology Procurement in Practice

Having examined the theoretical and policy implications of government technology procurement, as well as having described several cases, we now turn to the practical matter of implementing government technology procurement as part of a coherent technological infrastructure policy. We examine issues of timing, functional requirements, buyer coordination, developing buyer competence, organization of design work, and issues related to internationalization

4.1 The Phases of Technology Procurement

Because the development of new products and systems is often very time-consuming, government technology procurement is normally a long-term process. Often the process continues over several election periods, making it vulnerable to political determinations regarding continuation or termination of the project. The long maturation process also leads to the risk that the boundary between the purchaser and the supplier becomes obscured, a situation which is problematic for the management and control of the procurement process.

These factors advise the conceptual partition of the process into phases, from the formulation of the product-concept that embodies the technology to the ultimate delivery of the product or system. The purposes for dividing the procurement process into steps include controlling costs and estimating (and distributing) risks. It is most convenient to divide the technology procurement process into four phases:

- study phase,
- project phase,
- design phase, and
- production phase.

Of course, certain time overlaps can occur, but the sequence is useful (Linder 1982, 7).

During the *study phase* the need to be satisfied or the problem to be resolved by the procurement is identified. At this point the specific objective is articulated through the identification of the function which the system or product is to perform. During the *project phase* this function is specified. This function specification must be done very thoroughly. During the *design phase* the solution is worked out in detail and a prototype is often produced. Finally, serial production follows during the *production phase*.

A clear distinction must be made between the first two phases and the last two; the buyer must be responsible for the identification of needs and specification of the functional demands while the supplier takes responsibility for product development and production. These two roles must be clearly separated from each other.

4.2 Functional Requirements and Buyer Competence

Technology procurement requires that the buyer can clearly identify the need to be satisfied or the problem to be resolved by the procurement; the objective of the procurement must be clear. With this condition as a starting point, the buyer must be able to formulate and articulate his requirements concerning function and performance characteristics of the product, equipment, or system being procured. The needs to be satisfied consequently must be translated into functional demands and these clearly communicated to the supplier.

In government technology procurement the functional requirements are formulated by the national, regional, or local public purchaser. As illustrated in the case of Sweden's X2000 rapid train, the formulation of these specifications and functional demands can be extensive and very detailed. This is especially the case if the specifications include the requirements concerning real operating conditions and focus on the total or life-cycle cost, an approach which is becoming increasingly common because of its substantial advantages. It is thus important that the buyers are knowledgeable and exacting, capable of formulating precise and rigorous functional requirements. At the same time, these requirements must not become so detailed that the buyer does the design job himself.[14] In other words, *buyer competence* is absolutely central.

It is by no means self-evident that government organizations have the necessary buyer competence which is needed to formulate the functional demands. This is especially the case if the buyer is a small unit or if procurement of the specific commodity is uncommon for the buyer. These problems tend

[14]It can sometimes be difficult to find engineers who understand the importance of formulating specific functional demands, but avoid doing the design work themselves, i.e., provide the solution.

to be greater for local and regional administrations than for large national government agencies. This can require coordination between buyers, as discussed in the following section. At the same time, this problem also argues for systematically developing buyer competence, as will be discussed in section 4.4.

4.3 Buyer Coordination

Different types of government technology procurement call for different organizational forms. These organizational forms can revolve around development work (see section 4.5) or the initiation of the procurement process, i.e., the demand itself. Here we address coordination of potential users.

As has already been mentioned, procurement tends to be more difficult when buyers are fragmented. The simplest case is when the procurer is a large government agency (e.g., SJ or Televerket), while problems are the greatest when end users are individual consumers. Intermediate cases include county authorities and municipalities. There are strong reasons to coordinate procurement for intermediate-size buyers. This could be achieved through existing organizations such as the Federation of County Councils and the Association of Local Authorities. Further initiatives can, however, be justified.[15] Increased coordination between the technology procurement of various county councils could lead to exciting developments in the field of medical technology, for instance. Alternatively, more temporary coordination forms can also be useful. For example, the many medium-sized (around 100,000 inhabitants) cities in Sweden which are interested in an energy efficient and environmentally friendly system of passenger transportation could create a temporary organization to prepare and carry out a procurement project.

Earlier we saw that the NUTEK-organized procurement of energy-efficient technology was largely a matter of coordinating many small and fragmented end-users who individually neither represented a large enough demand nor possessed enough buyer competence to initiate a technology procurement process. NUTEK's coordination was a necessary precondition for the procurement. There are certainly many other opportunities where similar organizational forms can be developed and applied to other areas of technology development.

[15]The County Council's fund for technology procurement and product development (LFTP) has existed since 1982, but the problems with getting a coordination between different county councils have shown themselves to be rather large.

The observation that fragmentation of organizations leads to technology procurement being more difficult has significant import for current Swedish policies of decentralization and privatization. For instance, when the National Swedish Board for Public Building is organizationally split into several separate companies and regions, technology procurement is made more difficult. The same is true for the transformation of state-owned agencies into companies, as has been taking place in Sweden in the early 1990s. This can mean that the boundary between state and private activities becomes blurred within many areas, complicating the distribution of responsibility regarding planning and implementing technology planning efforts which are motivated by long-term societal interests (Andersson et al. 1993, 68)

Similarly, by transforming the state agency Televerket into the company Telia, the government no longer has a natural coordinator for the continuous build-up of Swedish infrastructure in the field of information technology. There are also plans to transform the Post Office (Posten), the State Railway Agency (Statens Järnvägar), and parts of the road works (Vägverket) and rail works (Banverket) agencies into companies. This has diminished the strength of the public sector, a tendency which will continue as long as the process of privatization continues. This process will weaken the capacity to pursue a far-sighted and long-term government technology procurement policy with the dual objectives of solving societal problems and indirectly strengthening Sweden's competitiveness.

Government technology procurement as part of a technological infrastructure policy requires forceful public actors who can identify needs, formulate demands, and thereby influence technical change. An active technological infrastructure policy which includes government technology procurement as an important tool, thus requires either a break in the trend toward decentralization and privatization, or an alternative strategy that would establish new procurement coordinating frameworks. One area where such an organization can be justified is in infrastructure for information technology. In this field Televerket once had over-arching responsibility, which it applied in a visionary and well-administered fashion, as evidenced by the initiative to collaborate with Ericsson on the AXE project. Though the need for long-range and visionary planning is even greater now, it is being neglected because no institution has been assigned the responsibility.

These examples demonstrate that user coordination and other organizational aspects of technology procurement are important components of a technological infrastructure policy. The specific conclusion in the case of Sweden is that some type of coordinating organization with responsibility for creative long-range planning and pilot projects ought to be created.

4.4 Support for Development of Buyer Competence

We have seen that the existence of buyer competence is a precondition for effective government technology procurement. Generic procurer competence can greatly enhance technological capability at the firm level and thereby strengthen competitiveness of firms. Government's role in supporting the development of these competencies is an intrinsic part of a technological infrastructure policy.

There are several ways in which the government can support the generation of procurer competence. The most direct is to subsidize the employment of competent engineers in those public administrations and agencies which carry out technology procurement. A similar approach is to require every such organization (government department, regional or local administration) to appoint a person to be responsible for technology procurement.[16]

A third way to support the development of buyer competence is to systematically educate those who are responsible for technology procurement in the various public administrations. Such training is important, since lack of appropriate project leadership is often a limitation to well-planned technology procurement. A large part of this education can consist of diffusing the experience and knowledge from those public administrations that have successfully undertaken substantial programs of technology procurement.

The organization of research projects as part of the study and project phases of technology procurement is a fourth method of developing of buyer competence. These projects could focus on special "needs and technology areas" that have previously been identified as having exceptional potential for technical, social, and economic achievement (see Annex I for examples). Targeting specific needs or problems which could be satisfied or solved by technical development, such projects should ideally include interdisciplinary research teams and operate over relatively long periods. The interdisciplinary composition – including behavioral scientists, economists, engineers, natural scientists, and innovation researchers – is critical for synthesizing the social and technical aspects of technology procurement. The long-term nature of the research projects is essential for participants to develop their own long-term perspectives for identifying needs and potentials.

[16]Technology procurement must then be separated from procurement of standard products.

Drawing on the diverse expertise of the professional staff, the research projects should investigate a wide array of issues, including questions of what products and systems are socially desirable, technically feasible, and economically viable. The results should form the basis for the formulation of specifications and functional requirements. By explicitly anticipating future needs, research will be directed toward identifying how industrial competence can be developed to satisfy those needs. The long-term perspective implies that those involved in the project can – and should – consider factors which normal market actors disregard.[17]

Representatives of potential suppliers should not participate in these research groups since competition between potential producers is a necessary part of government technology procurement. Too close a relationship between potential suppliers and those involved in the specification of the technology to be procured is undesirable from both the innovation policy and the industrial policy perspectives. On the other hand, having mentioned the advantage domestic firms may have due to the existence of national networks and early access to information, it may be justified to strengthen this advantage by inviting potential domestic suppliers to take part in reference groups of the research projects. However, it is critical that potential suppliers not be given influence over the project, and that open competition in bidding not be compromised.

Conversely, potential users ought to be encouraged to take part in the research projects. While during the identification of appropriate areas for preparatory research projects needs and problems ought normally to have the initiating role, nothing should stop industry representatives or researchers from suggesting projects themselves. The key criterion is that needs and potential technological solutions are matched in the process.

4.5 Organization of the Development Work

When the objective of the procurement is clear and the functional demands clearly specified, the product or system can be developed and designed. When it concerns the organization of the development phase, there are reasons to show openness and flexibility. It was earlier argued that the design phase

[17]Government intervention shall always be a complement to the market and never replace it. Intervention can only be justified if the result can be expected to be better with it than without it. The division of labor between markets and governments is discussed in detail in Edquist (1994a).

should be clearly separated from the formulation of functional requirements. In the simplest case the supplier develops the product on the basis of the functional demands, but still completely independently. However, when the expected product or system is complicated, it is advantageous for the design phase to also be clearly separated from the production stage. The end result of the design phase should then be a functioning prototype.

In the cases of high-tension current transmission and rapid trains the organization of the procurement was relatively simple; the buyers and the suppliers simply negotiated with each other. However, each individual procurement project requires its own suitable organizational forms. For example, in the case of AXE, a joint company (ELLEMTEL) was created, despite the fact that the procurement involved only one purchaser and one supplier. ELLEMTEL is an example of a specially modeled organizational solution. The case of NUTEK and energy-saving technology procurement of refrigerators illustrates a more complicated specially modeled organizational solution where the role of the government agency was to be a catalyst and a coordinator of potential users.

4.6 Technology Procurement, Internationalization, and European Integration[18]

It is often alleged that government technology procurement is becoming increasingly difficult because of the ongoing processes of internationalization and European integration. From the perspectives of both innovation policy and industrial policy, this claim is unwarranted.

Since its entry in the European Union in January 1995, Swedish public procurement has been subject to new regulations. In principle, these changes make Swedish procurements of substantial size more open to European suppliers.[19] The purpose is to increase competition and thereby reduce prices.

[18]This section is partly based on a conversation with Hans Sylvén at the Swedish Committee for Technology Procurement.

[19]Specifically, every procurement tender which is larger than 1.9 million Swedish crowns and all construction projects larger than 47 million crowns must be announced in the EU's Supplement to the Official Journal of the European Communities. Bids can be received within 52 days and the advertisement in the journal must be written in one of the EU's nine official languages. The procurer can, however, request that the bids shall be rendered in Swedish. In Denmark the request that the bids be rendered in Danish has lead to considerably decreased interest from foreign suppliers. In many EU countries – especially in southern Europe – a practice has developed not to follow all of the EU rules in detail.

According to various estimates it is expected that local Swedish authorities will be able to save 10-15 billion crowns per year as a result of these new procurement procedures.

Previous Swedish laws on government procurement (i.e., requirements for procurement on business lines, for competition, and for nondiscrimination) remain in effect. With regard to basic principles, the differences between the two regimes are not especially great. On the other hand, the new rules have changed the forms for procurement and have enlarged the category of procurement units to which the rules are applicable.

With regard to government technology procurement as an *innovation policy* instrument, the EU rules are only an advantage. Their effect is to increase the number of potential suppliers, as well as to strengthen openness and competition among them. This can only enhance the ability of government technology procurement to satisfy societal needs and solve specific problems.

The effects of the new procurement rules on government technology procurement as an *industrial policy* instrument are more complicated. As noted earlier, technology procurement as an industrial policy instrument requires openness in tenders and competition between potential suppliers, including foreign firms. The procurement law therefore forbids that Swedish-controlled companies are favored intentionally in government procurement.[20] The law also prohibits requirements from the buyer that a foreign supplier shall carry out the development work or the production operations in Sweden.[21] This however does not prevent the foreign bidders from offering to carry out production in Sweden. To the extent that the new rules enhance the relative competitive position of non-Swedish EU firms in Swedish technological procurements, the industrial policy role of those procurements is diminished.

Recall though, that we previously argued that it can be expected that Swedish suppliers will naturally be overrepresented in Swedish government technology procurement. To the extent that this advantage is not completely overwhelmed by the increased openness to foreign bidders, this means that technology procurement can still function as an industrial policy instrument from the Swedish point of view.

[20]This, however, does not apply to technology procurement of defense material which involves the country's security and which does not have any civilian use.

[21]Since the Swedish law is general, it even concerns non-EU suppliers. This means that procurers in a country outside the EU – e.g., the U.S. or Japan – can demand that Swedish suppliers carry out local production there, but that Swedish procurers can not require the same from a company controlled from that country.

Additionally, the law even allows indirect discrimination of foreign companies in a number of ways. As mentioned earlier, it can be required that the bid be rendered in Swedish. Similarly, the entire detailed basis for the bid (demand specifications, technical descriptions and other conditions) can also be provided in Swedish. It is also possible – and even natural – that information from the buyer is given selectively before the formal procurement process begins. Swedish procurers can, for instance, inform Swedish manufacturers of their plans for the next three or five years.[22] Another compensating advantage is that employees sometimes move between Swedish authorities and Swedish companies. Specialists in different government agencies and companies are also inclined to exchange information informally. These are natural processes and underscore the argument that domestic manufacturers are often favored in government technology procurement.[23]

Government technology procurement can thus continue to be an important industrial policy instrument at the national level in an integrated Europe because of the geographical and cultural closeness between domestic procurers and suppliers. In light of these observations, we conclude that the rules that are associated with EU membership will not significantly decrease the potential to use government technology procurement as either an innovation or industrial policy instrument. The earlier Swedish rules did not constitute an obstacle for effective government technology procurement and neither will the new ones.

However, it appears that many European Union countries are currently too small for handling technology procurement in certain technological areas. For example, there is no need for 15 rapid train systems or commercial aircraft designs in Europe. There are, therefore, strong arguments for using technology procurement at the EU level. If the EU, in this way, is seen as one unit, technology procurement certainly has the potential to serve as an innovation policy and industrial policy instrument for the community as a whole. Earlier experience at the national level is of great value in an attempt to develop such a new policy option for the EU.

[22]On the other hand, a procurer cannot advertise in any other journal than Official Journal before the same day as the advertisment is sent to this journal. The Official Journal has to publish the advertisement within 12 days. Normally it is published within 5-6 days. Swedish bidders can thus have an advantage if the buyer advertises in a Swedish journal at the same time as the announcement is sent to Official Journal.

[23]In addition, procurement within the water, energy, transportation, and telecommunications sectors are subject to special rules, implying that domestic suppliers can be favoured to a larger extent. These sectors account for a large part of the government technology procurement.

5. Summary and Conclusions

What roles the market and the government respectively ought to play within innovation activity varies between countries, regions, sectors, and functions. In this chapter we have discussed how state, county, and local authorities can use their orders for products and systems to influence innovations and technological change through government technology procurement. We have shown that the government activities can be used powerfully and to long-term advantage in this respect.

Through technology procurement the government sector plays an important industrial policy role by providing important support to the development of the competitiveness of Swedish companies. The government sector can also play an important innovation policy role by using its demand to influence technological development so that it can help to satisfy important societal needs, by solving, for example, environmental or social problems.

The competence to specify the functional characteristics of products and systems and the technological capabilities which are generated as a result of government technology procurement can be considered to be important parts of the technological infrastructure (TI) in a country. The protodemand which the public procurer articulates has had a market-building effect. This requires, though, that the procurer has "a vision of the future" and can translate these into functional characteristics. This may require organizational innovations, whereby the government creates public organizations where procurer competence is generated and recreated. In this way government technology procurement involves generating infrastructures which, in turn, create capabilities at the firm level and support the competitiveness of firms. Technology procurement can, in other words, be an important part of a technology infrastructure policy (TIP).

Annex: Possible Future Procurement Projects

- Infrastructure for information technology: information superhighways.
- Interactive television systems.
- IT-based security alarm systems for elderly and disabled.
- "Electronic assistants" for intellectually retarded people.
- Energy-effective and environmentally sound urban traffic systems.
- "The clean car".
- Systems for renovation of old apartments and houses.
- Systems for renovation of old water and sewage networks.
- New energy conversion technologies.
- Software for administrative systems.
- Specific kinds of medical technology.

References

Andersson, H., B. Atlestam, G. Dreborg, Gunnel, and L. Stenberg. 1993. Teknikpolitik för tillväxt – Åtgärder för bättre utnyttjande av teknologisk kompetens (technology policy for better exploitation of technological competence), in Swedish. Published in *Nya Villkor för Ekonomi och Politik*, Bilagedel 1, Ekonomikommissionens Förslag, SOU 1993:16.

Andersson, P.G. 1982. Snabbtåget X2 – Teknikupphandling (The rapid train X2 – technology procurement), in Swedish. Published in IVA.

Arrow, K. 1962. Economic welfare and the allocation of resources for invention. In *The Rate and Direction of Inventive Activities*. Princeton University Press.

Carlsson, B. and R. Stankiewicz. 1991. On the nature, function and composition of technological systems. *Journal of Evolutionary Economics* 1: 93-118.

Cohen, W.M. and O.A. Levinthal. 1990. Absorptive capacity: A new perspective on learning and innovation. *Administrative Science Quarterly:* 128-151.

Edquist, C. 1993a. Innovationspolitik för förnyelse av svensk industri (Innovation policy for renewal of Swedish industry), in Swedish. *Tema T Rapport* Nr 33, 1993, Tema Technology and Social Change, Linköping University.

——. 1993b. Systems of innovation - a conceptual discussion and a research agenda. Paper presented at workshop no. 3 "Globalization Versus National or Local Systems of Innovation" of the EUNETIC Network at BETA in Strasbourg, March 11-12, 1993.

——. 1993c. Technological unemployment and innovation policy in a small open economy. Paper written for the OECD and presented at the conference on Technology, Innovation Policy and Employment, Organized by the OECD and the Finnish government, Helsinki 7-9 October, 1993. Published as Tema T Working Paper No 124, 1993.

——. 1994a. Technology policy - the interaction between governments and markets. In *Technology Policy – Towards an Integration of Social and Ecological Concerns*, edited by G. Aichholzer and G. Schienstock. Berlin: Walter de Gruyter.

——. 1994b. Offentlig teknikupphandling för behovstillfredsställelse och konkurrenskraft (Government technology procurement for needs satisfaction and competitiveness), in Swedish. Published as Tema T Report 34.

Edquist, C. and B.Å. Lundvall. 1993. Comparing the Danish and Swedish systems of innovation. In *National Systems of Innovation – A Comparative Analysis*, edited by R. R. Nelson. Oxford University Press.

Edquist, C. and M. McKelvey. 1994. The Swedish paradox: High R&D intensity without high tech products. Paper presented at the European Association for Evolutionary Political Economy (EAEPE) Conference, Copenhagen, October 28-29. To be published in the proceedings from the conference.

Fridlund, M. 1993. Tekniska utvecklingspar – En länk mellan sociotekniska system och innovativa nätverk (Technology development pairs – a link between socitechnical systems and innovative networks), in Swedish. Paper presented at the seminar "Teknisk och industriell utveckling", arrangerat av Chalmers, Handelshögskolan i Stockholm och NUTEK, Särö, 17-19 November.

Glete, J. n.d. High technology and industrial networks. Some notes on the cooperation between Swedish high technology industries and their customers. Mimeo.

von Hippel, E. 1988. *The Sources of Innovation*. Oxford University Press.

IVA. 1982. Teknikupphandling. Dokumentation från ett symposium (Technology procurement. documentation from a symposium), in Swedish. IVA-Rapport 233, Ingenjörsvetenskapsakademin, Styrelsen för teknisk utveckling.

Kline, S.J. and N. Rosenberg. 1986. An overview of innovation. In *The Positive Sum Strategy – Harnessing Technology for Economic Growth*. edited by R. Landau and N. Rosenberg. Washington D.C.: National Academy Press.

Linder, G. 1982. Teknikupphandling – Något om förutsättningar och problem (Technology procurement – something about preconditions and problems), in Swedish. In IVA.

Lundvall, B.Å. 1985. Product innovation and user-producer interaction. *Industrial Development Research Series No. 31*. Aalborg University Press.

Lundvall, B.Å., ed. 1992. *National Systems of Innovation – Towards a Theory of Innovation and Interactive Learning*. London: Pinter Publishers.

McKelvey, M. 1991. How do national systems of innovation differ? A critical analysis of Porter, Freeman, Lundvall and Nelson. In *Rethinking Economics: Markets, Technology and Economic Evolution*, edited by G. Hodgson and E. Screpanti. Aldershot, England: Edvard Elgar Publishing Ltd..

—. 1994. National systems of innovation. In *The Elgar Companion to Evolutionary and Institutional Economics*, edited by M. Too, W. Samuels and G. Hodgson. Edvard Elgar Publishing.

Nelson, R.R. 1959. The simple economics of basic research. *Journal of Political Economy* 67.

Nelson, R.R., ed. 1993. *National Systems of Innovation: A Comparative Analysis*. Oxford: Oxford University Press.

Teubal, M., T. Yinnon, and E. Zuscovitch. 1991. Networks and market creation. *Research Policy* 20: 381-392.

Vedin, B.A. 1993. Så blev AXE ett sjumilakliv (This is the way AXE became a great leap forward), in Swedish. In Daedalus, 1994. Tekniska Museet, Stockholm.

II. TIP for Leading-Edge Industries

The Design of High-Technology Consortia: Lessons from SEMATECH

Peter Grindley, David C. Mowery, and Brian Silverman*

Since its establishment in 1987, SEMATECH (the Semiconductor Manufacturing Technology consortium) has received considerable attention from U.S. policymakers and managers. The consortium has been endorsed by the Clinton Administration as a model of government-industry cooperation in supporting research in "critical technologies".[1] Other industry-led consortia have been proposed or established in areas ranging from electric automobiles (the Advanced Battery Consortium) to high-definition displays and other semiconductor industry products (the proposed multi-chip module foundry consortium) to aircraft development and production (the proposed Aerotech consortium). Contrary to the predictions of Cohen and Noll (1992), SEMATECH has been followed by a number of other U.S. technology development consortia funded from public and private sources.[2]

This paper attempts to derive some lessons from SEMATECH for the design of such consortia in other industries. Although SEMATECH had few precedents in the U.S., both the Japanese and Western European governments have sponsored civilian technology consortia since the 1960s. With and without

* An earlier version of this paper appeared in the *Journal of Policy Analysis and Management* (1994). Research for this paper was supported by the Alfred P. Sloan Foundation through the Consortium on Competitiveness and Cooperation, by the U.S. Air Force Office of Scientific Research through the U.S.-Japan Industry and Technology Management Program, and by the Canadian Institute for Advanced Research. We are grateful to Dr. William Spencer for his invaluable assistance with this research, and to two anonymous referees and Dominique Foray for suggestions and comments.

[1]"SEMATECH, an industry consortium created to develop semiconductor manufacturing technology...can serve as a model for federal consortia funded to advance other critical technologies. Programs will be encouraged in the development of a new automobile, new construction technologies, intelligent control and sensor technologies, rapid prototyping, and environmentally conscious manufacturing" (Clinton & Gore 1993).

[2]The analysis by Cohen and Noll (1992) builds on their previous study of federal civilian technology programs (1991). But Cohen and Noll appear to overstate the importance of the results-focused research agenda of publicly funded civilian technology programs in arguing that they are fundamentally incompatible with the distributive politics of the U.S. Congress. Moreover, Cohen and Noll may understate the distributive politics of large-scale publicly funded basic research programs in the U.S. (Mowery 1992a).

M. Teubal et al. (eds.), Technological Infrastructure Policy, 173–216.

federal funding, consortia have sprung up in other U.S. industries since 1980. We compare SEMATECH with these other domestic and international consortia to develop some insights into the relevance of collaborative R&D programs in other U.S. industries.

Changes in SEMATECH's role and research agenda since its foundation have important implications for this discussion. Since its foundation, SEMATECH has shifted from *horizontal* research cooperation to *vertical* collaboration between its members – major U.S. users of semiconductor process equipment and materials and the U.S. suppliers of these goods – a structure that may be applicable to other U.S. industries. In many respects, SEMATECH now resembles an industry association, diffusing information and best-practice techniques, setting standards, and coordinating generic research. Like many Japanese cooperative research projects, SEMATECH is concerned as much with technology diffusion as with the advancement of the technological frontier (Hane 1994 makes a similar point). The SEMATECH research agenda emphasizes near-term results, a characteristic that it shares with other industry-led consortia in the U.S. and elsewhere.

Nevertheless, SEMATECH's vertically oriented research agenda has not met the goals of all of its original members, several of whom have left the consortium. Moreover, SEMATECH's efforts have not prevented several equipment firms from exiting the industry. The weakness of U.S. equipment firms' competitiveness in certain key areas that have received substantial attention and funding from SEMATECH (such as lithography) underscores the point that improved technology alone is not sufficient to reverse the fortunes of weakened firms.

1. Design and Policy Issues for R&D Consortia

The economics of collaborative research have received considerable attention in recent research (Katz and Ordover 1990; Mowery and Rosenberg 1989; Bozeman et al. 1986). The hypothesized advantages of collaboration in research include the ability of participating firms to lower costs and spread risks, reduced duplication in their R&D investments, and the exploitation of economies of scale in the R&D process. Research consortia also may be able to appropriate more of the returns to their R&D investments and internalize more of the interfirm spillovers that discourage R&D investment by individual firms; the hypothetically greater ability in appropriating returns favors collaboration in basic research.

These hypothesized advantages of collaborative research have not been subjected to detailed empirical scrutiny, however, and may be overstated. The output of cooperative research must be absorbed by the participant firms and transformed into commercially relevant knowledge. In order to exploit externally performed research – whether this research is performed in a multifirm consortium, a federal laboratory, or a university – participant firms must invest in the creation of in-house expertise (Mowery 1983; Cohen and Levinthal 1989). Some duplication of the in-house research investments of firms thus is likely even among participants in research consortia.[3]

In addition, any reduction in duplication of research projects may come at a cost. Diversity in research projects provides an important hedge against the possibility that any single research project will be fruitless (Nelson 1961). Yet if all or most firms in an industry participate in a cooperative research scheme, reducing the number of independent lines of inquiry, such diversity will be reduced. Although it is not a necessary result of research cooperation, centralization and reduced R&D diversification could lower the productivity of an industry's or a nation's research investment.

How important are the economies of scale in R&D that are widely cited as a source of efficiency gains through research collaboration? The average costs of research experiments at large, specialized research facilities, such as wind tunnels or advanced instrumentation, will decline with more intensive use. But no empirical evidence suggests that all industrial R&D is characterized by such economies of scale (Fisher and Temin 1973), nor are there indicators that enable one to identify the industries or technologies characterized by R&D scale economies.

Finally, economic analyses of cooperative research often overlook the extent to which appropriability issues manifest themselves within a consortium. Even a precommercial R&D consortium typically involves the sharing of sensitive information and know-how among competing firms. The same fear of knowledge spillovers that often causes firms to underinvest in R&D in the

[3]Moreover, to the extent that basic research requires a larger investment in this in-house "absorption mechanism", participants in cooperative research may react to the high costs of exploiting collaborative basic research by shifting the collaborative research agenda away from basic research (Mowery and Rosenberg 1989). The experience of the industry-funded Microelectronics and Computer Technology Corporation (MCC) illustrates this tendency. Since its foundation, MCC has shifted its research agenda from basic to near-term research, and now relies more heavily on member firms for the performance of this research. The shift in MCC's research to smaller-scale, shorter-term projects reflects the desire of members to show measurable returns on their contributions, as well as the difficulties of translating basic research into products (Lineback 1987a and 1987b).

absence of a consortium may also lead them to resist knowledge-sharing within a consortium (Evan and Olk 1990). Concerns over appropriability and spillovers, coupled with tensions arising from overall differences and conflicts among the objectives of consortia members, may impair the performance of consortia, increase their costs, and contribute to their dissolution.[4]

The complexity of consortium design and management is attributable to at least three challenges: definition of the research agenda and choice of projects; transfer of research results to participants; and adaptation to change in the economic and technological environment. Defining a consortium's research agenda is difficult because of appropriability concerns and divergent member-firm objectives. Firms may be unwilling to entrust the development of their core technologies to the consortium or to risk the leakage of key proprietary technological and strategic information. These problems, particularly divergence among members' objectives, become more severe as the number of firms in a consortium increases.

Consortia in the U.S., Western Europe, and Japan have developed different solutions to this challenge. Those involving a large number of participants (more than 30), including such European consortia as ESPRIT and Alvey, often adopt a decentralized structure and research agenda, funding a number of independent projects that each involve a small number of firms, perhaps as few as two or three.[5] This structure facilitates control of the research by individual firms and reduces knowledge spillovers among participants. The U.S. MCC consortium began with a complex research agenda that emphasized selective collaboration within a broad umbrella organization, but has evolved toward a decentralized structure that in some respects resembles that of ESPRIT.[6] Another U.S. consortium, NCMS, has developed a similar structure.

Recent Japanese consortia, such as VLSI, also display tensions between sharing and appropriating the results of research in their blend of firm-specific, individual projects and joint research facilities (Fransman 1990; Hane 1994). But large Japanese project consortia with fewer participants (less than 10) exhibit greater centralization in the establishment of their research strategy than the Western European consortia. This organizational design was palatable

[4]"Consortiums are the wave of the future, but they sure are hard to manage," according to former SEMATECH CEO Paul Castrucci (quoted in Smith 1989).

[5]A summary listing of the characteristics of some recent R&D consortia in the U.S., Western Europe, and Japan may be found in Table 1. For a listing of acronyms used in our discussion, see Appendix.

[6]This change in its structure occurred concurrently with expansion in MCC's membership.

to the members of previous Japanese consortia because all of the participants were some distance behind the world technological frontier. But the improved technological capabilities of these firms have made cooperative research agreements among them more difficult (Odagiri and Goto 1993).[7]

A related issue in the definition of the research agenda for R&D consortia is the time horizon of the research. Despite their frequent promotion as a solution to myopic R&D management, consortia that do not include universities or national laboratories rarely focus on long-term, fundamental research. Instead, successful consortia in the U.S., Western Europe, and Japan typically pursue relatively short-term research agendas. Indeed, the temporary nature of most Japanese consortia makes it difficult for them to undertake long-term research.

The second challenge facing consortia design involves the transfer of technology to participants. This process operates more successfully when cooperative research activities are complemented by significant in-house investments in related research by program participants. The size of this parallel investment and managerial commitment to the projects will depend on the selection of projects that are central to the firms' broader commercial and research interests. Technology transfer also is aided by rotation of member firms' personnel through collaborative or external research establishments.

A third important challenge for consortia is adjusting to change in their environment or research agenda. The difficulties of negotiating a consortium research agenda in the first place mean that gaining agreement among members on its revision may be impossible. Even if agreement can be reached, the consortium's original structure or facilities may prove unsuited to the new agenda. The organizational structure of a consortium also affects the feasibility of revisions in its research agenda – a strong administrative structure, rather than a weak secretariat, appears to be necessary. Some consortia, such as the Alvey and ESPRIT programs in the U.K. and Western Europe, have largely avoided flexibility in the research agenda. For SEMATECH, such flexibility has been an important strength, albeit one that has come at some cost.

[7]See Samuels (1987) and Fransman (1990, 59), who notes that "...the perceived threat from IBM as the dominant player in the international computer industry, helped to galvanize a significant degree of solidarity and cooperation in Japan. It was obvious that Japanese computer producers were under far greater competitive pressure from IBM than they were from each other. The dominance of IBM therefore helped create the conditions for a closer degree of cooperation between the Japanese companies."

2. The Foundation and Structure of SEMATECH

2.1 Structure and Objectives

The origins of SEMATECH can be traced to late 1986 and early 1987. Alarmed by Japanese firms' growing domination of the global market for semiconductor memory chips, a 14-member committee of the Semiconductor Industry Association formulated a proposal for a research consortium.[8] These 14 companies, which accounted for over 80 percent of U.S. semiconductor component manufacturing capacity, became the founding members of SEMATECH.[9] In addition to paying dues totalling $100 million per year (which are matched by $100 million from ARPA), the member firms contribute roughly two-thirds of SEMATECH's 300-member research staff through temporary (usually 2-year) rotation of assignees at the consortium.

Concurrently with the foundation of SEMATECH, U.S. semiconductor materials and equipment (SME) suppliers formed SEMI/SEMATECH to facilitate linkages between U.S. SME suppliers and SEMATECH. SEMI/SEMATECH currently has 135 members (out of roughly 800 U.S. firms in the SME industry) who account for more than 85% of U.S. SME sales. The organization, which has eleven full-time employees headquartered in a wing of SEMATECH's Austin facility, is funded by membership dues and by fees generated through sponsorship of conferences.[10] The president of SEMI/SEMATECH sits on the board of SEMATECH, and the boards of SEMI/SEMATECH and SEMATECH meet jointly once each quarter.[11]

[8]Concerned by the implications for national defense of U.S. dependence on foreign semiconductors, the Defense Science Board (an advisory committee within the Department of Defense) developed a competing proposal that recommended creation of a manufacturing facility jointly owned by government and industry to produce semiconductor components (McLoughlin 1992).

[9]SEMATECH's founders included the following firms: Advanced Micro Devices, AT&T, Digital Equipment Corporation, Harris Corporation, Hewlett-Packard Company, Intel Corporation, IBM, LSI Logic, Micron Technology, Motorola, National Semiconductor, NCR, Rockwell International, and Texas Instruments.

[10] Dues are based on company revenue, and start at less than $5,000. SEMI/SEMATECH's 1991 operating budget was approximately $1.6 million.

[11]SEMI/SEMATECH should not be confused with the trade association SEMI (Semiconductor Equipment and Materials International). SEMI has over 1500 member firms, of which only 800 are U.S.-based. While the members of SEMI/SEMATECH are generally also members of SEMI, the two organizations are separate and have distinct roles and objectives.

2.2 Evolution of SEMATECH Objectives

SEMATECH's original objectives – to enable member firms to improve their semiconductor manufacturing process technology – underpinned the decision to build a large-scale fabrication facility in Austin. Fab One, completed in 1989 in 32 weeks, was a triumph of collaborative effort (U.S. General Accounting Office 1989). But SEMATECH had difficulty developing a research agenda that could exploit this research facility. Members questioned whether the development of advanced manufacturing processes was an appropriate objective for an industry consortium. Process technology expertise is central to the competitive advantage of individual semiconductor manufacturers, and member firms were reluctant to share such sensitive information. SEMATECH member firms also differed considerably in their level of manufacturing technology sophistication, raising the danger that some firms could "free ride" on the contributions of technology leaders.

In response to these controversies, SEMATECH altered its research agenda to one that sought to improve the technological capabilities of U.S. suppliers of semiconductor manufacturing equipment (Katz and Ordover 1990; U.S. Congressional Budget Office 1990), and to strengthen vertical cooperation between U.S. suppliers and U.S. users of semiconductor process equipment.[12] This research focus potentially benefits all members without threatening their proprietary capabilities, and thereby may reduce the appropriability problems associated with the original research agenda. The new research agenda has shifted the consortium's focus from the development of a state-of-the-art production process in its Austin facility to knowledge diffusion and technology transfer. In the words of William Spencer, current SEMATECH CEO, "We can't develop specific products or processes. That's the job of the member companies. SEMATECH can enable members to cooperate or compete as they see fit" (Burrows 1992, 58).

This revised research strategy is illustrated by SEMATECH's announcement in January 1993 that it had achieved its goal of manufacturing 0.35 micron line-width integrated circuits. Rather than developing a "recipe" and associated processes for the production of specific components that could be readily adopted by its member firms, the consortium had succeeded in developing and integrating a process and equipment that could produce a variety of 0.35 micron integrated circuits. In other words, the 0.35 micron

[12]According to Levy and Samuels (1991, 123), 80% of collaborative research agreements in Japanese industry, few of which are publicly financed, involve vertical relationships.

line-width process technology was a generic industry technology, rather than a firm-specific one. According to a spokesman for AMD, "What SEMATECH has done is an outstanding achievement, but it's not realistic to think a member of SEMATECH could take the process in its entirety and produce product with the same equipment and the same alignment" (Cassell and Dunn 1993).

The shift in its research agenda was associated with changes in SEMATECH's intellectual property policies. Originally, the results of SEMATECH research were licensed to member firms on an exclusive basis for two years, after which they were to be made available to U.S. firms at nominal royalty rates. Its new emphasis on the supplier industry led SEMATECH in 1991 to relax its prohibition on the disclosure of SEMATECH-funded equipment advances to nonmember firms; instead, member firms now have priority in ordering and receiving such equipment (U.S. General Accounting Office 1992a). Once SEMATECH member demand is satisfied, equipment suppliers can sell to any customer.[13] The change in disclosure policies benefited U.S. equipment firms and addressed criticism from some semiconductor manufacturers that SEMATECH was exploiting general tax revenues to benefit member companies at the expense of nonmembers. This shift in policy means that equipment firms participating in SEMATECH projects now can sell products embodying the results of this work to all customers, including foreign firms. Still, SEMATECH participants retain important advantages because of their lead time and greater familiarity with new equipment developed under SEMATECH sponsorship. Assignees gain familiarity in the use of new technology and, by influencing the evolution of equipment technology, may improve its fit with their needs.

The resulting R&D agenda and associated distribution of costs and benefits of SEMATECH R&D provide an example of public financial support (matched by industry contributions) for research whose results are *excludable*, but also *nonrival*, in Romer's terms (1993). Relaxation of SEMATECH's licensing restrictions has decreased the excludability of its research results. Nevertheless, many SEMATECH firms thus far appear to view the returns to their hands-on involvement with equipment suppliers as sufficient to justify their continued investment in the consortium. Their behavior suggests that a significant element of excludability still characterizes SEMATECH research results.

SEMATECH provides a central funding and testing organization for

[13]For some sophisticated equipment, a supplier may only be able to produce a handful of units per year. In such a case, SEMATECH members are likely to purchase most or all of the available product.

equipment development and qualification (demonstration of their performance specifications), which is less expensive than individual prototyping or testing. By reducing duplication among semiconductor manufacturers in the development and qualification of new tools, the consortium can lower the costs of adopting new process equipment.[14] Decreasing the fragmentation of product designs and increasing knowledge of equipment maintenance, software, and operation can facilitate the adoption of U.S. manufacturing equipment by SEMATECH members.[15]

The shift in the SEMATECH research agenda has not eliminated all conflicts among its members. ARPA, the federal agency funding SEMATECH, expressed some concern over the shift in the research agenda away from long-term, more fundamental research issues to equipment development (Katz and Ordover 1990), and the goals of SEMATECH member firms and Pentagon policymakers have not always been aligned throughout the consortium's brief history.[16] Three firms have withdrawn from the consortium since its foundation (Micron Technology, LSI Logic, and Harris Semiconductor withdrew in 1992, 1992, and 1993 respectively). For both Micron Technology and LSI Logic, smaller semiconductor producers with weaker capabilities in process technology, withdrawal reflected dissatisfaction with the consortium's decision not to pursue development of an advanced manufacturing process that all members could apply to the production of chips.[17] At least two current members

[14]Leyden & Link (1992) present some preliminary estimates of resource savings associated with similar activities performed by the National Institute of Standards and Technology. They find that these standardization activities yield social returns exceeding 100%. Interviews with some managers of SME firms, however, suggested that a number of SEMATECH member firms still require extensive in-house testing and qualification of new equipment.

[15]SEMATECH also actively participates in industry efforts to develop "road maps" of anticipated technological change, to support university research (primarily through contributions to the Semiconductor Research Corporation), and to increase communication between member firms and SME suppliers. For a detailed overview of SEMATECH's current operations, see Grindley et al. 1994.

[16]These divergent views of SEMATECH's agenda and role are likely to surface in other Pentagon-funded programs in dual-use technologies, and suggest that integrating the defense and civilian industrial bases and improving the economic competitiveness of U.S. high-technology industries through DoD R&D policy may prove to be difficult (Alic et al. 1992). As we note below, SEMATECH announced in October 1994 that it will no longer seek federal block funding after 1996.

[17]As the U.S. General Accounting Office (1992a, 12) reported, both LSI and Micron "were more interested in pursuing SEMATECH's initial strategy to improve yields and reduce costs through better manufacturing methods than in working with suppliers to develop next-generation equipment."

of the consortium, AT&T and Rockwell, have stated that they are evaluating their commitment to SEMATECH.[18]

2.3 Distinctive Structural Features of SEMATECH

The original aims of SEMATECH resembled those of many Western European cooperative R&D programs (characteristics of some of which are summarized in Table 1), in extending government R&D assistance to an industry under pressure from foreign producers. Like these programs as well as those of Japan, SEMATECH pursued a precompetitive research agenda, avoiding commercial development activities. But SEMATECH's development of an initial research agenda that won the support of its members, as well as its shift in this agenda, depended on some structural features that differ from those of many European programs.

Its original design called for SEMATECH to carry out research in a central facility. This approach differed from those of the European programs, most of which consist of a collection of independent projects performed in-house by groups of member firms under an umbrella directorate or secretariat.[19] Rather than filling a purely administrative role, as did the European directorates (for example, the Alvey Directorate did not staff project selection committees but simply oversaw the disbursement of funds), SEMATECH managers were closely involved in their development projects. SEMATECH's approach also differs from that of many Japanese research associations, which conduct a great deal of their R&D in member firms.[20]

[18]In late December 1993, AT&T submitted notification to SEMATECH that it wished to retain the option of leaving the consortium in two years (Dorsch 1994). AT&T's action does not constitute a formal notification of a decision to withdraw from SEMATECH, but it indicates that senior management is giving the possibility serious consideration. Rockwell submitted a similar notification to SEMATECH in late 1992.

[19]See Ouchi (1989) for the distinction between centralized *operating-entity* consortia and *umbrella* or *secretariat* consortia.

[20]Japan's VLSI Project, in which joint research accounted for less than 20% of the total research budget, was the first Japanese research association to utilize joint research facilities. Even in the joint facilities, research was primarily conducted between pairs of firms rather than among all consortium members. Almost 90% of the Japanese consortia surveyed by Aldrich and Sasaki (1993, 15) conducted the bulk of their research in member firms, well above the 44% of U.S. consortia reported to do so in the survey.

Table 1: Major Research Consortia in Japan, U.S.A., and Europe

Name	Country	Dates	Budget[a] $m	Govt share	Structure	Role of Managers	Central R&D site	Members	Projects	Description of research focus
VLSI	Japan	1976-80	350	40%	centralized	proj mgmt	yes	6	-	Semiconductor R&D, manuf.
OMCS	Japan	1979-85	90	100%	centralized	proj mgmt	yes	-	-	Opto-electronics R&D
VHSIC	USA	1980-89	900	100%	centralized	admin.	no	9[f]	-	Military semiconductor R&D
Supercomp.	Japan	1981-89	130	100%	centralized	proj mgmt	no	-	-	High speed computer R&D
FED	Japan	1981-90	40	100%	centralized	proj mgmt	no	-	-	Semiconductor R&D
5G	Japan	1982-91	426	100%	centralized	proj mgmt	yes	10	-	Computer software, AI
Alvey	UK	1983-88	500	50%	umbrella	admin.	no	127	309	Computer software, VLSI
ESPRIT I	Europe	1984-89	1800	50%	umbrella	admin.	no	500	226	IT, software
ESPRIT II	Europe	1988-93	3800	50%	umbrella	admin.	no	-	-	IT, software
Eureka	Europe	1985-96	7700	50%	umbrella	admin.	no	-	297	Electronic technology
RACE	Europe	1985-96	3000	50%	umbrella	admin.	no	-	-	Telecommunications
JESSI	Europe	1989-96	4000	50%	umbrella	admin.	no	-	70	Semiconductor manuf. R&D
MCC	USA	1983-	80 pa[b]	0%[c]	mixed[e]	proj mgmt	yes	50	-	Semiconductor basic R&D
NCMS	USA	1986-	150 pa	50%[d]	umbrella	proj mgmt	no	180	-	Manufacturing processes
SEMATECH I	USA	1987-92	1000	50%	centralized	proj mgmt	yes	11	-	Semiconductor manuf. R&D
SEMATECH II	USA	1993-	200 pa	50%	centralized	proj mgmt	yes	11	-	Semiconductor manuf. R&D

[a] Total amounts (government plus industry).
[b] 1981-86.
[c] pa: per annum. Reflects the most recent annual budget figure.
[d] While MCC gets no lump sum government funding, the consortium competes for grants.
[e] Combination lump-sum and competitive grant funding.
[f] Began centralized, has evolved toward umbrella form.
[g] VHSIC participants total 14 firms with joint contractors.

Note: VLSI Budget: ¥73.7bn (MITI contribution: ¥29.1bn); OMCS: ¥16bn; Supercomputer: ¥Y23bn; FED: ¥7.6bn 1981-1986 only (Fransman, 1990, p. 182); 5G ¥24bn [$133m] 1982-87 only (Fransman, 1990, p. 228)]; Alvey: £350m; ESPRIT I: ECU 1.5bn; ESPRIT II: ECU 3.2bn; Eureka: ECU 6.5bn (Watkins, 1991); JESSI: ECU 3.5bn original plan, since reduced to about ECU 2.5bn (Watkins, 1991); MCC: $300m 1983-90; RACE: ECU 550m first 5 years only. All ECU translated at $1.20 to ECU 1; Yen at ¥210 to $1 for VLSI and ¥180 to $1 for more recent Japanese consortia. Japanese budget data taken from Fransman (1990); European budget data from Watkins (1991); U.S. budget data from annual reports of relevant programs and personal communications.

Another important difference between SEMATECH and many European consortia is the fact that the design of SEMATECH and much of the impetus for its creation came largely from industry, rather than from the national and regional governmental organizations that have spawned many European programs.[21] Substantial industry involvement (the participation of senior managers and their commitment of corporate funds) in SEMATECH's formation, management, and staffing increased the likelihood that the consortium would focus on problems viewed by industry as central to its near-term survival.

This commitment by U.S. firms to SEMATECH, along with its centralized management structure, gave the consortium sufficient autonomy to revise its research agenda in response to changing developments in the semiconductor industry. This flexibility contrasts with the fixed agenda of the European consortia, such as ESPRIT and Alvey, which committed the bulk of their funds to a fixed menu of projects at an early point in their development (Grindley 1988a). In this sense SEMATECH more closely resembled Japanese consortia such as VLSI, which relied on committees made up of member-company vice presidents and departmental managers to select research topics and allocate resources (Sakakibara 1983), and to revise these decisions, over the life of the consortium.

SEMATECH also differs from many Western European consortia in the underlying strength of its target industry, especially in product technology and design. Indeed, this underlying strength facilitated the commitment of resources by member firms to SEMATECH (in contrast to many U.S. SME firms, which lack the necessary human or financial resources). Although serious, the competitiveness problems of the U.S. semiconductor industry were less severe than those of the European or U.K. information technology industries at the inception of ESPRIT or Alvey (Grindley 1988b). SEMATECH could focus on a limited array of issues in process technology and equipment; Alvey and ESPRIT, on the other hand, faced a need to virtually rebuild entire industries that had lost important complementary capabilities, such as marketing or firm-financed R&D. SEMATECH has had serious difficulties in segments of the U.S. equipment industry that require much more than access to cooperative R&D.

[21]Although ESPRIT's design and structure incorporated the advice of a succession of European industry advisory groups, the impetus for the program came from senior EC policymakers, such as Etienne Davignon.

SEMATECH's organizational structure contrasts with that of its major privately funded counterpart in the U.S. electronics industry, MCC. SEMATECH was more successful at getting member firms to provide high-quality assignees than was MCC, a staffing policy that has facilitated the transfer of research results to SEMATECH member firms. The research agenda of MCC and SEMATECH also differ, inasmuch as SEMATECH focuses on process development, rather than basic research, and makes its research results available to all members.[22]

3. Evaluating SEMATECH

3.1 Defining Evaluation Criteria

Is SEMATECH a success? Evaluating this consortium is difficult, for reasons that are common to many public-private technology partnerships – agreement on the goals of such programs and on evaluation criteria are lacking. Another problem is that the relevant time horizon for determining the effects of SEMATECH extends well beyond the consortium's five years of operations – the competitive effects of intrafirm R&D are realized only after a considerable lag, and it is reasonable to expect that consortia similarly take time to affect the economic fortunes of an industry. Even if the relevant quantitative data could be assembled, it is difficult to specify the counterfactual case against which one should compare the results of the public investment in SEMATECH. For example, what would have been the effects of $500 million in federal grants to individual firms for the support of research on semiconductor manufacturing processes and equipment? Alternatively, should SEMATECH be compared against a situation in which no such investment in civilian semiconductor manufacturing research was made?

SEMATECH's adaptability yields another problem that is likely to arise in future consortia: many of its original goals are no longer relevant to its evaluation. Flexibility in the SEMATECH research agenda is an important strength, the absence of which has impaired the performance of many European technology programs. The consortium's ability to revise its goals and research agenda nevertheless create difficulties of accountability and evaluation that

[22]MCC projects often are selected and sponsored by groups of member firms, with research performed in-house or in member firms. In return for this effort the project sponsors have the rights to any innovation for a year after its development; other MCC members then are granted access for an additional year, after which the rights are available for licensing outside MCC.

cannot be ignored in a program that receives substantial public funds.[23] We cannot provide definitive answers to these questions, but the widespread view of SEMATECH as a model for U.S. technology policy demands some consideration of its effects.[24]

3.2 SEMATECH's Impact on the Semiconductor and SME Industries

SEMATECH has been widely credited with improving the competitiveness of U.S. semiconductor producers and equipment firms.[25] The turnaround in the competitive performance of the industry has coincided with the formation of SEMATECH. Since 1990 the U.S. share of the world market for semiconductor components has been increasing, and in 1992 the share returned to equality with that of Japanese firms (Figure 1).[26] U.S. semiconductor equipment suppliers have also improved their global market performance

[23]This tension is aptly stated in a U.S. General Accounting Office report on SEMATECH: "While generally agreeing with our lessons learned, SEMATECH officials stated that our matters for Congressional consideration neglect the necessity for flexibility in a consortium's response to changing environments, citing its decision in 1989 to alter its R&D program to give crucial attention to equipment suppliers. While we agree that a consortium needs flexibility, we continue to believe that a consortium needs to develop an initial operating plan with realistic objectives and milestones on the basis of a comprehensive assessment of the industry." (1992b, 13).

[24]Recent attempts to evaluate SEMATECH have examined its impact on the total R&D investment of its member firms (Irwin and Klenow 1994; Horrigan 1994). This method, while a creative approach to evaluating SEMATECH, raises some concerns. Methodologically, it is difficult to specify an appropriate model which can control for all relevant differences between members and nonmembers. In addition, without intrafirm line-of-business data on R&D spending, scholars must choose between including member firms such as AT&T and IBM, whose R&D is diversified far beyond semiconductors (the approach taken by Horrigan), and limiting the sample to undiversified firms, which both reduces sample size and risks introducing bias into the results (see Irwin and Klenow 1994). Both papers take as their criteria the influence of SEMATECH on member firms' R&D spending, but the connection between increases in spending on R&D – an input – and the outputs – technology-related public goods or competitiveness – remains unclear.

[25]For a representative comment, see the *Washington Post* 1992.

[26]Dataquest figures for 1993 estimate U.S. firms' share of world semiconductor component sales at 43.7%, slightly larger than the 40.4% share of Japanese firms (Siegman 1993). These market-share data measure only reported sales of components and therefore exclude the so-called captive production for internal use of firms such as IBM. Indeed, were these figures computed in terms of total output, thus including captive production of such U.S. firms as IBM and AT&T, U.S. firms would be shown to have maintained their dominance through the 1980s.

since 1991 (Figure 2), reversing a 20-year decline in their share of the world equipment market. In 1992, U.S. firms regained the leading positions in sales of both semiconductor components (Intel) and semiconductor equipment (Applied Materials) (see Tables 2 and 3).

This improvement in the performance of the U.S. semiconductor manufacturers and equipment suppliers may be based on gains in manufacturing performance that are attributable to SEMATECH-supported work on process technologies and SEMATECH support for improved equipment design, operation, and maintenance.[27] But there is little credible evidence that SEMATECH is directly responsible for such gains. The measured resurgence in U.S. semiconductor producers' market share, which is based on value of shipments, largely reflects the intense price competition during much of this period in the commodity segments of the industry, such as memory components, that has resulted from entry into memory production by large South Korean producers.[28] The strong performance of U.S. firms in the more profitable microprocessor market has shielded their revenues somewhat from the effects of stiff price competition elsewhere in the industry. Relatedly, the competition in memory chip production, combined with the broader Japanese business downturn, has led Japanese semiconductor firms to reduce their capital investment, while the entry of Taiwanese and South Korean firms has enhanced foreign demand for non-Japanese equipment. A portion of the apparent improvement in U.S. SME firms' market share appears to reflect this shift in the global distribution of new equipment demand, and may imply nothing about the competitiveness of these firms in markets in which they had previously lost market share to foreign firms.[29]

[27]According to VLSI Research, SEMATECH has improved yields of U.S. equipment to the point where the Japanese yield advantage has shrunk from 15% in 1985 to 9% in 1991 (U.S. General Accounting Office 1992a). A survey of SEMATECH member firms found that in 1991 they expected to purchase 40% of their new fab equipment from U.S. producers; in fact, however, members actually purchased 70% of their equipment from U.S. suppliers, which may also reflect SEMATECH-supported cooperation (SEMATECH 1991).

[28]Fiscal year 1992 profits for Japan's four largest memory chip producers – NEC, Fujitsu, Toshiba, and Hitachi – declined 72%, 85%, 67%, and 44%, respectively, from their 1991 levels (Young 1992; Doe 1992). Korean firms have garnered 20 percent of the 1-megabit DRAM market, and in 1992 Samsung replaced Hitachi as the world's largest producer of 4-megabit DRAMs (Young 1992). In addition, some experts predict severe overexpansion of 16-megabit DRAM capacity by early 1997 (*Business Week* 1994).

[29]In 1993, U.S. SME suppliers held 38% of the South Korean market for equipment, as compared to 20% of the Japanese market. However, this share of the Japanese market did represent an improvement over the U.S. suppliers' 15% share in 1990 (*Fortune* 1994).

Figure 1: Worldwide Semiconductor Market Shares (by company base)

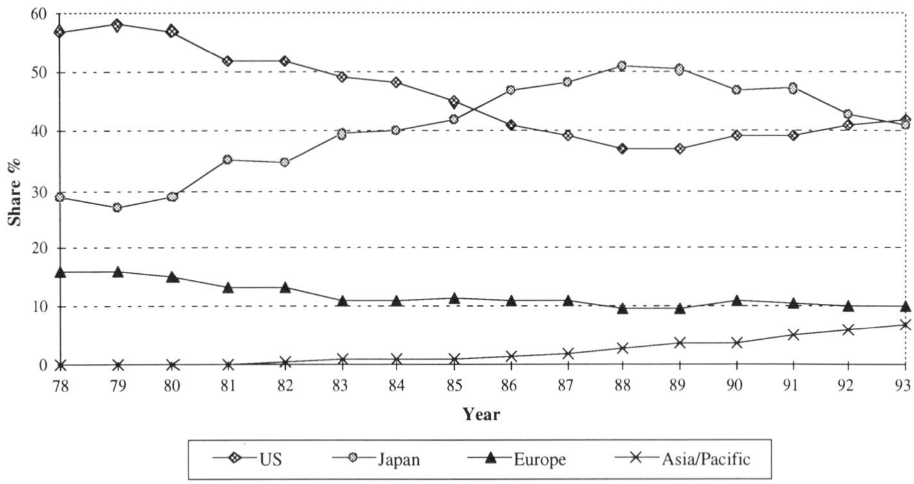

Source: Dataquest.

Figure 2: Worldwide Semiconductor Manufacturing Equipment Production Shares (by producer base)

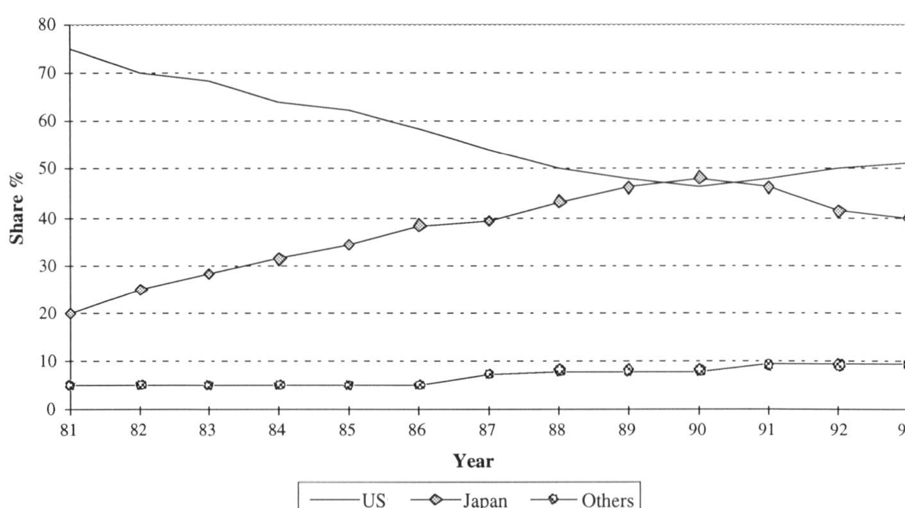

Source: VLSI Research.

Table 2: Top 10 Worldwide Merchant Semiconductor Firms (1980-93)

	1980			1990			1993		
Rank	Company	Sales ($m)	Mkt. Share (%)	Company	Sales ($m)	Mkt. Share (%)	Company	Sales ($m)	Mkt. Share (%)
1	Texas Inst.	1580	12.2	NEC	4952	8.6	Intel	7950	9.6
2	Motorola	1110	8.5	Toshiba	4905	8.5	NEC	6173	7.4
3	Philips	935	7.2	Hitachi	3927	6.8	Motorola	5973	7.2
4	NEC	787	6.1	Motorola	3692	6.4	Toshiba	5754	6.9
5	National	747	5.7	Intel	3135	5.5	Hitachi	5038	6.1
6	Toshiba	629	4.8	Fujitsu	3019	5.3	Texas Inst.	4003	4.8
7	Hitachi	622	4.8	Texas Inst.	2574	4.5	Samsung	3047	3.7
8	Intel	575	4.4	Mitsubishi	2476	4.3	Fujitsu	2931	3.5
9	Fairchild	566	4.4	Matsushita	1945	3.4	Mitsubishi	2804	3.4
10	Siemens	413	3.2	Philips	1932	3.4	IBM	2510	3.0
	Others	5036	38.7	Others	24943	43.4	Others	36817	44.4
	Total	13000	100	Total	57500	100	Total	83000	100

Source: Dataquest.

Table 3: Top 10 Worldwide SME Suppliers (1979-93)

	1979			1989			1993		
	Company	**$m**	**HQ**	**Company**	**$m**	**HQ**	**Company**	**$m**	**HQ**
1	Fairchild	111	US	Tokyo Electron	634	Japan	Applied Materials	1160	US
2	Perkin-Elmer	101	US	Nikon	582	Japan	Tokyo Electron	889	Japan
3	Applied Materials	54	US	Applied Matls.	523	US	Nikon	688	Japan
4	GCA	54	US	Advantest	399	Japan	Canon	528	Japan
5	Teradyne	53	US	Canon	384	Japan	Advantest	397	Japan
6	Varian	51	US	General Signal	354	US	Lam	372	US
7	Tektronix	39	US	Varian	335	US	Varian	321	US
8	Eaton	38	US	Hitachi	210	Japan	Teradyne	314	Japan
9	Kulicke & Soffe	37	US	Teradyne	200	US	Schlumberger	268	Europe
10	Balzers	34	Europe	ASM Intl.	187	Europe	SVG	255	US
Market share:									
	US:		76%			48%			54%
	Japan:		16%			41%			38%

Note: Part of change in relative market shares due to Yen appreciation over period.

Source: Tokyo Business Today, 1990; Davis, 1990; VLSI Research.

Sales by American- and Japan-based SME manufacturers broadly followed equipment-purchase trends in their respective national markets, trends that reflected contrasting growth rates in the capital spending of U.S. and Japanese semiconductor producers (Table 4). U.S. equipment purchases in 1992 rose by 3% over their 1991 level, matching the 3% rise in U.S. SME firms' sales during this period. Japanese equipment purchases in 1992 fell by more than 25% from their 1991 level, while Japanese SME suppliers' revenues declined by 13%. (U.S. semiconductor manufacturers' capital expenditures declined by 9% while Japanese capital expenditures declined by 29% during the 1991-1992 period). The increase in U.S. worldwide SME market share thus is explained largely by the temporary shrinkage of the Japanese market for equipment rather than by some long-term improvement in the performance of U.S. equipment firms.[30] In fact, given the global shifts in demand for semiconductor manufacturing equipment, Japanese SME firms appear to have performed at least as well as U.S. firms. Therefore, once the Japanese economy recovers, so may the sales and market share of Japanese SME firms.

In one important segment of the U.S. SME industry, lithographic *steppers*, SEMATECH programs have not prevented erosion in U.S. firms' competitiveness. This class of equipment accounts for a substantial share of SEMATECH spending on equipment development and improvement.[31] GCA, the inventor of the wafer lithographic stepper in 1970 and a major U.S. supplier, steadily lost market share from 1985 on, largely to foreign firms. GCA was an important participant in SEMATECH's equipment development programs, which supported roughly 20% of the costs of developing the firm's advanced XLS stepper (Dunn 1993a).[32] SEMATECH also purchased 14 GCA ALS 200 I-line steppers and leased them to semiconductor manufacturers for beta-site testing and qualification, at a cost of $19 million (Zipper 1990; U.S. GAO 1992a). GCA technology was considered excellent, but the firm's business problems led to a March 1993 decision to suspend stepper production. Following lackluster bidding by domestic suitors, GCA was shut down in May 1993 (Dunn and Robertson 1993).

[30]Historically, Japanese semiconductor manufacturers have purchased 80% of their equipment from Japanese suppliers (Thompson 1993, 123).

[31]Between 1988 and 1992, SEMATECH devoted 38 percent of its $287 million contract research budget to lithography (U.S. General Accounting Office 1992a).

[32]SEMATECH reportedly devoted roughly $70 million of joint development funds to GCA (Dunn 1993b).

Table 4: Changes in Semiconductor Capital Expenditure, Equipment Purchases, and SME Market Share (1991-92)

		1991		1992		Growth 92/91*
		$bn	Share (%)	$bn	Share (%)	%
Capital Expenditure:	Japan	5.6	43	4.0	36	-29
	USA/Canada	3.8	29	3.5	32	-8
	Europe	1.2	9	1.1	10	-8
	Asia/Pacific	2.4	18	2.4	22	0
	Total	13.0	100	11.0	100	-15
Equipment Purchases:** (by market)	Japan	4.1	46	3.0	37	-25
	USA/Canada	2.7	30	2.8	34	3
	Europe	0.9	11	1.0	13	11
	Asia/Pacific	1.2	14	1.2	15	2
	Total	8.9	100	8.1	100	-9
SME Sales:** (by HQ of supplier)	Japan	4.8	45	4.2	41	-13
	USA/Canada	5.0	47	5.2	51	3
	Europe	0.6	5	0.6	5	-5
	Asia/Pacific	0.3	3	0.3	3	-1
	Total	10.7	100	10.2	100	-5

* Growth 92/91 = [$bn 1992 - $bn 1991] / [$bn 1991] (using unrounded data)

** Equipment Purchases include semiconductor equipment only. SME Sales include equipment and process materials.

Source: Capital Expenditure - Dataquest; Equipment Purchases - Semiconductor Equipment and Materials International (SEMI) and Semiconductor Equipment Association of Japan (SEAJ); SME Sales - VLSI Research.

SEMATECH and its members are now engaged in an attempt to reinvigorate the only remaining U.S. producer of leading-edge lithographic equipment, SVG Lithography (SVGL). In May 1993 SVGL announced that it would enter a joint venture with Canon of Japan for the development and production of the Micrascan stepper, another project that had received substantial SEMATECH support.[33] In a departure from its previous restrictions on foreign firms' access to its research projects, SEMATECH announced that it would continue to support SVGL's development of the Micrascan.[34] This stance was not warmly received by Congressional and other U.S. policymakers, who pressured SEMATECH and its members to do more to save SVGL.[35] In the event, SVG and Canon proved unable to resolve all of the details of their proposed joint venture agreement, and in October 1994 the two parties abandoned their plans for collaboration. At the same time, SEMATECH announced a large, multistage plan to support SVGL's development of the next generation of Micrascan product (Clark 1994). Central to this plan is an advance commitment by several SEMATECH members to purchase a fixed number of Micrascan III steppers over the next several years, a commitment that was conspicuously missing from SEMATECH's attempt to save GCA.[36]

Although the current attempt to save SVGL may yield better results, SEMATECH's efforts in this important segment of the U.S. SME industry

[33]According to Dunn (1993c), SVGL received $30 million directly from SEMATECH, and roughly $20 million from IBM in support of Micrascan project development. IBM also contributed approximately $80 million through equipment purchases.

[34]At the time, William Spencer, the SEMATECH CEO, noted, "This is another step in the evolution of realizing that markets are no longer just the U.S., and if we're going to have strong companies, they have to sell worldwide" (Dunn 1993c).

[35]In response to SEMATECH's May 1993 announcement that it would continue to work with SVGL despite its alliance with Canon, the House Armed Services Committee criticized "the rationale of Congressionally initiated investments if these investments do not result in jobs and industry in the U.S. In essence, in many respects the Department of Defense has become a research laboratory for aggressive foreign investors and cartels. It seems pointless to provide public funding primarily for the benefit of foreign firms" (Robertson and Dorsch 1993).

[36]Under the current arrangement, over the next four years SVGL will devote $36 million of its own R&D funds to development of the Micrascan III. SEMATECH will contribute $30 million in several stages, conditional upon the meeting of predetermined milestones. Several SEMATECH members, including Intel, Motorola, and AMD, will commit to purchasing a sufficient number of steppers that SVGL will be able to manufacture at full capacity over the next two years. Finally, these member firms have agreed to collectively buy $30 million of equity in SVGL if the federal government will contribute $40 million in R&D funding to SVGL over the next four years (*Solid State Technology* 1994).

have not prevented the outright failure or financial decline of major U.S. producers. Indeed, its inability thus far to reverse the decline in the U.S. stepper industry highlights the argument that technology alone is rarely sufficient to restore the competitiveness of firms or industries that lack critical, complementary personnel, marketing, managerial, and financial resources.

If the evidence on SEMATECH's direct contribution to improved economic competitiveness in semiconductor manufacturing or equipment is ambiguous, an assessment of the consortium's technical performance suggests a more positive verdict. SEMATECH has met most of its revised objectives in the development of process technology and equipment. A 1992 survey of SEMATECH members (U.S. General Accounting Office 1992a) found that most reported positive returns on their investments in SEMATECH, based on reductions in their costs of purchasing, operating, and maintaining U.S. equipment (Burrows 1992). The consortium also set out specific technical goals calling for scheduled demonstrations of the capability to produce smaller line-widths and larger wafers. These have also been accomplished essentially as planned. Phase 3, which demonstrated the production of 0.35 micron line-width integrated circuits on 8-inch wafers, was achieved in late 1992 (SEMATECH 1992). But this new capability has brought U.S. semiconductor manufacturing technology to parity with Japan, rather than enabling the U.S. semiconductor industry to surpass its Japanese competition, as SEMATECH's members originally expected (U.S. General Accounting Office 1992a).

The ambiguity of the link between SEMATECH as cause and improved competitiveness as effect will bedevil policymakers and evaluators of other public-private consortia. The experience of the U.S. lithographic stepper industry, however, suggests that collaborative support for technology development and improvement will not by itself suffice in industries populated by small, undercapitalized firms with severely constrained stocks of managerial and technical talent, especially when these firms face larger competitors with deeper pockets.

3.3 SEMATECH and the Role of Vertical Research Relationships within Consortia

Although SEMATECH's emphasis on infrastructure has reduced appropriability problems among its members, some of these problems have been shifted upstream to the SME industry. Just as SEMATECH's original agenda for research on manufacturing-process technology raised concerns by chip manufacturers over sharing information with competitors, sharing between SME suppliers and SEMATECH member firms has raised concerns among suppliers that proprietary information will be disclosed to competitors. One SME supplier, Invax, has sued SEMATECH for allegedly revealing sensitive technical information to a competitor (Burrows 1992).[37]

SEMATECH's role in supporting equipment development and qualification also gives it considerable power to strengthen individual firms or technological alternatives in an industry faced with great uncertainty.[38] This power means that important alternative paths of technological development may be ignored – as we noted earlier, consortia can reduce diversity in technology development. Nevertheless, unlike earlier federally supported efforts to support the development of technologies with commercial applications (e.g., the SST), SEMATECH is largely managed by the customers for these technologies. Its members have strong incentives to maintain a diversified portfolio of technological alternatives, and to revise their choices in response to new information. Thus far, SEMATECH also has diversified its risks in many technologies, pursuing contracts with more than one supplier and more than one engineering approach.

SEMATECH has begun to assign considerable weight to firms' financial strength in contract awards in its equipment development program. This policy may benefit firms with greater resources, and thereby affect the structure of the U.S. SME industry. The consortium's centralized qualification procedure for new equipment means that SME firms working on equipment development or improvement contracts require SEMATECH approval to sell prototype tools, a policy that favors larger SME firms with the financial resources to

[37]In interviews, managers from some supplier firms expressed concern that if they fail to meet the equipment performance and schedule requirements associated with a SEMATECH project, the consortium might transfer sensitive information produced by the project to competitors in an effort to develop equipment that can meet performance goals. Another supplier feared that its proprietary technology would be leaked by SEMATECH member firms to a competitor through long-established procurement relationships between this competitor and specific member firms. There is no evidence, however, of any such leakage of sensitive information (Interviews with SEMATECH, member firm, and supplier personnel, May-October 1993).

[38]Although SEMATECH has been accused of favoritism in awarding equipment development and improvement contracts (Dunn 1989), SME firms generally seem satisfied with SEMATECH's contract award process (Rice 1989; Burrows 1992). The ability of SEMATECH to avoid major disputes with suppliers thus far may be due in large part to its exhaustive contract proposal review procedures, which include meetings by SEMATECH personnel with each bidding supplier to provide a detailed critique of their proposal's strengths and weaknesses.

weather the waiting period. Although these features of its programs should increase concentration in the U.S. SME industry, SEMATECH's effects on industry structure thus far are uncertain. The number of mergers in the U.S. SME industry has grown since SEMATECH revised its research agenda,[39] but this trend reflects many factors other than the presence of SEMATECH. Indeed, many recent mergers and acquisitions within the SME industry reflect the divestiture of their SME divisions by established U.S. firms. High levels of restructuring thus coexist with a large number of small firms and continued fragmentation in industry structure. The pace of acquisition of U.S. SME suppliers by foreign firms also appears to have increased since SEMATECH changed its focus.[40]

SEMATECH accepts equipment development and improvement proposals from firms that are not members of SEMI/SEMATECH, and has on occasion awarded a contract to a nonmember firm, a policy that has increased tensions between the consortium and SEMI/SEMATECH.[41] As a result, some SME suppliers have questioned the value of membership in SEMI/SEMATECH (see Dunn 1989). Other equipment suppliers argue that SEMATECH occasionally acts as a barrier, rather than as a support, to communication and information exchange between equipment suppliers and manufacturing firms. These equipment-firm managers prefer to work directly with individual firms, rather than the consortium, in equipment development projects, arguing that this policy would lower the costs and duration of these projects.

The efficiency gains from SEMATECH testing and qualification of equipment also have been questioned by some industry managers, who argue that individual purchasers of equipment still must conduct extensive debugging and beta-site activities within their production facilities. Some of this activity undoubtedly reflects differences among the manufacturing processes of member firms, as well as concerns by members over disclosure of sensitive

[39]Between 1987 and 1989, there were fewer than 60 acquisitions, mergers, or dissolutions in the U.S. SME industry. Between 1989 and 1991, there were more than 100 (Davis 1990; SEMATECH 1991 and 1992).

[40]Foreign firms acquired 28 U.S. SME suppliers between late 1987 and the end of 1989 (Advisory Council 1990). Between January 1990 and June 1992, foreign firms acquired 53 U.S. suppliers (Interview with Miller Bonner, SEMATECH, July 1992).

[41]SEMATECH's willingness to entertain proposals from SME suppliers that are not members of SEMI/SEMATECH may be driven in part by political considerations. Some suppliers (not members of SEMI/SEMATECH) bristle at the idea of being required to pay a membership fee in order to have the right to bid for government-funded contracts (Dunn 1989).

aspects of their processes to other members. Other SME firms complain about SEMATECH's bureaucratic processes for setting standards and the length of time SEMATECH takes to develop and award contracts. But these drawbacks must be balanced against SEMATECH's stabilization of equipment demand, which simplifies operations for SME suppliers.

SEMATECH faces a dilemma between its members' desire to capture the benefits of SME improvements and innovations and its objective of strengthening U.S. SME suppliers. As was noted above, the consortium has reduced its restrictions on sale to nonmembers of equipment developed with SEMATECH funding. SME suppliers opposed the restriction because it reduced their potential market, revenues, and profits. Relaxation of the restriction means that SEMATECH-financed equipment advances diffuse throughout the global semiconductor industry more rapidly, increasing the benefits of SEMATECH research for domestic and foreign SME suppliers and nonmember chip-makers.

Perhaps the greatest challenge for SEMATECH in structuring and managing this vertical research agenda results from the significant differences between its members and SME firms in financial and managerial resources. Many of the SME firms are small, financially pressed, and staffed by few engineers and managers. Such firms may not have the resources to assign engineers to work at SEMATECH or at a member firm on a 6-month equipment improvement project. In some instances, equipment suppliers have been unable to absorb and apply sophisticated equipment design or experimental modeling techniques developed in SEMATECH projects in which they had participated. Moreover, interviewees repeatedly stressed the point that the critical weaknesses of many equipment firms were financial and managerial, as much as technological, in nature.[42] Fundamental project management or marketing skills often were lacking.

SEMATECH has instituted training programs in quality management for equipment firms, but more than training may be necessary to revive or sustain some equipment firms. Many of SEMATECH's management and personnel policies, as well as its technology objectives, have been revised to address the very different nature of the technology development and transfer tasks when equipment firms, particularly small enterprises, are involved. But much

[42]SEMATECH managers have on occasion been confronted with serious financial crises in equipment firms, and have been forced to consider financial bailouts for these enterprises. SEMATECH's mix of public and private financing, however, imposes serious restrictions on the consortium's use of funds for such purposes and reduces its flexibility.

remains to be done. These problems of collaboration between large user firms and small suppliers are likely to affect efforts at vertical collaboration in other U.S. industries, such as aerospace or automobiles.

3.4 Is Exclusion of Foreign Firms Feasible or Desirable?

Restrictions on foreign membership in SEMATECH are difficult to sustain in a global industry like semiconductors. Tensions between national and international competitive realities are apparent among both equipment suppliers and semiconductor manufacturers, and SEMATECH members have become increasingly involved in joint ventures with large foreign rivals.[43]

These foreign alliances have complex implications for U.S. suppliers. Although SEMATECH member firms control any flow of SEMATECH-developed technology to foreign partners within these alliances, international joint ventures weaken the effectiveness of the exclusion of foreign firms from SEMATECH. But these joint ventures also provide U.S. SME suppliers a means to gain access to foreign markets. If the American member of an alliance can persuade its foreign partner to use U.S. equipment in their venture, then the foreign chip-maker may continue to buy from the U.S. suppliers. The SME firms in SEMI/SEMATECH also have entered into a number of international joint ventures.[44] To the extent that these ventures among equipment firms encourage outward technology flows that reduce the lead-time advantages that otherwise would accrue to SEMATECH member firms, member-firm benefits from participation in the consortium could decline. These international alliances influenced SEMATECH's May 1993 relaxation of its restrictions on indirect participation by foreign firms.[45]

The exclusion of foreign firms from a publicly supported technology development consortium also has created some difficulties for U.S. policy toward other governments' subsidization of high-technology industries.

[43]For example, AMD-Fujitsu; AT&T-Mitsubishi; AT&T-NEC; IBM-Toshiba; Motorola-Toshiba; Texas Instruments-Hitachi.

[44]For example, Applied Materials-Komatsu; Eaton-Sumitomo; Lam Research-Tokyo Electronic; Varian-Tokyo Electronic. Many of these collaborations appear to cover marketing and distribution in foreign markets, rather than product development.

[45]Any significant change in this policy, such as opening SEMATECH to membership by foreign (particularly Japanese) semiconductor manufacturing firms, would raise questions about future federal funding. However, SEMATECH's recent announcement of its intention to end requests for federal funding after 1996 may eventually lead to changes in its membership policies.

Although the U.S. government supports funding for technology development activities within the SEMATECH consortium, it opposes technology and product-development subsidies extended by European governments to the Airbus consortium. The apparent inconsistencies between the U.S. government's position on SEMATECH (or more recently, flat-panel display technologies) as a matter of technology policy and its position on subsidies as a matter of trade policy reflect broader problems of coordination between trade and technology policies that have become prominent in industries other than semiconductors (Mowery 1992b).

4. Lessons of SEMATECH

As we noted in the introduction, SEMATECH is likely to serve as a model for federal support of industrial technology in other sectors of the U.S. economy.[46] The internationalization of such industries as aerospace or automobiles, in which large firms such as Boeing, General Electric, or Ford pursue extensive international collaborations with foreign counterparts and suppliers, has intensified competitive pressure on U.S. supplier firms in these and other industries. Many of these supplier firms, like U.S. SME firms, are relatively small and have serious technological, managerial, and financial weaknesses. In aerospace, many of these firms also are important sources of defense, as well as civilian, components. Can the vertical-collaboration model of SEMATECH strengthen U.S. suppliers and sectoral competitiveness in these and other U.S. industries? This section considers some lessons of SEMATECH's structure, agenda, and management.

[46]U.S. joint public-private consortia established since SEMATECH include the Multichip Module (initial budget, funded from public and private sources, $60 million for an unspecified time period); Amtex ($25 million from the Department of Energy and $25 million from the U.S. textiles industry over 5 year); and the Advanced Battery ($260 million from public and private sources over 4 years) Consortium. More than a dozen dual-use technology consortia are included as part of a $550 million Defense Technology Conversion initiative announced in 1993. Proposals include consortia to support the development of huge memory chips and parallel computers (Robertson 1993). Cohen and Noll's observation (1994) that public funding for research is declining is valid on its face, but they do not differentiate between defense R&D, which has fallen significantly, and civilian R&D, which has remained level or increased. The defense-nondefense mix of U.S. R&D funding has shifted slightly toward nondefense R&D, and within this category, additional consortia are likely to receive funding.

4.1 Short-Term Focus

Like other industry consortia in Japan and the U.S., SEMATECH now focuses on technology development, rather than fundamental research. The shift in SEMATECH's research agenda from leading-edge, proprietary technology to near-term support for the industry's supplier infrastructure resolved conflicts over the research agenda, at the cost of losing several smaller members. SEMATECH's experience supports the argument that consortia are most effective in supporting the improvement and adoption of technology, and rarely focus on long-term research.

Its near-term focus, however, means that SEMATECH-based R&D cannot substitute for the longer-term R&D that government, universities and industry have long performed. SEMATECH deals with the technological exploitation of fundamental research, and especially with the adoption of technological advances. But it cannot substitute for broader support for the scientific and technological infrastructure of the U.S. economy. Nor, in the nature of the case, is much of this support likely to flow from private sources.

SEMATECH's efforts to improve the semiconductor equipment industry are insufficient by themselves to restore the competitiveness of this industry, but they have aided equipment and manufacturing firms with strong financial and management resources. This issue is especially important in SEMATECH's efforts to revive or strengthen U.S. SME firms, which often lack the in-house resources to exploit the results of SEMATECH projects. For example, strong firms are better able to provide the assignees that are essential to SEMATECH's operations. Where an industry's problems reflect broader weaknesses, which may be the case in the European Information Technology industry (and, possibly, the U.S. lithographic stepper industry), incremental programs such as SEMATECH may be insufficient. In fact, SEMATECH may accelerate rationalization or even a shakeout within the U.S. semiconductor equipment industry.

4.2 Freestanding Organization

Formulating and implementing a technology strategy, and especially revising this strategy, are more feasible in a centralized organization than in an umbrella consortium of independent projects. As we noted earlier, SEMATECH's relatively centralized structure appears to have enhanced the flexibility and responsiveness of its research agenda, especially by comparison with Western

European consortia. But a centralized structure with many members may not be well suited to the development of proprietary technology. SEMATECH proved incapable of pursuing a research program in the manufacturing process technologies that its members felt were essential to their competitive performance. Collaboration in these more competitively sensitive areas may require projects with fewer members. This consideration, along with the force of distributive political motives, may account for the umbrella structure of European programs.

4.3 Informed Management and Industry Support

SEMATECH's ability to shift its research agenda depended crucially on industry involvement in program management. Such involvement was ensured by industry's role in initiating the SEMATECH program and in financing the consortium. The importance of industry cofunding and participation has long been emphasized in evaluations of U.S. and other nations' technology programs, and the SEMATECH experience supports this conclusion.[47]

In contrast, programs such as Alvey and ESPRIT were motivated more by broad economic and political concerns than by industry initiative, and had equivocal support at best from industry – although they did enlist industry funds, and ESPRIT was a response by the EC Commission to pressure from European information technology firms. Weak industry involvement in strategic planning meant that projects were selected according to obscure criteria and priorities were difficult to change. As we have noted, maintaining broad involvement by its member firms in project selection and oversight is an important continuing objective of SEMATECH.

4.4 Is or Was Government Sponsorship Necessary?

SEMATECH has recently announced its intention to stop accepting federal matching contributions at the end of 1996 (Kehoe 1994). But its role as a model for other consortia means that the case for initial federal funding, as well as the justification for the particular form of federal financing (a year-to-year block grant) both merit scrutiny. The eight years of federal support for

[47]For an early evaluation of federal demonstration projects that stresses the importance of financial commitments from industry, see Baer et al. (1977).

SEMATECH can be justified on several grounds. Public funding strengthened the credibility of the SEMATECH proposal, and thereby may have strengthened the commitment of individual U.S. semiconductor firms to the broad notion of collaborative R&D. Public financing also provided some guarantee of continuity for early years of the consortium's operation and, at least implicitly, indicated government cooperation on other policy issues. By improving cooperation among manufacturing firms and (even more important) between semiconductor manufacturers and their U.S. suppliers of equipment and materials, SEMATECH had an important catalytic effect on this U.S. industry. SEMATECH's role within the U.S. semiconductor industry was itself catalyzed by the federal financial contribution. But this catalytic function of federal funding is an argument for a short-term, rather than a permanent, block funding mechanism.

A second argument for federal financing of SEMATECH, as we noted above, flowed from the changed relationship between defense-related and commercial technological developments in the U.S. semiconductor industry. In order to maintain a strong defense-technology base in semiconductors, this argument asserted, defense agencies had no choice but to support the advance of commercially relevant technologies, because these constituted the source of military applications. This argument has been applied recently to other programs, such as flat-panel displays, despite the absence of much evidence that the technologies needed for military and civilian applications really are generically similar. In the case of SEMATECH, the industry-dominated research agenda has favored projects with relatively near-term benefits that in some cases yield little "spin-on" to military applications, negating a primary justification for funding of the consortium by the Defense Department. Moreover, the defense agency responsible for overseeing SEMATECH's financing and operations, ARPA, has disagreed with the industry members over various aspects of the consortium's research agenda.

In other words, the evidence from SEMATECH suggests that whatever their merits as mechanisms for supporting advances in commercially relevant technologies, industry-led consortia may have limited utility for the dual-use or spin-on technology policies that have received considerable attention in recent years. Moreover, the disagreement between SEMATECH's defense sponsor and its member firms over research priorities indicates that the extent of generic similarity between defense and nondefense dual-use technologies may be overstated by the proponents of such programs.

A final justification for public financing of SEMATECH and similar consortia, flows from the public-goods characteristics of many of the outputs of R&D consortia. As Nelson (1992) and Romer (1993) have argued, technology-related public goods that are based on knowledge span much more than just basic research. SEMATECH appears to have produced some important industry-wide knowledge that has benefited both member and nonmember firms, although the near-term economic effects of this knowledge are uncertain. But many of the consortium's current activities, such as equipment certification and testing, resemble those carried out in other industries by trade associations or standard-setting bodies that receive no federal funding. There is little a priori basis for arguing that these other activities do not merit greater public support, and Ergas (1987) notes that the U.S. is unusual in its historic lack of support for them. Nevertheless, the economic case for federal financial support for SEMATECH and other consortia is strongest in research areas characterized by high uncertainty, lengthy time horizons, and substantial spillovers and external effects. Ideally, some partitioning of the activities of SEMATECH and other consortia might allow for more generous public support of these higher-risk activities, without precluding the possibility of more modest support for the equipment certification and standard-setting activities.

Government financial support for the creation of public goods has long been justified by the presence of free-rider problems that might otherwise undercut their private provision. Even the results of SEMATECH's lower-risk activities benefit more than just its member firms, and therefore appear to be susceptible to free-rider problems. The severity of these free-rider problems has been reduced somewhat by the fact that SEMATECH's members account for the bulk of the U.S. semiconductor industry's output. Although several of the original members withdrew from the consortium, SEMATECH has passed a critical market test of viability, as 11 of the original member firms thus far have maintained their contributions.

The public-goods argument, however, does not provide a strong justification for the particular funding mechanism chosen by the federal government. SEMATECH's federal funding was not based on any formal competition with other proposals for industry-led consortia (in contrast to the Commerce Department's Advanced Technology Program) or peer evaluation of its research performance. But however desirable competitive funding of R&D consortia may be, reliance on contracts from federal agencies for SEMATECH R&D would enhance federal agencies' control over SEMATECH's research agenda and reduce the responsiveness to industry needs of the consortium's research agenda. Nonetheless, some mechanism to encourage competition among proposals for initial and continuing institutional support from the federal government for industry-led consortia is desirable.

Does the proposal by Romer (1993) for an industry-wide tax for the competitive support of generic industry research (including R&D consortia) provide a superior solution to these various dilemmas, while balancing the objectives of competition and industry control? In view of the apparent reluctance of consortia to undertake longer-term R&D, the Romer mechanism ia likely to expand support for only relatively short-term R&D.[48] Smaller U.S. semiconductor firms also might well oppose a plan that mandated contributions from all industry members, since they currently reap some benefits from SEMATECH at little cost. Further, Romer's advocacy of a decentralized system for the expenditure of these tax revenues conflicts with SEMATECH's relatively centralized structure for agenda-setting and administration of its research activities. Nevertheless, the Romer proposal does provide an intriguing mechanism for retaining substantial industry control over the provision of industry-specific public goods, while maintaining competition among various research entities.

Although 11 of the 14 original members of SEMATECH remain active, we noted earlier that 2 of these remaining member firms have announced their intention to review their future involvement. Combined with the consortium's decision to forego federal support in the future, any withdrawal of additional firms from the consortium would mean that a smaller number of semiconductor manufacturers will have to underwrite the costs of research benefiting the entire industry (including, in some cases, the non-U.S. semiconductor industry). Should this possibility be realized, SEMATECH might be forced to develop a more proprietary policy toward disclosure of research results. But such restrictions would undercut the benefits of SEMATECH to the U.S. equipment industry.

An alternative to increases in membership costs and shrinkage in the member-firm population is expansion in SEMATECH's domestic membership,

[48]Romer suggests various voting schemes to determine industry desire for such a tax. Romer also proposes a highly decentralized mechanism for project selection, under which each firm can devote its contribution to whatever topics it chooses, subject only to the requirement that projects deal with "common property resources that benefit the entire industry" (1993, 349). At least two problems remain to be worked out: how to define the boundaries of the industry (i.e., who is subject to the tax), and how to determine whether projects deal with common property resources.

which would require a reduction in membership fees.[49] At the limit, this mechanism bears a strong resemblance to the Romer proposal, and might reduce the conflicts between the financial burdens of membership and disclosure of research results. But SEMATECH's centralized structure could prove unwieldy in the face of the substantial growth in membership that might result from this change in policy. This system might well result in a consortium that resembled the Western European programs, consisting of a number of independent projects under the sponsorship of a central secretariat. Yet another potential response would open SEMATECH to foreign firms, expanding the number of member firms willing to underwrite the current membership fees and able to benefit from the current distribution of SEMATECH research results and benefits. The admission of foreign firms to SEMATECH also could provide significant benefits to U.S. equipment suppliers.

None of these shifts in SEMATECH policy appears likely as long as the consortium continues to receive federal financial support. In the long run, however, and especially after the end of federal financial support, it is difficult to see how one of these scenarios can be avoided, if SEMATECH is to remain robust. The eventual admission of non-U.S. manufacturers to the consortium surely is among the more likely possibilities.

[49]When SEMATECH was established, the consortium adopted rules under which a new member would have to pay all past dues that it would have paid had it joined at SEMATECH's inception. This policy was designed to discourage potential members from adopting a wait-and-see attitude during SEMATECH's early years. SEMATECH's Board of Directors has agreed in principle to eliminate its back-dues requirement after 1992 (U.S. General Accounting Office 1992a). Nevertheless, few nonmember firms have the resources to make membership in SEMATECH worthwhile as long as the $1 million minimum annual dues requirement remains.

One way to increase membership, as suggested by the U.S. General Accounting Office (1992b, 9), would be to establish a lower-cost associate membership through which smaller firms could participate in a subset of SEMATECH activities. This is similar to the structure adopted by MCC in 1984 in an attempt to expand its membership. As of July 1993 MCC had 47 associate members in addition to its 22 shareholder members (in March 1993 MCC established "small business associate" and "university affiliate" memberships as well; as of July 1993 the consortium had 15 of the former and 11 of the latter). A reduced-fee option might attract new members to SEMATECH – Cypress Semiconductor CEO T.J. Rodgers acknowledges that "if [SEMATECH] cut their dues in half, we would take a very hard look at joining" (Burrows 1992) – but SEMATECH has not yet altered its dues structure.

4.5 Is SEMATECH a Model for Consortia in Other U.S. Industries?

SEMATECH has inspired numerous proposals for collaborative R&D in other U.S. industries. The feasibility and ultimate success of such collaboration in these other sectors will be influenced by three sets of factors: (1) the structure of the consortia; (2) the political or economic expectations imposed on the consortia by sponsors; and (3) the fit between the activities of consortia and the competitive problems within an industry.

SEMATECH provides some clear guidelines for structuring and managing consortia in the U.S. context, especially when compared with consortia in other industrial economies; these structural characteristics have been extensively discussed above. SEMATECH's experience suggests that consortia can support collaboration between users and suppliers that focuses on relatively near-term technology development, technology adoption, and improved communication and cooperation between users and suppliers. But this agenda focuses on incremental improvement and diffusion, rather than so-called frontier research. Public and private sponsors of industry-led consortia must recognize that these entities are not well suited to long-range research, and adjust their expectations accordingly. Public and private sponsors also must recognize the critical economic importance of foreign partners, markets, and suppliers to the U.S. firms that are likely to participate in such consortia. These economic realities mean that exclusion of foreign participants from domestic U.S. R&D consortia is likely to be infeasible, counterproductive, or both.

As we noted above, SEMATECH has been more successful in strengthening U.S. SME and semiconductor firms that were relatively strong. The consortium has been least effective in dealing with the financial or managerial weaknesses in the supplier tier of the U.S. semiconductor industry. The demands of collaboration between relatively small suppliers and well-established users of equipment differ substantially from those associated with collaboration among large, research-intensive firms. In other words, vertical collaboration in other industries is likely to prove more effective for supplier firms with significant in-house management, financial, and technological strengths. Indeed, consortia in other industries will have to address financial or managerial, as well as technology-related issues, in order to deal with the broader competitive weaknesses of the supplier firms in many of these sectors.

5. Conclusion

SEMATECH set out in 1987 to elevate U.S. semiconductor manufacturing to a world leadership position, without specifying the means to achieve this objective. The consortium's ability to define and modify its research agenda was aided by its structure, which supported adaptive behavior. Flexibility was crucial to SEMATECH, but its successful exploitation rested on the involvement of experienced industry managers in the consortium's operations. The involvement of industry executives in SEMATECH management meant that problems with the consortium's research agenda were recognized and effectively communicated to its members, and SEMATECH's autonomy facilitated its response to these problems. SEMATECH managers were sufficiently acquainted with the industry to develop an alternative agenda to partially reach its initial goals. Continued industry involvement is essential to maintaining SEMATECH's effectiveness in developing and transferring technology to member and SME firms.

By contrast, the coordination of projects in most European umbrella consortia more often than not has been purely administrative. Political and budgetary constraints also have tended to reduce flexibility in the research agenda of these programs. Nevertheless, the flexibility and autonomy that have aided SEMATECH's operations create dilemmas for public officials charged with (partial) funding and oversight of consortia. The broad goals of such undertakings, such as the "achievement of world leadership," generally are affected by so many influences other than collaborative R&D, and require so much time for their achievement, that they are unrealistic yardsticks for evaluation. In addition, a concept like "industry competitiveness" is difficult to measure objectively. An alternative evaluation criterion focuses on the specific tactics adopted to achieve broad goals. In the case of SEMATECH, these might include its technical goals, as well as improved cooperation between users and suppliers. If these tactical goals are agreed upon at the inception of such a program, evaluation can focus on their attainment. But if these tactics themselves are subject to significant shifts, evaluation becomes much more difficult. Equally important is the possibility that these tactical goals may not be well connected to the achievement of the broader goals and concerns of the political sponsors of such programs. Evaluation problems are intensified still further when the goals of public and private funders of a consortium (for SEMATECH, respectively ARPA and SEMATECH's members) diverge somewhat.

Perhaps because SEMATECH's research agenda originated from, and was subsequently modified by industry personnel, many of its tactical aims have been achieved. SEMATECH sought to strengthen a relatively narrow array of capabilities in which the U.S. semiconductor industry was weak in the late 1980s, rather than attempting to reconstruct an entire industry. This approach is reminiscent of Japanese consortia, which aim at specific areas of industry weakness but leave the rest to the firms. SEMATECH's narrower focus also contrasts with European consortia, many of which have attempted to solve an abundance of problems in weak industries. But SEMATECH has not been equally beneficial to all of its original or current members, nor have its activities strengthened all segments of the U.S. semiconductor equipment industry. Programs like SEMATECH can complement the technical and management strengths of member firms, but they cannot substitute for the complete absence of these strengths. Industry-led consortia in industries with little or no industry-funded R&D are likely to encounter severe problems.

Another element of similarity between SEMATECH and many Japanese research consortia is this consortium's focus on near-term projects of a generic character; sensitive, proprietary technologies are developed by firms privately and in limited alliances. By contrast, the European programs that focus on shorter-term projects have become more proprietary. SEMATECH, MCC, and Western European research consortia all have responded to industry demands for short-term results that are closer to the market and easier to transfer to the firms. All of these cooperative research efforts have shifted away from long-term research, which may not be appropriate for consortia and better left to individual firms, university-industry cooperative programs, or public research institutes. SEMATECH seems likely to continue to move in this direction, and increasingly stresses near-term deliverables in its aim to improve the returns on members' financial contributions. From a national perspective, industry-led consortia are insufficient by themselves to maintain the infrastructure of high-technology industries. Their research efforts must be supplemented by longer-range research and training activities, that by their very nature can be funded only from public sources and conducted in relatively open institutional environments that emphasize disclosure and dissemination (See Dasgupta and David 1994).[50]

[50]SEMATECH retains links to academic research institutes through its university-based "centers of excellence" and through the Semiconductor Research Corporation, so as to maintain contact with frontier research.

The U.S. semiconductor industry remains technologically strong, especially in product innovation in products that depend heavily on software for their functionality and commercial success. SEMATECH sought to improve an area in which U.S. semiconductor manufacturing firms lagged behind Japanese competitors, viz., the weakness of vertical user-supplier relationships. Its research agenda has more to do with catching-up with global best practice than with leading the next generation. SEMATECH may be a model for other catch-up situations, but its history suggests that such a program can only work in an industry that retains strength in other technical and nontechnical areas.

Consortia Abbreviations

Japan:	
5G	Fifth Generation Computer Project
FED	R&D Association for Future Electronic Devices
ICOT	Institute for New Generation Computer Technology
MITI	Ministry of International Trade and Industry
NFE	New Function Elements Project
OEIC	Opto-Electronics Integrated Circuits Project
VLSI	Very Large Scale Integration
UK:	
DTI	Department of Trade and Industry
MOD	Ministry of Defence
SERC	Science and Engineering Research Council
Europe:	
AIM	Advanced Informatics for Medicine
BCR	Meteorological and Chemical Measurement Research
BRIDGE	Biotechnology Research
BRITE	Basic Research in Industrial Technologies
ELDO	European Launcher Development Organization
ESA	European Space Agency
ESPRIT	European Strategic Program for Research and Development in Information Technologies
ESRO	European Space Research Organization
EURAM	European Research in Advanced Materials
EUREKA	European Research Technology Program
FAR	Fisheries Advanced Research
FLAIR	Food Technologies Research
JESSI	Joint European Submicron Silicon project
STAR	Special Telecommunications Action for Regional Development
RACE	R&D in Advanced Communications for Europe
USA:	
DARPA	Defense Advanced Research Projects Agency
DOD	Department of Defense
MCC	Microelectronics and Computer Technology Corporation
NCMS	National Center for Manufacturing Sciences
SDI	Strategic Defense Initiative
SEMATECH	Semiconductor Manufacturing Technology Corporation
SEMI	Semiconductor Equipment and Materials International
SIA	Semiconductor Industry Association
SRC	Semiconductor Research Corporation
VHSIC	Very High Speed Integrated Circuit Project

References

Advisory Council on Federal Participation in SEMATECH. 1990. *SEMATECH 1990: A Report to Congress.* Washington D.C.

Aldrich, H.E. and T. Sasaki. 1993. Governance structure and technology transfer management in R&D consortia in the United States and Japan. In *Managing Technology Development and Deployment in Manufacturing: U.S. and Japan Comparisons*, edited by J. Liker, J. Ettlie, and J. Campbell (forhtcoming).

Alic, J.A., L.M. Branscomb, H. Brooks, A.B. Carter, and G.L. Epstein. 1992. *Beyond Spinoff: Military and Commercial Technologies in a Changing World.* Boston: Harvard Business School Press.

Baer, W., L.L. Johnson, and E.W. Merrow. 1977. Government-sponsored demonstrations of new technologies. *Science*: 950-977.

Bozeman, B., A. Link, and A. Zardkoohi. 1986. An economic analysis of R&D joint ventures. *Management and Decision Economics*: 263-266.

Burrows, P. 1992. Bill Spencer struggles to reform SEMATECH. *Electronic Business*, 18 May, 57-62.

Cassell, J. and P. Dunn. 1993. SEMATECH, AMD, and HP alight on 0.35 micron. *Electronic News*, 25 January, 4.

Clark, D. 1994. SVG, Canon call off partnership in chip-making technology. *Wall Street Journal*, 28 November, B5.

Clinton, W. and A. Gore. 1993. *Technology for America's Economic Growth: A New Direction to Build Economic Strength.* Washington D.C.: White House Press Office.

Cohen, L. and R. Noll. 1991. *The Technology Pork Barrel.* Washington D.C.: Brookings Institution.

——. 1992. The federal role in research joint ventures. Stanford University. Mimeo.

——. 1994. Privatizing public research. *Scientific American* (Sept): 72-77.

Cohen, W. and D. Levinthal. 1989. Innovation and learning: The two faces of R&D. *Economic Journal* 99: 569-596.

Committee on Industry and Trade. 1929. *Final Report.* London: His Majesty's Stationery Office.

Dasgupta, P. and P.A. David. Forthcoming. Toward a new economics of science. *Research Policy.*

Davis, D.B. 1990. U.S. semiconductor equipment makers must move the earth to survive. *Electronic Business*, 14 May, 44-49.

Defense Science Board. 1987. *Report of the DSB Task Force on Defense Semiconductor Dependency.* Washington D.C., U.S. Department of Defense.

Doe, P. 1992. Recession walloped Asia's electronics industry too. *Electronic Business*, October, 143-145.

Dorsch, J. 1994. AT&T ponders SEMATECH/SRC exit. *Electronic News*, 7 February, 1.

Dunn, P. 1989. Gear firms question regime for sales to SEMATECH. *Electronic News*, 16 January, 1.

—. 1993a. GCA to halt stepper production. *Electronic News*, 29 March, 1.

—. 1993b. Litho question: To be or not to be: GCA shutdown seen imminent as managers buy service group. *Electronic News*, 17 May, 1.

—. 1993c. Litho question: To be or not to be: SVGL-Canon agreement marks new worldly SEMATECH attitude. *Electronic News*, 17 May, 1.

Dunn, P. and J. Robertson. 1993. Stepper maker GCA closes: SEMATECH to prop up SVGL. *Electronic News*, 24 May, 1.

Electronic News. 1993. U.S. equipment vendors gain. 19 April, 10.

Ergas, H. 1987. Does technology policy matter? In *Technology and Global Industry: Companies and Nations in the World Economy*, edited by B. Guile and H. Brooks. Washington, D.C.: National Academy Press.

Evan, W. and P. Olk. 1990. R&D consortia: A new US organizational form. *Sloan Management Review* (Spring): 37-46.

Fisher, F.M. and P. Temin. 1973. Returns to scale in research and development: What does the Schumpeterian hypothesis imply? *Journal of Political Economy*: 56-70.

Fransman, M. 1990. *The Market and Beyond: Cooperation and Competition in Information Technology in the Japanese System.* New York: Cambridge University Press.

Grindley, P. 1988a. The UK software industry: A survey of the industry and evaluation of government policy. Report Series, Centre for Business Strategy, London Business School.

----. 1988b. Assessing the Alvey software engineering programme. In *Information Technology: First Report.* House of Commons Paper 25-II, pp. 352-357.

Grindley, P., D.C. Mowery, and B. Silverman. 1994. SEMATECH and collaborative research: Lessons for the design of high-technology consortia. *Journal of Policy Analysis and Management* 35 (4): 723-758.

Hane, G.J. 1994. The real lessons of Japanese research consortia. *Issues in Science and Technology* (Winter): 56-62.

Heaton, G.R. Jr. 1988. The truth about Japan's cooperative research. *Issues in Science and Technology* (Fall): 32-40.

Horrigan, J.B. 1994. The benefits of consortia: The impact of MCC and SEMATECH on members' R&D spending. LBJ School of Public Affairs, University of Texas at Austin. Unpublished manuscript.

Irwin, D.A. and P.J. Klenow. 1994. High tech R&D subsidies: the effects of SEMATECH. Graduate School of Business, University of Chicago. Unpublished manuscript.

Jorde, T.M. and D.J. Teece. 1993. Rule of reason analysis of horizontal arrangements: Agreements designed to advance innovation and commercialize technology. *Antitrust Law Journal* 61: 579-619.

Katz, M.L. and J.A. Ordover. 1990. R&D cooperation and competition. *Brookings Papers on Economic Activity*, 137-203.

Kehoe, L. 1994. Chip research consortium to shun federal funding. *Financial Times*, 6 October, 4.

Kelly, K. 1993. Besting Japan. *Business Week*, 7 June, 26-28.

Levy, J. and R.J. Samuels. 1991. Institutions and innovation: Research collaboration as technology strategy in Japan. In *Strategic Partnerships and the World Economy*, edited by L. Mytelka. London: Frances Pinter.

Leyden, D. and A. Link. 1992. *Government's Role in Consortia.* Boston MA: Kluwer Academic Publishers.

Lineback, J.R. 1987a. Can MCC survive the latest defections? *Electronics*, 22 January, 30.

——. 1987b. It's time for MCC to fish or cut bait. *Electronics*, 25 June, 32.

McLoughlin, G.J. 1992. SEMATECH: Issues in evaluation and assessment. Congressional research service. Science Policy Research Division, #92-749SPR. Washington D.C.

Mowery, D.C. 1983. Economic theory and government technology policy. *Policy Sciences*: 27-43.

——. 1992a. The U.S. national innovation system: Origins and prospects for change. *Research Policy* 21: 125-144.

——. 1992b. The challenges of international trade to U.S. technology policy. In *Linking Trade and Technology Policies*, edited by M.C. Harris and G.E. Moore. Washington, D.C.: National Academy Press.

Mowery, D.C. and N. Rosenberg. 1989. *Technology and the Pursuit of Economic Growth.* Cambridge: Cambridge University Press.

Nasko, H. 1983. European Common Market. *IEEE Spectrum* (Nov): 72.

Nelson, R. 1961. Uncertainty, learning, and the economics of parallel R and D. *Review of Economics and Statistics* 43: 351-364.

——. 1992. National innovation systems: A retrospective on a study. *Industrial and Corporate Change* 1 (2): 347-376.

Odagiri, H. and A. Goto. 1993. The Japanese system of innovation: Past, present, and future. In *National Innovation Systems: A Comparative Analysis*, edited by R.R. Nelson. New York: Oxford University Press.

Okimoto, D. 1987. Regime characteristics of Japanese industrial policy. In *Japan's High-Technology Industries*, edited by H. Patrick. Seattle, WA: University of Washington Press.

Olson, M. 1965. *The Logic of Collective Action*. Cambridge: Harvard University Press.

Ouchi, W. 1989. The new joint R&D. *Proceedings of the IEEE* 77: 1318-1326.

Pollack, A. 1991. SEMATECH starts to make progress. *New York Times*, 19 April, D1.

Randasseze, L.P. 1994. GCA and domestic lithography: Managing procurement initiatives in SEMATECH and vertical consortia. RAND Project Memorandum PM-199-OSTP (not cleared for open publication).

Rice, V. 1989. What's right with America's IC equipment makers? *Electronic Business*, 15 May, 28-34.

Robertson, J. 1993. Dual-use funding would spread $550m. *Electronics News*, 15 March, 1.

Robertson, J. and J. Dorsch. 1993. GCA revives as U.S. drafts Corning, AT&T. *Electronic News*, 16 August, 1.

Romer, P.M. 1993. Implementing a national technology strategy with self-organizing industry investment boards. *Brookings Papers on Economic Activity*: 345-390.

Sakakibara, K. 1983. From imitation to innovation: The Very Large Scale Integration (VLSI) Semiconductor Project in Japan. Unpublished manuscript.

Samuels, R.J. 1987. Research collaboration in Japan. Presented at the annual meeting of the Association of Asian Studies, Boston, MA, 1 April 1987.

SEMATECH. 1991. *1991 Annual Report.*

SEMATECH. 1992. *1992 Annual Report.*

SEMI/SEMATECH. 1991. *SEMI/SEMATECH Status Report 1991-1992.*

Siegman, K. 1993. An American tale of semi-success. *San Francisco Chronicle*, 20 December, B1.

Sigurdson, J. 1986. Industry and state partnership in Japan: The VLSI Project. Discussion Paper No. 168, Research Policy Institute, University of Lund.

Smith, L. 1989. Can consortiums defeat Japan? *Fortune*, 5 June, 245-254.

Stokes, B. 1993. To `techies, free trade looks costly. *National Journal*, 24 (April): 988-989.

Teece, D.J. 1986. Profiting from technological innovation: Implications for integration, collaboration, licensing, and public policy. *Research Policy* 15: 285-305.

Thompson, T. 1993. The Americans are back in the semi equipment business. *Electronic Business*, July, 121-126.

Tokyo Business Today. 1990. Japan's semiconductor equipment makers challenge U.S. lead. February, 16-18.

U.S. Congressional Budget Office. 1990. *Using R&D Consortia for Commercial Innovation: SEMATECH, X-Ray Lithography, and High-Resolution Systems*. Washington, D.C.: U.S. Government Printing Office.

U.S. General Accounting Office. 1989. *The SEMATECH Consortium's Start-Up Activities*. Washington, D.C., July.

——. 1990. *SEMATECH's Efforts to Strengthen the U.S. Semiconductor Industry*. Washington D.C., September.

——. 1992a. *SEMATECH's Technological Progress and Proposed R&D Program*. GAO/RCED-92-223BR, Washington D.C., July.

——. 1992b. *Lessons Learned From SEMATECH*. GAO/RCED-92-289, Washington D.C.

Washington Post. 1992. When industrial policy works. 29 August, A18.

Watkins, T.A. 1991. The EC and institutions of cooperative industrial R&D. Harvard University. Mimeo.

Young, L.H. 1992. Japanese face slowing growth and shrinking profits. *Electronic Business*, November, 83.

Zipper, S. 1990. Four SEMATECH Members to test GCA steppers. *Electronic News*, 21 May, 34.

Biotechnology and Advanced Technological Infrastructure Policies: The Example of the UK's Protein Engineering Club

Margaret Sharp

This chapter examines a policy experiment in the UK – the Science and Engineering Research Council's Protein Engineering Club – which in many respects encapsulates the ambitions for *advanced technological infrastructure policies*. Technological infrastructure policies are aimed at building capabilities which are industry relevant and therefore different from straight science capabilities. They embody both human capital and physical capital. In particular, technological infrastructure policies:

- develop *generic* capabilities geared to sectoral needs rather than those of individual firms;
- promote *multidisciplinarity* in that the capabilities involved combine science and engineering skills;
- are *precompetitive* and generate only indirect economic value in that further investments are necessary before firms can commercialize a new product or process; and
- involve *economies of scope*, bringing together a critical mass of capabilities, rather than the traditional economies of scale associated with normal production activities.

Advanced infrastructure policies pertain to the application of new and untested technologies. As a result, there is a high degree of uncertainty about both what the technology itself can achieve and whether there are market needs that can be satisfied. Capabilities may have to be created *before* users can define their needs. In other words, the evolution of capabilities runs side-by-side with the identification of market potential; the two are interdependent – one cannot be identified without the other.

The Protein Engineering Club was set up in the mid-1980s with the express purpose of increasing awareness in both the academic and industrial communities to important new technology developments emerging from scientific research. It was not established in any sense as part of an explicit advanced technological infrastructure policy; rather, the Club was an

M. Teubal et al. (eds.), Technological Infrastructure Policy, 217–246.

experiment born of the tight funding climate for research in Britain in the mid-1980s. It sought simultaneously to pull together two important strands of the British research community – crystallographers and molecular geneticists – and to raise industrial awareness of scientific research which seemed to have major implications for routes to new drug discoveries and other novel developments. Monetary constraints dictated the simultaneous pursuit of industrial and public funding. The relatively small number of participants gave rise to a proactive organization in which the high level of interaction between university and industry partners helped shape the emerging research trajectory. Serendipity led to a workable solution and created a policy model which was subsequently copied in other Research Council initiatives. Like any experiment, it had its elements of success and failure, but on this occasion the successes were more enduring than the failures.

Because the Protein Engineering Club manifested the four key characteristics of advanced technological infrastructure policies – a *generic* technology at a *precompetitive* stage of development, applied in building *multidisciplinary* and *industrially relevant* capabilities – it therefore serves as a good example of how the concept may be put into practice. This chapter explains the main features of the Club in more detail and seeks to identify some of the lessons to be learned from the experience.

1. Biotechnology and Protein Engineering[1]

Biotechnology by its broadest definition is "the application of biological organisms, systems and processes to manufacturing or service industries" (ACARD 1980). In this sense biotechnology has been around since the New Stone Age when humankind first learned the arts of cross breeding plants and animals, and of using yeast to leaven bread and ferment alcohol. For many centuries broad empiricism sufficed as technology, but by the beginning of the twentieth century this was replaced by a more systematic attempt to screen and categorize the role and variety of micro-organisms existing in the natural environment and to exploit those that had useful application – penicillin being a prime example. This was the so-called second generation of

[1] The developments in biotechnology described in this section are covered in detail in a number of texts. Most comprehensive yet readable are the general texts written by OTA to explain the background and importance of successive developments in biotechnology. See OTA 1981, 1987 and 1991.

biotechnology and great hopes were pinned on what might be achieved in areas such as enzyme chemistry. The aspirations of microbiologists and biochemists in the early part of the 20th century were, however, to be overtaken by developments in hydrocarbon chemistry. The focus of innovation switched to petrochemicals and remained dominated by that sector until the 1960s. Today there is again considerable interest in some of these second-generation techniques – for example, using micro-organisms for energy production (biomass energy) and for waste management. There is also renewed interest in using biological catalysts in chemistry for complex biotransformations. But the main focus of attention has shifted to so-called third-generation techniques involving engineered micro-organisms.

The new or third-generation biotechnology dates from the early 1970s when two breakthroughs in molecular biology – the discovery of a mechanism by which part of a foreign gene could be inserted into another and thereby change its characteristics (recombinant DNA), and techniques for fusing and multiplying cells (hybridomas) – heralded the coming of genetic engineering. Using the methods of recombinant DNA, the genes of micro-organisms such as yeast could be reprogrammed to produce useful proteins, such as insulin. As the micro-organism multiplied itself, so it became, in effect, a minifactory manufacturing the protein.

The applications of these radical new techniques were rapidly appreciated. By the early 1980s genes had been cloned into foreign host cells to produce proteins of recognized therapeutic value such as insulin, human growth hormone, and Factor VIII (for the treatment of hemophilia) and scale production was underway. This heralded the emergence of a whole new range of protein drugs based on naturally occurring proteins in the body's immune system. These new biopharmaceuticals are only today emerging on the market as full-fledged drugs.

Applications for biotechnology outside the pharmaceutical industry also rapidly became apparent. In agriculture, genetic engineering had application to both animal husbandry and plants. In plants it led to the rapid development of transgenic plant species incorporating such desirable characteristics as resistance to fungi, pests, and even to particular types of herbicide. Growing public concern over the use of powerful chemical herbicides and pesticides also stimulated research into biological alternatives, for example the use of microbial pesticides, and many of these could be strengthened and made more effective by implanting new genes via recombinant DNA technology.

Figure 1 illustrates the three generations of biotechnology and identifies the five main pathways or *trajectories* of development from genetic engineering: monoclonal antibodies, pharmaceuticals, plant biotechnology, animal biotechnology, and chemicals. Over time the different trajectories have grown apart, each developing along its own pathway, building up its own specialized technology and applications. Nevertheless, there are important commonalities in developments. For example, although the pharmaceutical and agricultural trajectories are now distinct pathways, both took off with the discovery of methods of implanting foreign genes into the basic genome (recombinant DNA), and both trajectories are much influenced by developments in protein engineering. The diagnostics trajectory, built upon monoclonal antibodies, has seen an interesting crossover with pharmaceuticals as developments in antibody engineering bring monoclonals back into mainstream therapeutics while areas such as gene probes become a central feature of diagnostics.

Protein engineering emerged in the early 1980s as an important new development in the search for ways of manufacturing therapeutic and other useful proteins. Proteins are large, complex biological molecules which fulfill a wide range of functional tasks in living cells – they act as natural catalysts, sensors, transmitters, receptors, building blocks, and defence mechanisms. They are formed of a chain (sequence) of amino acids, but it is the way this chain is folded – its three dimensional structure – which, together with the chemical sequence, determines the behavior and functioning of the protein.

It is this three-dimensional structure that determines the active sites of the protein which bind, through a series of chemical receptors and ligands, to the cell wall. In turn, these chemical interactions affect the way the protein functions. Changing its chemical structure and shape can therefore change the way the cell functions. For example, eliminating toxic side effects may mean eliminating a series of receptors which in turn means giving the folded protein a different shape. The challenge to scientists is to be able to predict with confidence the effect on a protein of alterations to its primary gene sequence and to use this knowledge to design proteins to fulfill a specified function.

Figure 1: The Three Generations of Biotechnology and the Main Trajectories of Development in the New or Third Generation Biotechnology

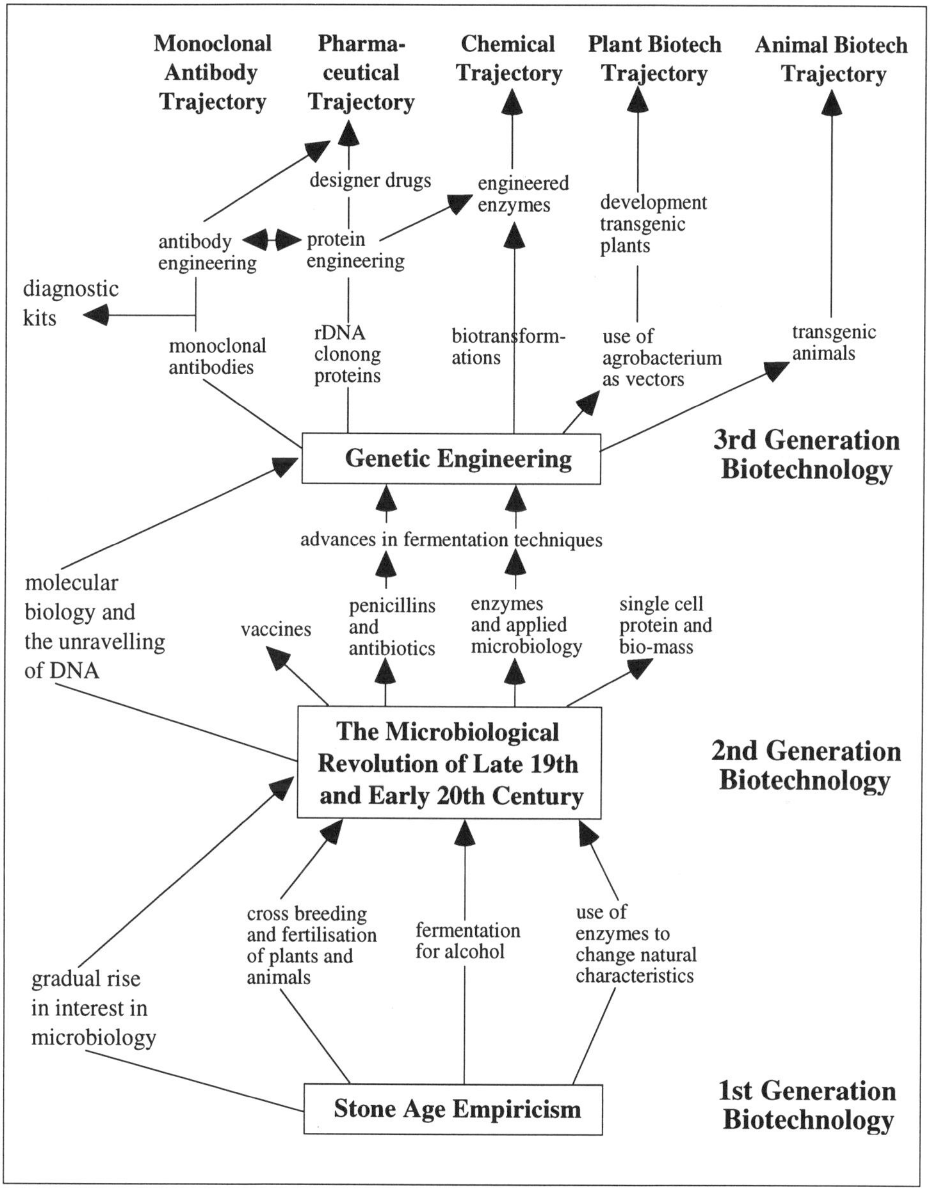

Source: Sharp 1991.

The process is illustrated in Figure 2. Starting at the center of the diagram, scientists can only alter the sequence of the protein (by the process of site-specific mutation) if the active site of the protein is known; this requires knowledge of its three-dimensional structure and the underlying amino acid sequence which determines this structure. Hence, the process of sequencing (bottom circle) and determining the crystallographic structure (top circle) go hand-in-hand. Once the two are known, an appropriate modification to the sequence can be made and the new cloned gene expressed and tested. In this regard, protein engineering brings together the skills of two distinct sets of scientists: the biophysicists and crystallographers (top circle), and the molecular geneticists and molecular biologists who dominate sequencing (bottom circle). In its early days, the whole process was estimated to take over 6 man-years. Today, although shortcuts have been discovered, it is still a labour-intensive activity.

Figure 2: The Protein Engineering Cycle

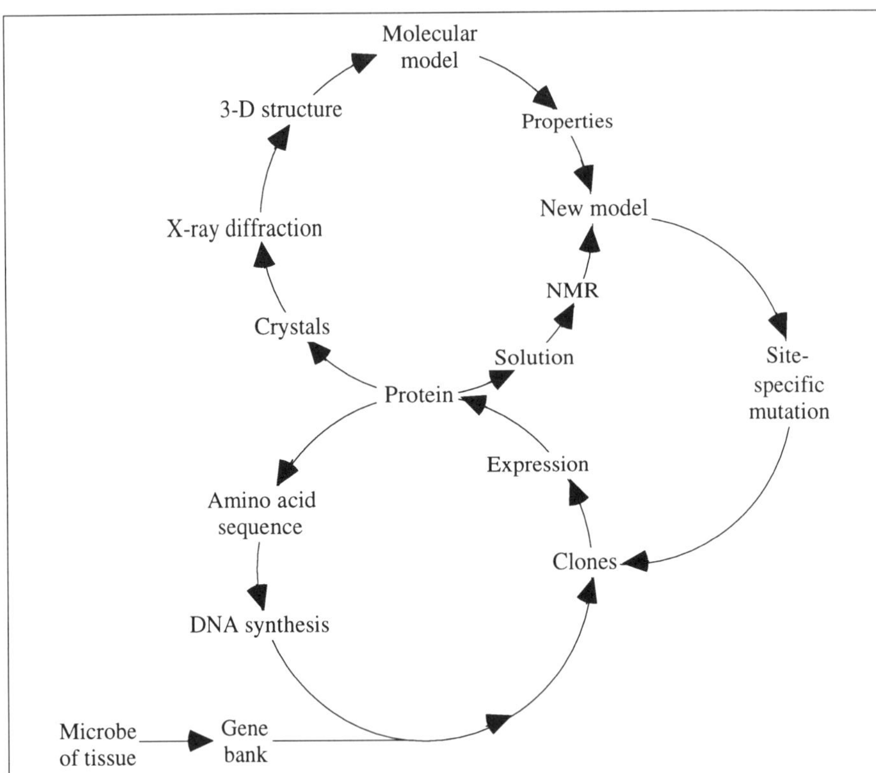

The protein enginnering cycle, adapted from Steffen Peterson, manager of Novo Industri's Department of Protein Engineering. Getting through the cycle took about 6.3 man-years in 1985.

The rapid emergence of protein engineering is well illustrated in Figure 3 which charts the growth of protein chemistry publications. The first attempts to modify proteins through site-directed mutagenesis were those of Alan Fersht and Greg Winter at the laboratory of Molecular Biology in Cambridge in the early 1980s. It is clear from this diagram that the period of take-off came in 1983-84 when the rate of publications began to grow exponentially. Its emergence marked a new phase in the development of the new biotechnology.

Figure 3: Protein Chemistry Publications in the 1980s*

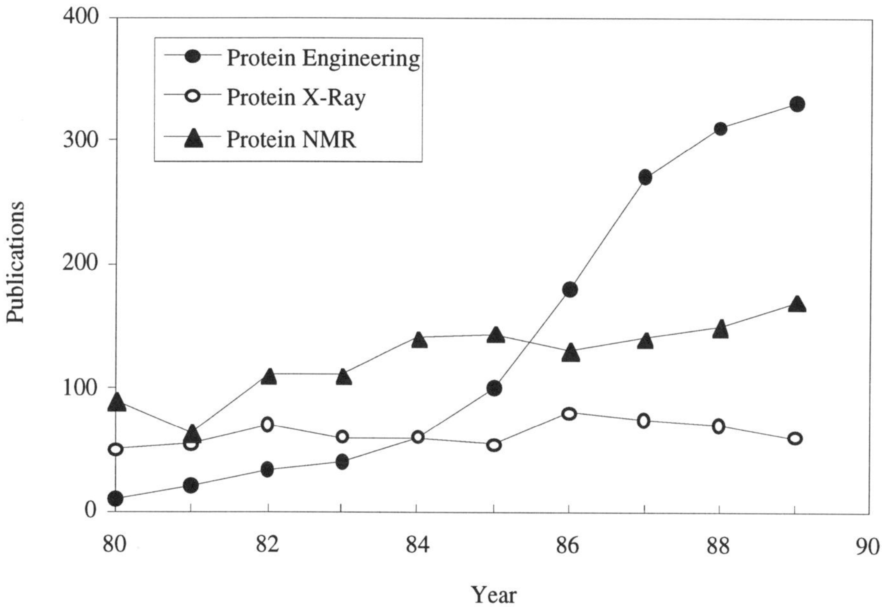

Source: Chemical Abstracts. For protein engineering the key words used were *protein engineering* and *site-directed mutagenesis*.

* I am indebted to Robin Leatherbarrow at Imperial Colledge for this diagram.

The first phase of biotechnology was characterized by the use of genetic engineering to manufacture relatively simple, large-molecule therapeutic proteins such as insulin. Most of the new biopharmaceuticals being launched during the current (second) phase are the product of second-generation technology introduced since 1984 – modified versions of these macromolecules, engineered to enhance their efficacy and make them more compatible with the body's immune system. The next (or third) generation of products currently under development by biotechnology and pharmaceutical companies goes one step further – from modifying existing proteins to designing, *ab initio*, novel proteins whose structure and function is based on rules deriving from protein engineering.

2. Biotechnology as Advanced Technological Infrastructure

Using the taxonomy developed by Tassey (1992) protein engineering emerges as an important *generic technology*, "the first phase in technology research whose objective is to show that the concept for an eventual market application works in a laboratory environment and thus reduces the typically large *technical* risks before moving on to the more applied phases of R&D" (98; Tassey's emphasis). Tassey distinguishes this from scientific research: "It is the organization of scientific principles into a *functional technical concept*", while "use of the word technical implies that technical performance is the target of this phase of R&D rather than economic or market viability." "The research objective is to show that the technological concept works in a laboratory environment" (99). This definition of a generic technology sets it clearly apart from *basic research*, which Tassey defines as "not leading, or being intended to lead, to an application", and from *applied research* which he defines as "the conversion of generic technology into specific prototype products and processes" where "the research output is a commercial prototype". This contrasts with generic technology where the "end point [is] a laboratory prototype providing proof of the technical concept" (95-96).

Protein engineering can thus be defined as an important *generic technology* within biotechnology, one among a number of crucial developments of technique which have helped translate advances in science into a new technology and given reality and viability to that new technology. In 1983-84, however, protein engineering was more science than technology. Fersht and Winter had published their seminal paper in 1982, and Fersht was working on altering the structure of subtilisin, an enzyme used in biological washing powders to give it greater resistance to extremes of temperature and detergent acidity. There were still, however, major uncertainties. In retrospect it is possible now to identify 1984-85 as the years when the major pharmaceutical and chemical companies shifted from a highly skeptical stance towards biotechnology and began investing earnestly in in-house teams and facilities.[2] Just as the turning point in microelectrics came in the 1970s when companies recognized that there were successive generations of chips and computers to be exploited, so biotechnology took off once it became clear that protein engineering opened the way to successive generations of new proteins.

The economic characteristics of this new technique were as follows:

- *pervasive uncertainty:* although the Fersht/Winter experiments had indicated that protein engineering might be used to modify protein structures, it was at this stage quite unclear how far their experiments would hold for other proteins and in other environments. Clearly, potential uses could be identified, but it was impossible to tell in detail what markets might emerge until further experimentation revealed how robust the technique was to prove;
- *a generic technology*: both in the sense Tassey discussed and in the sense that the potential application of the technology was a concept which could be applied across a range of sectors – chemical, pharmaceutical, and agricultural. It was also clear that more needed to be known about the technique and its capabilities before firms were in a position to move into specific product development. In particular, far more work was needed on the mapping and classification of protein structures and on the linkage between (gene) sequence and (protein) structure – essentially precompetitive rather than competitive research;
- *multidisciplinarity*: the capabilities involved in developing the technology needed to be drawn from a wide spectrum of the sciences. Biophysicists, crystallographers, and NMR experts led the work on protein structures; protein chemists and biochemists, molecular modelers and molecular biologists, and molecular geneticists dominated the work on gene sequencing. The need to build and work with huge databases meant also bringing in information scientists and systems experts; scaling-up meant working with biochemical engineers. In other words, protein engineering meant building a substantial multidisciplinary team.

These characteristics fit almost exactly the requirements for an advanced TIP set out at the beginning of the chapter: the generic nature of the technology, its multidisciplinarity, and the precompetitive nature of its development. Perhaps most noticeable were the uncertainties that surrounded it – critical

[2]This is examined in detail in Sharp and Galimberti 1993.

parameters were unknown, as was the way it might relate to performance. As required by the TIP model, capabilities needed to be created before users could define their needs. Indeed, many countries, recognizing its importance, introduced programs of one sort or another aimed to promote the new technology. The British Science and Engineering Research Council's Protein Engineering Club was one such model.

3. The Biotechnology Directorate and the Protein Engineering Club

The emergence of the Protein Engineering Club is connected to some of the wider developments in British science in the early 1980s. The Biotechnology Directorate was established by the Science and Engineering Research Council (SERC) in late 1981 as one of a series of initiatives aimed at promoting university-industry links in the new emerging field of biotechnology. The Spinks Report (ACARD 1980), published the previous year, had identified the British academic base in the life sciences, especially molecular biology, as one of Britain's strengths but had lamented the lack of industrial interest and called for initiatives which would bring the two sectors closer together. The SERC had in fact been considering what it might do to promote biotechnology, whose interests cut across its own structures, biology coming under the Biological Sciences Committee (BSC) of its Science Board and biochemical engineering under its Engineering Board. A joint committee between the two boards had been considered, but the interest that focused on biotechnology as a result of the Spinks Report and its aftermath led the then-chairman of the SERC, Sir Geoffrey Allen, to suggest the establishment of a higher profile body, and it was decided to establish the Directorate (Senker and Sharp 1988, 7-21).[3]

The Biotechnology Directorate (BTD) was the third directorate to be established by the SERC, its two predecessors being in polymer engineering and marine technology. In both cases the technology had been deemed ripe for commercialization and the directorates' job had been to stimulate the academic research base to develop work which could provide an underpinning

[3]In fact it was the Government response to the Spinks Report, published in a White Paper in April 1981 (UK Government 1981) that caused problems, particularly amongst academics, by suggesting that the Government need do nothing and that the market would provide all necessary stimulus. See Senker and Sharp 1988, 13-14.

for industrial development, the expectation being that their role would be completed within a five-year span. The BTD was in a very different position. Its science and technology base, far from being mature and ripe for commercialization, was still highly experimental with new discoveries and breakthroughs being announced each week. Moreover, whereas the two earlier directorates had been responsible directly to the Engineering Board of the SERC, the BTD reported both to the Science Board and to the Engineering Board. The Science Board was, at least initially, very suspicious of the idea of a *directed* program; its traditional funding mode was to respond to grant proposals coming from researchers on the basis of peer review rating, those judged to be the best science being given priority funding. The science Board was extremely unhappy at the prospect of funds being diverted to projects that were less-highly rated for their scientific qualities but which might be judged to be industrially relevant.

As the term *directorate* implies, the BTD's function was, in part, to *direct* research funding towards areas judged to be of high priority. The BTD's Management Committee was organized so that half of its membership was made up of scientists from industrial research laboratories, and its terms of reference required it to:

- survey the strengths and weaknesses of the British academic base and, in the light of this, establish a prioritized program of research;
- promote multidisciplinary research and the formation of a research community;
- encourage industrial involvement in the program, including the dissemination and application of its research.

The Protein Engineering Club emerged as one of the relatively early initiatives of the Directorate. A preliminary meeting was held in November 1983 with academic and industrial participation, and sufficient interest was shown to encourage follow-up. Talks began with about a dozen firms who had shown some interest in the idea of forming a club of industrial sponsors, and in April 1894 representatives from eight of these joined discussions about the shape of the research program to be undertaken. During the summer of 1984 academic groups were invited to submit specific proposals in the different areas identified by this program. By autumn the first funding decisions were taken and the Club began to take shape, although the program did not officially start until April 1985.

4. The Protein Engineering Club 1985-1989[4]

4.1 Beginnings – 1984-86

The Club did not have an easy beginning. Many academics resented the degree to which the program aimed to manage research, while the firms involved were required to play a protracted game of "chicken and egg" in the formulation of the program. For their part, the companies did not wish to commit themselves to joining the Club until they saw how the final program shaped up, yet they felt inhibited in voicing their own views (and for that matter given the early stage of developments, did not really have strong views) until such decisions had been taken. The result was lengthy, and perhaps unnecessary, rounds of negotiations between the industrialists, the BTD, and the university groups involved. Only four firms – Celltech, Glaxo, ICI, and Sturge[5] – of an initial eight who participated in the early agenda-setting rounds, eventually joined the Club. Each committed itself to paying an annual subscription of £30,000 for four years towards the cost of the research program, making a total of £480,000. The SERC had committed £1.5 million, bringing the total funds initially available to just under £2 million.

The Club, as such, consisted of all those paying subscriptions – the four firms, and both the BTD and the BSC from the SERC. It was run by a Steering Committee consisting of two representatives from each of the four firms, and two representatives each from the BTD and the BSC. The Steering Committee was chaired by one of the industrial members. The Club manager, who was paid from Club funds and appointed by the Steering Committee, was an *ex officio* member of the Steering Committee. The university groups undertaking the research were not Club members *per se* but contractors to the Club.[6] Figure 4 illustrates the Club's organization.

[4]This section draws substantially on the two evaluation reports written about the Club, Senker and Sharp 1987, and Sharp 1993.

[5]Sturge was a medium-size specialist chemical producer whose main interest was the production of citric acid. In 1984 it had been taken over by the multinational RTZ who hoped to use Sturge to pioneer diversification into biotechnology. By 1988 it had changed tactics. Sturge was sold first to Rhône Poulenc and subsequently to Bayer. In the process, the initial interest in protein engineering was largely eroded.

[6]This caused some ill-feeling amongst the university groups who saw themselves as being cast as second-class citizens. In this sense they misunderstood the set-up of the Club – members were those who paid the subscription. But their unhappiness also reflected their dislike of the managed-research concept inherent in the Club (Sharp 1993, 24)

Figure 4: Protein Engineering Club

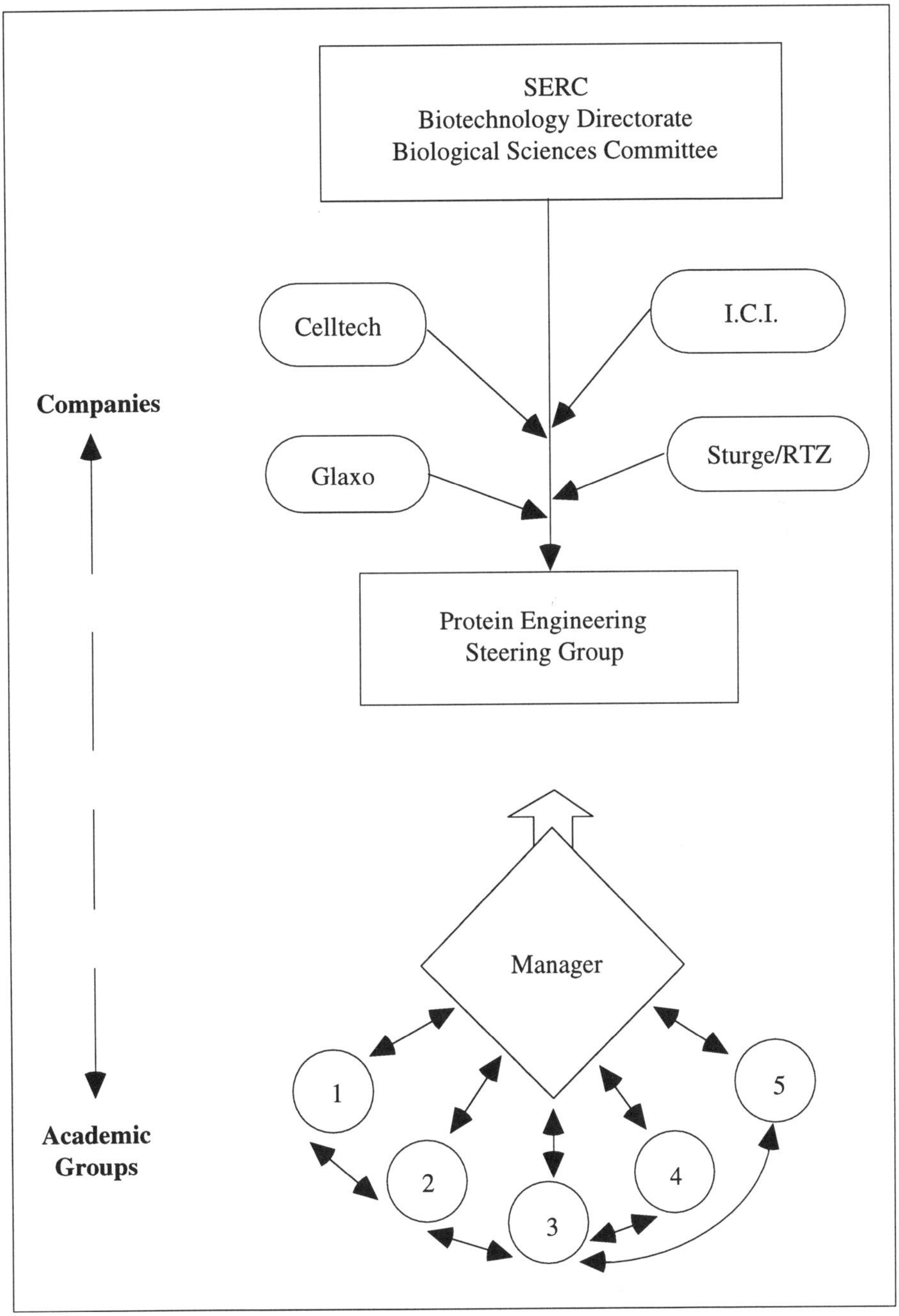

The initial round of funding decisions came in for considerable criticism. Only half the allocated funds were in fact committed and there appeared something of a bias in favor of crystallographic research groups, reinforcing the views of those critics who argued that the Club was merely an elaborate strategy of self interest to help secure more funding for these groups.[7] Moreover, instead of the four- or five-year program grants to two or three sizeable departments which the community had expected, the money was distributed largely as small project grants and split among 11 departments. It took several more rounds of funding and careful management from the Steering Committee and program manager to give the program greater coherence and acceptability. Ultimately it involved 23 separate project grants, mostly of 2-3 years duration, to groups at 11 different universities.

The strategy of using industrial funding to leverage more research funds from the public research budget paid off handsomely. A year after its foundation the Club was rewarded with an extra commitment of £1.5 million from the SERC's Science Board – bringing total funding for the Club program to £3.5 million over its four-year life. Of this, 85 percent came from the SERC and only 15 percent from industry. In addition, spurred in part by the unhappiness of the academic community at their heavy commitment of funds to a *managed* program, the SERC established what was termed an *underpinning program* to fund more basic research in protein engineering. Over the course of the six years, 1984-90, when the Club was in existence,[8] the total commitment of funds to the underpinning program was £3.6 million, matching precisely Club funding itself.[9]

[7]It is important to recognize the context of the funding decisions. The initiative came at a time when British universities were suffering from substantial real cuts in research funding and many groups were finding it difficult to obtain funding to keep research going. This helps to explain the jealousies and tensions between groups. See Balmer and Sharp 1993.

[8]The Club's official life spanned April 1985 to April 1989 – four years. However, as is clear, a good deal of work in setting up the Club took place in 1984, and since some of the grants extended beyond the April 1989 end point, the Club was de facto in existence for the six years 1984-90.

[9]It can be argued that, given the importance of protein engineering, much of this money would have been forthcoming irrespective of the Club initiative. In my evaluation of the Club program I suggest that the Club was responsible for accelerating the underpinning program by approximately two years and therefore that it legitimately can be claimed that the Club initiative brought £3.18m extra resources into Protein Engineering – £480k from industry, £1.5 million from extra Science Board commitments to the Club, and £1.2 million from the first two years of underpinning money. See Sharp 1993, 76-81.

4.2 Towards Maturity – 1987-88

It was two years before the Club to begin to meet some of the aspirations of its founders. By that stage most of the grants had been awarded and the shape of the program was well defined. As Table 1 shows clearly, Club expenditures went into three main areas. The most important category (43 percent of funding) was what was termed *enabling* technology.[10] This included two major database projects. The first, at Birkbeck, was a comprehensive database on protein structures built from the U.S. Brookhaven data[11] by cleaning up the raw data from that source and developing a set of software to provide easier access and analysis. The second database was developed at Leeds and again consisted of a comprehensive cleaning up and pulling together of existing data – this time for all the protein sequence data. The two databases were put together as the ISIS database (Integrated Structure: Interpreted Sequence) and were available in a relatively rough state to Club members from 1987 onwards.[12] In addition, the Club also funded as part of its enabling technologies research, work on crystallographic methods (at Imperial College and Sheffield University), on the use of NMR techniques in protein structure work (Oxford and Leicester), and service facilities in solution conformation and purification techniques at Newcastle and Leicester.

The two other main funding categories identified in Table 1 were the work undertaken on *model* proteins and that on the *target* proteins. The former were proteins where, at the start of the program, there was already considerable knowledge of sequence and structure and which, therefore, could be used for experimentation in different types of engineering. Target proteins were those for which there was little or no knowledge of sequence and

[10]The term *enabling technology* was used by the Club itself. Under the Tassey taxonomy discussed earlier in the chapter, these developments would be subsumed within the term infratechnology which Tassey defines as "not embodied in the product technology" but which "make possible the organization of core technology elements into efficiently functioning systems" (Tassey 1992, 100).

[11]Every scientist who derives three-dimensional coordinates for a protein structure is required to record them with the public domain Brookhaven database.

[12]They were both subsequently mounted at the SERC Daresbury laboratory as part of the SEQNET facility and made available to academics as an on-line facility. The Birkbeck structures database and its associated software is now marketed commercially through Oxford Molecular as its IDITIS package. See Sharp 1993, 42-56.

Table 1: Grants Commitments and Staff Posts by Functional Category

Category	University	Grant Awarded £k	% of Total	Posts
I Enabling Technology				
Structure/Sequence	Birkbeck	546.2	13.6	3 *
Database & Software	Leeds	497.1	14.8	3 +
Science of	Imperial	131.3	3.9	1
Crystallisation	Sheffield	69.8	2.0	1
NMR	Oxford	107.8	3.3	1
	Leicester	101.3	3.0	1
Solution Conformation	Newcastle	49.9	1.6	1
Purification Studies	Leicester	20.0	0.6	0
Subtotal		**1433.4**	**42.8**	**11**
II Model Protein				
Subtilisin	Imperial	105.9	3.2	2
Protease Inhibitor	Imperial	61.1	1.8	1
PGK	Bristol	168.4	5.0	2
Isocitrate dehydrogenase	Glasgow	123.7	3.7	2
Arom Enzyme	Glasgow	107.3	3.2	2
CAT	Leicester	132.9	4.0	2
DHFR	Leicester/			
	UMIST	158.9	4.7	1
Antibodies	Oxford	70.6	2.2	1
Myoglobin	York	35.0	1.1	1
Antibodies as enzymes	Sheffield	36.3	2.5	1
Subtotal		**1050.1**	**31.4**	**15**
III Target Protein				
Penicillin Acylase	York	338.8	10.1	3
	Newcastel	71.7	2.1	1
Glucose Isomerase	Imperial			
	- Blow	116.6	3.5	2
	- Harley	292.4	8.7	3
Methanol Oxidase	Leeds	45.6	11.4	1
Subtotal		**865.1**	**25.8**	**10**
Grand Total		**3348.6**	**25.8**	**36**

* Includes computer manager post.
+ Includes information officer post granted 1988, filled 1989.

Source: SERC Printout of Club grants, September 1990.

structure, but which were perceived to have commercial potential. As Table 1 shows, there were a large number of university groups engaged in one or the other program, with Imperial College, Bristol, and York Universities among the major players.

By 1987 the Club had begun to emerge from the infighting that had characterized its early days and to develop its own agenda. The annual two-day meeting and workshop, to which all participating university groups and firms were invited, took place in January 1987 at York University and attracted over 120 participants. Many of those attending felt it was at this meeting that for the first time the Club achieved some coherence and a sense of belonging.[13] The same sentiments were expressed after the annual meeting at Royal Holloway College in January 1988. They came particularly from the junior researchers involved with the university projects who seldom attended the specialist and international conferences held, for example, by the Biochemical Society. By contrast the senior researchers who were frequent participants at these international conferences found the workshops less valuable, partly because they were able to do their networking elsewhere. For many of the junior researchers the meetings and workshops run by the Club provided the only opportunity of getting away from the laboratory and meeting others working in the same field. It was also useful for them to meet some of the industrial scientists, although representation from the firms involved in the Club was fairly sparse and the main point of contact remained the firms' representatives on the Steering Committee.

By 1988 it became clear that a number of the projects undertaken by the Club would be highly successful in academic terms. The ISIS database developed by Leeds and Birkbeck, in spite of early difficulties, was widely regarded as an academic first, putting the UK ahead of its competitors at that time. Interestingly, it was also widely acknowledged that such work would not have been funded in the normal course of responsive-mode, peer-reviewed

[13]One interesting aspect of this achievement was that it made those in the protein-engineering community who were not involved in Club activities feel like outsiders, and there was considerable pressure on the SERC to open these workshops to other university groups. This was generally opposed by the industrial membership of the Steering Committee which argued that there were aspects of the research which might have commercial relevance in due course and which they did not wish to be openly discussed. They also felt that since they paid a subscription for Club membership, it ought to be worth something in terms of exclusivity. By contrast, the university groups participating – who were not, as explained, Club members but contractors to the Club – would have preferred to open up attendance at the annual workshop (Sharp 1993, 87-90).

research because it would not have been judged original, novel research.[14] Other developments regarded as highly successful in academic terms were the work at Imperial College on the crystal structure of arthrobacter glucose isomerase,[15] the Bristol work on PGK, and the Oxford work on antibody engineering, the latter breaking new ground which was later to become seminal in this area.

As the Club program developed, a strong bond was forged between members of the Steering Committee. Unlike some other club programs (for example those developed in the early 1980s in the UK by the Department of Trade and Industry where club membership really meant no more than being kept in touch with developments – i.e., a reactive, information-receiving role), academic and industrial members played an active role in reviewing and monitoring work on projects. Reports were presented at each of the biannual Steering Committee meetings with individual members being asked to take responsibility for commenting on the reports from specific groups. This often led to close contact with the groups concerned, which were followed up by site visits and further discussion. As a result of this proactive involvement, members of the Steering Committee felt they were involved in a worthwhile activity and their enthusiasm helped to create a sense of purpose and commitment amongst the wider membership of the Club (Sharp 1993, 81-86).

4.3 Problems – 1988-90

The final two years of the Club, however, were more difficult than anticipated. Two issues dominated Steering Committee discussions – the problems associated with the Birkbeck/Leeds ISIS database and the future of the Club.

[14]A similar situation arose later with the UK's Human Genome Program where the relatively routine work of identifying gene sequences, vitally necessary to the project, was nevertheless perceived as boring and dull science. It is interesting that in both cases the managed program of research recognized its importance and funded the research in order to provide the necessary scientific infrastructure. The value of this work is now recognized by the scientific community which accepts, in retrospect, that these projects were both eminently worth funding. This provides an interesting example of how funding can, in the longer run, succeed in changing what seem to be immutable attitudes and behavior. See Sharp 1993, 69-76; and Balmer 1993, Chapter 5.

[15]The team at Imperial College under Professor Blow maintained that it was wrong for the Club to take credit for this work which, had they not received Club finance, would have proceeded as it had begun, under College sponsorship. In principle, they are obviously correct. But the case illustrates how difficult it is to attribute success to one particular source of funding. It remains true that the Club and the associated underpinning program opened up new and important sources of funding for such groups and enabled them to complete research which might otherwise have taken much longer to yield results.

The ISIS database

The problems surrounding the database are interesting because they were fundamentally about issues of intellectual property and its commercialization. By 1988 it was clear to those who had used the database that it was potentially a very useful tool for all researchers involved with developments in protein engineering, enabling, for example, the rapid identification of homologous sequences and structures and enabling links to be made between structure and function. The Steering Committee took the view that priority should be given to making the data available to other academic groups, but that simultaneously the possibility of commercial exploitation should be pursued.

Under the terms of the original agreement, intellectual property rights (IPR) on any exploitable outcome of the research sponsored by the Club were vested in its main paymaster, the SERC, although the Club companies had rights both to favorable license access and to a share of any income deriving from exploitation. The SERC in turn looked to the British Technology Group (BTG), at that time a state-owned technology transfer organization which had first option rights on the commercialization of any government financed research.[16] Problems arose because the BTG, while recognizing the potential of the database, found it difficult to find an appropriate commercial sponsor and was anxious not to see the market "spoiled" by too easy an access for academic users before a licensee had been found. Commercialization also meant that the two groups involved (Leeds and Birkbeck) were required to spend a good deal of time verifying the work they had done and writing the software they had developed in a form which was accessible to the nonspecialist. For the research groups this meant giving valuable time, which might otherwise have been spent on using the database for developing their

[16]The BTG had previously been known as the National Research and Development Corporation (NRDC) and had been set up in the 1950s expressly to help academics and government scientists by relieving them of the problems associated with patenting and licensing their discoveries. In this capacity, it had a number of major successes to its name; for example, it owned the cephalosporin antibiotic patents which had yielded a steady flow of income over the years. However, it was not a popular institution with academics who complained that it frequently failed to recognize the potential in the work they brought to its attention (the notorious example here being the case of monoclonal antibodies where between them the MRC and the NRDC failed to patent) and, even when it did, they received minimal reward for their efforts. Under the Thatcher government of the 1980s it lost its first option rights, universities being given the right to seek patents for themselves, and it has subsequently (1993) been privatized. The Protein Engineering Club contract was drawn up in early 1985 just before the BTG lost its first option rights. For a full discussion of BTG and its role, see Dodgson 1991, 30-35.

own research, in what was essentially a time-consuming and rather tedious task for which they earned no academic accolades. They did not give it high priority and the task therefore dragged on longer than it might have otherwise.[17]

Moreover, as time passed there was also need to keep both databases up-to-date, a not insubstantial task. For its part, the BTG only belatedly recognized the need to provide the resources necessary to support this sort of development work in the academic groups, and for some time there was a good deal of bad feeling between the various parties on this account. The net result was that it was not until 1990 that the database was first mounted for on-line academic access (through the SERC's SEQNET facility at Daresbury) or that negotiations began with the eventual licensee (Oxford Molecular) over how best to commercialize the work. In the event, they chose only to exploit the Birkbeck structures database and associated software, marketing it as their IDITIS package. Nevertheless, this was described by the company in 1993 as "the most innovative piece of IPR in our portfolio."[18]

From club to LINK program

The other main issue that preoccupied the Steering Committee in the latter phases of the Club's life, was how it might best be continued. In the late 1980s the UK Government shifted the focus of its R&D support away from schemes such as the Support for Innovation (SFI) scheme which had helped companies meet the development costs of implementing new technologies (dubbed near-market R&D) towards support for R&D which was more distant from the market-place – precompetitive R&D. The prime vehicle used to support such R&D was known as the LINK program, essentially linking academic and industrial research. LINK rules initially required that at least two companies and one academic institution participate in any scheme that was approved for funding. The second stage of the Protein Engineering Club was seen as a prime candidate for a LINK program and proposals to this effect were first mooted in early 1988, the hope being that the project could be cleared in time for a smooth transition from Club to LINK program in April 1989 when the Club itself officially came to an end.

[17]There were a number of other problems which help explain the delays. It turned out that a good deal of the early work on the structures database at Birkbeck needed to be checked and reworked. In addition, there were complications with both groups over signing the Heads of Agreement document which assigned IPR to the BTG (Sharp 1993, 47-48).

[18]See comments on Oxford Molecular's views in Sharp 1993, 50-51.

Unfortunately, delays of one sort or another dogged the project and it was not until mid-1989 that the LINK scheme was approved. A meeting held in July 1989 showed a good deal of company interest, with some of those who had withdrawn from the early Club discussions expressing regret that they had not maintained their interest. Again, negotiations stalled over program content with exactly the same "chicken and egg" issue, companies being unwilling to commit themselves until they saw what was on offer, yet Club rules requiring that what was offered was substantially chosen by the companies involved.

The LINK rules added further complications, because they allowed not only for a Club format (in which several companies came together to help fund a common program of projects), but also for collaborations involving as few as two companies and one academic group. They also laid down that, whatever the format, companies had to contribute at least 50 percent of the funding – in other words, that it was the same price whether they went for the Club "table d'hôte" menu or chose to go "a la carte," picking off research groups individually. Perhaps, not surprisingly, they chose the "a la carte" menu. As attempts to put together a Club menu grew increasingly protracted, a number of companies began to negotiate with the individual academic groups on separate deals and, as a result, the Club bid disintegrated (Sharp 1993, 58-67).

In many respects this was rather a sad end to what was, in other ways, a successful experiment. For both the ISIS database and the attempt to take the Club forward under LINK rules, the story was one of delays, misunderstandings, and bad feelings on many sides which soured relations and overshadowed real achievements. It is summed up by the comment of one grant-holder who said, "What the Club needed was a managing director who would take a grip and stop these things from happening" (Sharp 1993, 67).

5. The Protein Engineering Club as an Advanced Technological Infrastructure Policy

It was argued earlier that protein engineering in 1984 represented a classic example of a technology ripe for the application of advanced technological infrastructure policy and that the SERC's Club represented just such a policy. It is therefore worth looking back at the experience of the Club to see what lessons can be learned from the experiment.

5.1 Coping with Uncertainty

The key characteristic differentiating advanced from basic technological infrastructure policies is that they deal with new technologies where uncertainty is pervasive. Such policies are particularly appropriate when nobody knows what the technology itself can achieve or whether there are market needs to be satisfied (see Justman and Teubal in this volume). In this case, the Protein Engineering Club provided a mechanism for flagging the importance of the technology and for carrying it forward, while at the same time keeping industry informed of new developments. The inability of the industrial members to identify what they wanted from the Club in the early negotiations over the program specification well illustrates their uncertainties at that stage as to where market potential lay. By contrast, six years later in 1990, the companies involved in the LINK negotiations had a much clearer idea of what they wanted from a research program and the difficulty was reconciling their different interests. Indeed, one of the reasons why the LINK initiative to continue the Club program failed was that in the course of its six-year life, protein engineering had moved from being precompetitive to being competitive research. By 1990, companies were much more interested developing specific projects with bilateral relationships with academic groups than they were in collaborating with their competitors.[19] The Club had enabled the companies participating to maintain a "window" on the technology, to follow developments in the UK and, through the informal networks with academics developed through Club participation, to keep abreast of what was happening worldwide. When the time came to invest in building up in-house competencies (as Club members Glaxo and ICI began to do in 1988) they knew where and how to tap into the requisite skill base.

There would appear to be three lessons from this experience regarding the uncertainty of technology which could be carried over more generally into technological infrastructure policies:

(i) Clubs should be considered an *especially appropriate* means of underpinning research and university-industry linkage in cases of newly emerging technologies *when the wide interaction between science and*

[19]In this regard, it is interesting that the LINK rules were changed in late 1990 to allow one-to-one company-academic linkage because, in areas such as protein engineering, initiatives were failing since companies did not wish to share research results with competitors. It was after this change that the LINK protein engineering program finally began to move forward, but not, as we have seen, with a club element to it. Most of the projects embraced within the program involved one-to-one or two-to-one partnerships (Sharp 1993, 63-64).

technology which they foster is needed to help shape the research agenda. They are not necessarily an appropriate form of support once a technology moves from being precompetitive to being competitive. This is particularly the case in areas such as chemicals and pharmaceuticals, where IPR is an important aspect of competition (and therefore where sharing IPR poses problems); perhaps less so in sectors involving engineering skills where that learning means that IPR plays a lesser role.

(ii) The precompetitive nature of the research fostered by clubs argues in favor of *flexibility in funding rules*. The Protein Engineering Club involved only a 15 percent contribution to costs from industry, and an 85 percent contribution from government. Had the 50 percent LINK rules applied in 1984, it is doubtful whether the Club would have emerged. When interviewed, participants made it clear that in 1984 they would not have been prepared to put up more funding because of the uncertainties surrounding the whole area at that time. Yet to lose industrial participation in the project risks slower implementation, less industrial relevance, and less-effective technology transfer. It is difficult to quantify such benefits and therefore to introduce explicit cost-benefit analysis in such cases, but little harm is done if greater flexibility is allowed in funding rules.

(iii) A relatively *proactive stance in setting the initial agenda* for such clubs is likely to save much management time and frustrations with little detriment to the usefulness of the program. Given the uncertainties inherent in the emergence of new technologies such as protein engineering, it is difficult for industry to define in advance what their needs will be. But this argues also for caution in committing all funds immediately; better, perhaps, to allow the program to develop along its own trajectory as uncertainties begin to clear. In retrospect, the much-criticized failure of the Protein Engineering Club to spend its initial allocation of funds may have been a blessing in disguise.

5.2 Industry's Involvement in Technology Transfer

One of the main objectives of advanced technological infrastructure policies is to speed the assimilation of new technologies, especially by the industrial sector. In this regard, the Protein Engineering Club aimed to achieve deep involvement by industry in club activities on the grounds that there were mutual benefits to be gained – industry was well informed of scientific progress in the field and, perhaps even more important, knew whom to contact if they needed advice; academics gained a better appreciation of industry's needs.

This philosophy underlay the proactive role expected of Steering Committee members. It was also recognized that an important source of technology transfer comes through the training of young researchers in academic laboratories who are then recruited into industrial research laboratories.[20] It is interesting to see how far the Club met these two objectives.

The depth of contact between industrial scientists in the Club and their counterparts in the universities varied considerably from group to group. There were some university groups that had little contact with any industrial members of the Club throughout its lifetime, whereas others had almost daily contact. The most frequent contact was between the industrial members of the Steering Committee and the main grantholders at the different universities. Only at the annual review meetings did more junior members of the academic and industrial groups get a chance to meet each other.

The most regular and successful meeting ground and source of stimulus to each side was the Steering Committee itself. The industrial members who devoted the most time to Club business benefited the most from: contacts and awareness of developments; privileged access to the ISIS database; knowledge and availability of other infrastructure services; and in the longer run, an opportunity to identify possible recruits. For the academics there were also benefits; six of the eleven groups subsequently received direct research contracts from industrial members, while the industrial credibility gained from working on the Club program helped win other research contracts. Moreover, in spite of initial skepticism, most academics participating in the Club program felt that they had benefited from closer knowledge and contact with industrial views (Sharp 1993, 81-86).

As regards training, over its lifetime the Club supported 36 postdoctoral research posts and 19 predoctoral studentships. 57 different individuals filled the postdoctoral research posts, of whom 14 were "birds of passage" who either gave up the post after a short period or filled it on a temporary basis when the original appointee had moved on. Of the 43 long-stay researchers, a substantial majority (33) were still active in academic research in 1993. Seventeen of the 19 doctoral students completed their doctorates by 1993, eight of these going on to postdoctoral positions in the UK, and another four to postdoctoral positions abroad. Overall, therefore, the program made a

[20]In their evidence to the recent Waldegrave White Paper (on the future of the science base in the UK), industry made it clear that from its point of view one of the most important functions of academic scientific research was to train a cadre of recruits for their laboratories (Lyall 1993, 37-45).

considerable contribution towards developing skills and capabilities in protein engineering, but not, it would appear, for industry's benefit. Indeed, the most notable feature of the statistics about the Club trainees was the very small number going into industry. Only eight out of the 43 long-stay, postdoctoral researchers and two of the doctoral students were working in industrial laboratories in 1993. There were three known cases of industrial offers having been rejected in favor of academic posts, but even allowing for this, the figures provide scant support for industry's claim that they need academic research as a source of training for recruits. Perhaps most disturbing was the very negative image of industry amongst the young researchers and graduate students; industry was seen as profit driven and exploitative, and industrial research as yielding little benefit to society (Sharp 1993, 90-97).

There would seem to be three main lessons to be learned regarding industry involvement:

(i) *A proactive approach to club membership is fully justified.* In general, the more involved the industrial members in shaping and running the collaborative research program, the more effective the technology transfer. By contrast, clubs which involve little more than information-sharing are likely to be least effective.

(ii) *Regular meetings are an essential feature of collaborations*, both more formal review meetings as well as smaller, specialist workshops on specific topics. At such meetings it is particularly important to encourage participation by junior staff from both academic groups and industry since they are less likely to meet on other occasions. Informal contacts are just as important as formal contacts in technology transfer and ways need to be found to promote such liaison.

(iii) *Industry should make particular effort to deepen their involvement where they participate in collaborations of this type*, for example by arranging for exchange visits amongst junior staff. Academic researchers remain poorly informed about the way research is used in industrial laboratories and full two-way collaboration can be of mutual advantage.

5.3 Multidisciplinarity

Advanced technological infrastructure policies can also fulfill the function of creating new capabilities which cut across older established disciplines. In science new ideas frequently come at the boundaries of disciplines, deriving from the cross-fertilization of knowledge systems. Protein engineering is a classic example of such cross-fertilization and one of the prime objectives of the Club was to bring together two of the major disciplines involved – the crystallographers and biophysicists who had pioneered work on protein structures, and the molecular geneticists who had developed the techniques of genetic engineering which underpinned the biotechnology revolution.

How far did the Club succeed in creating such a research community? Interestingly, the early developments of the Club did more to sow schism than harmony. The research community in Britain reacted strongly against the concept of directed research implicit in the establishment of the Biotechnology Directorate within the SERC, and the Protein Engineering Club proved to be the most controversial of its initiatives. In particular, it was attacked by the Medical Research Council (MRC) for two reasons. First, having supported the researchers who had pioneered the techniques of protein engineering, the MRC regarded the topic as its domain and resented the intrusion of the SERC. Second, the concept of directed and managed research geared to the needs of industry which was implicit in the Club idea was totally alien to the MRC's *scientistic* culture.[21]

However, as the Club matured an increasing sense of community developed. The annual review meetings and specialist workshops were particularly important in this regard, especially for the more junior researchers who did not have the links of their seniors to the national or international academic fraternity. The increasing acceptance of multidisciplinary research also reflected the growing recognition of the importance of biotechnology and its need to pull across disciplines. It is noteworthy that whereas many universities in Britain began the 1980s with separate departments for biochemistry, genetics, microbiology, and even sometimes botany and zoology, by the end of the 1980s these had frequently been amalgamated into departments of biological sciences. Funding, particularly the availability of funds from the Biotechnology Directorate, played a part in stimulating these changes, in the same way that funding of infrastructure projects which represented useful rather than novel science helped to change attitudes towards their

[21]The term *scientistic* is used in the sense that the MRC has traditionally put the highest emphasis on science for its own sake. It maintains that the best science is that which develops from the scientist's own search for knowledge and thus should be curiosity-led rather than directed by administrators or profit-driven by industry. Funding should therefore be provided essentially in response to the scientists' own proposals and judged purely in the light of whether it constitutes good science as judged by their peers. All that the Protein Engineering Club stood for violated the tenets of this philosophy. The MRC therefore refused from the beginning to cooperate in any way with the initiative and maintained its opposition throughout its lifetime (Balmer and Sharp 1993).

acceptability.[22] Given the fight put up by the MRC against the Protein Engineering Club, it is ironic to record that in 1994, the victory for the directed program of industrially relevant research is seemingly almost complete. The new era of research funding introduced by the Waldegrave White Paper of 1993 (UK Government 1993) makes the directed program the norm rather than the exception.

There are two lessons regarding multidisciplinarity to be drawn from this experience for advanced technological infrastructure projects:

(i) *Promoting multidisciplinary research is an important vehicle to foster change and innovation* in academic scientific institutions. Academic institutions are amongst the most conservative in society. They change slowly and only when considerable pressure is put upon them. But there has to be a dynamic – a way for new ideas to gain credence and acceptability. For all its merits, the danger with the peer review system is that becomes incestuous and reinforces conservatism. Programs such as the Protein Engineering Club have sometimes to seek deliberately to break the mold, and in doing so, inevitably get attacked by the traditionalists. But they can be successful ultimately in effecting change.

(ii) *Deliberately fostering a sense of community is valuable* within programs of this sort in order to promote interaction. Meetings, workshops, and newsletters can all play a part in doing so and should be seen as an integral, rather than peripheral part of the program.

6. Conclusions

There are many different ways of viewing the Protein Engineering Club experiment. One is to see it as the product of the funding climate in Britain in the 1980s. The Conservative government's squeeze on the universities and research funding at that time had created what John Ziman was to label "science in a steady state" (1987). Funds were limited, extra expenditures on new initiatives in one area meant less to spend in another, and the government itself was demanding more relevance from science. The Club was a brilliant way of squaring all these circles – it identified and flagged an important new area of development where Britain had relevant scientific expertise but needed to train a new generation of multidisciplinary researchers and awaken industrial interest. It not only brought extra funding in terms of the industrial subscriptions,

[22]See footnote 14 and in particular the comments recorded by one of the Club participants quoted in Sharp 1993, 72.

but by demonstrating this industrial relevance, it also unlocked extra government resources as a reward.

It would be unfair and wrong, however, to see the Club only as a cynical manoeuvre by some well-placed academics to secure funding for favored projects. It was also an experiment aimed both at technology transfer and at shifting the focus of research from the traditional disciplinary basis towards new multidisciplinary areas. As this chapter has demonstrated, it succeeded in many aspects of this objective – it raised industrial awareness and brought academic and industrial interests together; it developed important aspects of the research infrastructure; it trained a new generation of researchers; and by the end of its span, it succeeded in changing the orientation and goal of multidisciplinary projects. Along the way there were some minuses to set off against these pluses – it had, rightly or wrongly, succeeded in alienating a number of key players in the field; attracted only marginal industrial interest; and bungled important aspects of IPR management. It was not all a bed of roses.

This chapter has sought to abstract from the British context and from the immediate pros and cons of the Club itself to view the Club mechanism as policy tool, one that is appropriate to technological infrastructure policies, and particularly *advanced* infrastructure policies. In this context, the Club mechanism emerges as highly relevant for the following reasons:

- Industrial participation ensures *industrial relevance.*
- It is concerned with building *generic technology* capabilities in the sense that it seeks to develop laboratory rather than commercial prototypes.
- When applied to an emerging area of technology such as Protein Engineering, the program is by its nature *multidisciplinary.*
- Collaboration between firms ensures that it concentrates on *precompetitive* research where the interaction between industrial and academic players is an important element in shaping a new trajectory and establishing the critical parameters of the new technology.
- It provides an important *mechanism for handling the uncertainties* inherent at this stage of development.
- The continuing interaction between science and technology implicit in such a program provides an *important constituent of the continuing dynamic.* As Rosenberg has pointed out, "[the] natural trajectory of certain specific technological improvements serves to identify and define the limits to further improvements which, in turn, serves as a focusing device for science" (Rosenberg 1992).

It is tempting at this point to move from the positive (clubs of the Protein Engineering Club type clearly fit into the category of advanced technological infrastructure policies) into the normative (what then is the best way of organizing such a club). A number of general guidelines which emerge from the Protein Engineering Club experience were discussed earlier in this chapter relating to issues as diverse as the funding formula and the need for regular meetings involving junior as well as senior researchers. Perhaps the most important of these relates to not being inhibited about breaking new ground. There are always considerable forces for conservatism which resist new developments. Change only comes as a result of challenging this built-in inertia in society. Beyond this, there are some fairly obvious points:

- Involvement by industrial members suggests that large groupings, say over a dozen, would be cumbersome.
- Given that such a Club spans the period between scientific discovery and commercial application, its life span is, by definition, limited – five to ten years seems appropriate but this will vary from case to case.
- The precompetitive nature of the research means that there is unlikely to be much output for which IPR are important. Clubs should not therefore be judged by their IPR. Equally, it is sensible to lay down clear guidelines in advance as to how IPR, should it arise as an issue, should be handled.
- Clubs need firm management and will not be self-organizing.

The message, however, should be to beware of generalizations. Each sector, each technology, bears its own characteristics; each country, its own culture and institutions. In some sectors, at some stages of development, clubs of this sort may be a useful way of carrying a new technology forward. Each will need to be adapted to its own set of circumstances.

References

Balmer, B. 1993. Mutations in the research system?: The Human Genome Mapping Project as science policy. Ph.D. thesis, submitted to the University of Sussex.

Balmer, B. and M. Sharp. 1993. The battle for biotechnology: Scientific and technological paradigms and the management of biotechnology in Britain in the 1980s. *Research Policy* 22: 463-578.

Dodgson, M. 1991. *The Management of Technological Learning: Lessons from a Biotechnology Company*. de Gruyter.

Lyall, C. 1993. The 1993 White Paper on Science and Technology: Realising our potential or missed opportunity? M.Sc. Thesis, submitted to the University of Sussex.

OTA. 1982. *Genetic Technology: A New Frontier.* Washington D.C.: U.S. Congress. Office of Technology Assessment.

---. 1987. *New Developments in Biotechnology: US Investment in Biotechnology*. Washington D.C.: U.S. Congress. Office of Technology Assessment.

—-. 1991. *Biotechnology in a Global Economy.* Washington D.C.: U.S. Congress. Office of Technology Assessment.

Rosenberg, N. 1992. Science technology in the twentieth century. Chapter 2 in *Technology and Enterprise in a Historical Perspective*, edited by G. Dosi, R. Giannetti, and P. Toninelli. Oxford.

Senker, J. and M. Sharp. 1988. The Biotechnology Directorate of the SERC: Report and evaluation of its achievements. Mimeo, obtainable from Science Policy Research Unit, University of Sussex, Brighton BN1 9RF.

Sharp, M. 1987. The Protein Engineering Club: The first two years. Mimeo, obtainable from Science Policy Research Unit, University of Sussex, Brighton BN1 9RF.

—-. 1991. Pharmaceuticals and biotechnology: Perspectives for the European industry. Chapter 13 in *Technology and the Future of Europe*, edited by Freeman, Sharp and Walleer. London: Pinter Publishers.

—-. 1993. The Protein Engineering Club: An assessment of the Club as an experiment in promoting a programme of strategic research. Mimeo, obtainable from Science Policy Research Unit, University of Sussex, Brighton BN1 9RF.

Sharp, M. and I. Galimberti. 1993. Europe's chemical giants and the assimilation of biotechnology. FAST Coherence and Diversity of the Innovation System of Europe, FOP 350, June.

Tassey, G. 1992. *Technology Infrastructure and Competitive Position*, Kluwer Academic Publishers, Dordrecht.

UK Government. 1981. *Biotechnology.* Cmnd 8177, London, HMSO.

—-. 1993. *Realising our Potential: A Strategy for Science, Engineering and Technology.* CM 2250, HMSO (The Waldegrave White Paper).

Ziman, J. 1987. Science in a "steady state": The research system in transition. SPSG Concept Paper No. 1, Science Policy Support Group, London.

Generic R&D Collaboration Between Firms: The Israeli Experience

Dan Kaufmann and Tamar Yinon*

In June 1992 the chief scientist of Israel's Ministry of Industry and Trade launched a program aimed at strengthening and expanding the country's technological infrastructure. The MAGNET Program[1] was established as the only Israeli framework, private or public, for the support of collaborative *generic research* (GR).[2] Magnet also facilitates the diffusion of technology by subsidizing advanced technology users clubs. The Israeli initiative was not novel; similar programs had been initiated more than a decade ago by most OECD governments. What Magnet has in common with programs established earlier elsewhere is the recognition of the growing importance of GR and an awareness of the importance of a technological infrastructure policy that promotes collaborative technology development.

The relatively late penetration of collaborative GR into Israeli industry provides us with an opportunity to analyze this technology policy while utilizing the rich experience accumulated elsewhere. Such an analysis offers two advantages often absent in so-called real-time studies:

- information on the evolution and success of other programs provides a point-of-reference regarding the performance of the Israeli collaborative GR program;
- the tools needed for studying the various phenomena have become rather focused.

The Israeli case also makes an attractive study because Israeli industry is relatively young and small, shedding additional light on some components of

* This study is based on the authors' investigation of the MAGNET program, undertaken during 1993-4 on behalf of Israel's Chief Scientist at the Ministry of Industry and Trade. This research was followed by a complementary investigation of similar programs in Europe, including in-depth interviews of key participants in these programs. The research results have been documented and reported to the Chief Scientist and remain the property of his office.

[1]MAGNET is the Hebrew acronym for "Generic R&D and Technological Infrastructure".

[2]Generic research can be defined as being "beyond the stage of fundamental research, but be such that at the conclusion of the project it will still need further R&D to produce marketable products or processes" (BRITE Information Package). Tassey defines generic technology as "the first step in a process by which basic scientific knowledge is made increasingly more usable for market application" (1991, 356).

M. Teubal et al. (eds.), Technological Infrastructure Policy, 247–267.

industrial infrastructure which do not yet exist or have not matured, but which are crucial for collaborative GR.[3]

In this chapter we evaluate Israel's Magnet program, and in doing so, we show that the effective development of collaborative generic research requires a complementary set of institutional and policy changes. Israel's failure to implement a complementary technology infrastructure policy is an important reason for the restricted success of the Magnet program, in particular, its failure to be widely used. We begin by briefly discussing the reasons for the growing importance of generic research and examine some of the associated market failures that have impeded the advancement of collaborative GR. Following a brief description of how GR has been supported in other industrialized countries, we turn to the main characteristics of the Israeli program. Using insights gained from the international experience in launching and facilitating successful collaborative GR, we examine:

- the requirements for complementary technological infrastructure;
- the need for other policy mechanisms such as intellectual property rights, legal support, and assistance with partner selection;
- the importance of pecuniary assistance; and
- the relative advantages of neutral versus selective policy approaches.

Based on the identification of the complementary infrastructure elements important in the promotion of collaborative GR, we conclude by offering some policy guidelines to policy makers.

1. Generic Research and Technological Infrastructure Policy

Generic research has become increasingly important in technological development for several reasons. The gradual shortening in the life cycle of products has made it much more difficult to recoup investments in applied R&D. As a result, firms tend to shift some of their efforts from incremental innovative research to activities that have the potential for more radical technological change. When successful, this reorientation of research may lead to several new applications with longer lead time. Another reason for conducting GR is the steadily increasing complexity of the technology embedded in industrial products (Chesnais 1988). Increasingly, high-tech

[3]For example, the absence of well-functioning industrial technology centers in Israel helped us to identify their central role in the initiation and execution of GR projects.

products directly incorporate results from basic-science research (Soete 1991). An additional motivation for GR is the globalization of industrial R&D and production, as illustrated by the rapid rise of the newly industrializing countries (NICs). The NICs proved that an effective technological infrastructure policy, based on purchasing or imitating existing technologies, can generate a major competitive advantage. This advantage is especially significant in those countries where high-skilled, low-wage labor is available. In developed countries GR promotes the capitalization of academic research capabilities by creating a useful link between universities and industry. But promoting collaborative GR in Western market economies is made difficult by the existence of a number of obstacles.

1.1 Market Failures Associated with Collaborative GR

Market failures associated with collaborative GR can be divided into two groups: those inherent to GR and those inherent to collaboration within a competitive market.

Similar to any applied research, GR suffers from the typical market failures associated with R&D activity: uncertainty, indivisibility, and inappropriability (Arrow 1962). However, these market failures are more extreme for GR than for other forms of applied research. GR is closer to basic research, and therefore further removed from the market (Soete 1991); accordingly, the risks and costs involved in GR exceed those for applied research. As to inappropriability, patent systems are typically not designed to effectively protect property rights of GR output (Katz and Ordover 1990; Ouchi 1989). This is because GR results are less tangible than the results of other forms of applied research, while patent systems are aimed at protecting products or processes for which at least prototypes have been developed.

Market failures may also arise because GR generally contributes to several commercial applications simultaneously. Often firms will lack the complementary assets needed to exploit many potentially profitable opportunities in different markets. In such a case investments in GR may not be fully commercially exploited by an individual firm and the social optimum is not reached. Free-rider possibilities may also deter firms from conducting GR individually because research results may strengthen other industries that did not pay for the research.

These arguments support the case for conducting GR in a collaborative environment. Interfirm collaboration is conducive to raising the considerable sums of money required, to spreading the risks, and to exploiting the GR results in all relevant markets. Interfirm collaboration is also efficient because of the multidisciplinary nature of GR. However, obstacles to interfirm cooperation may also exist, at both the public and the firm levels.

Collaboration is generally regarded undesirable according to neoclassical economic theory which believes competition to be imperative to the optimal functioning of the market. Institutionally this finds its expression in the formulation of strict antitrust laws which can limit interfirm collaboration (Jorde and Teece 1990). On the other hand, at the level of the firm, competition is part of corporate culture and organization. The orientation toward competition and rivalry is such that market forces will not lead to interfirm collaboration, even when the survival of companies is threatened. Implementing interfirm collaboration requires a paradigmatic shift in both the private and public perception of the source of competitiveness in modern industrial organization. In such conditions, forming consortia – especially among market-place competitors – may involve high transaction costs. Both the search for collaborators and the formulation of contractual agreements are acts which are new, and even more importantly, contradict existing market perceptions.

1.2 Programmatic Responses

Recognizing the potential offered by GR and collaborative research, most OECD governments responded by developing support programs more than a decade ago. The Japanese Government was the first to recognize the potential benefits of collaborative GR, introducing the VLSI (Very Large Scale Integrated Circuits) program in 1976. The perceived major success of this program, and their intention to launch the Fifth Generation Computer Program employing the same organizational approach, inspired imitative efforts both in Europe and in the U.S. In 1982 the 10 major U.S. microelectronics companies initiated the first American collaborative-GR program, the Microelectronics and Computer Technology Corporation (MCC); "the recognition of accelerated foreign competition was clearly a factor contributing to the formation of MCC" (Dove 1989, 1365). In contrast to the Japanese case, MCC was a private initiative. In 1984, and for similar reasons, the British Government inaugurated the ALVEY program, intended to enhance research in the field of information technology. In the same year the European Commission launched the first Framework Program, concentrating on research in the fields of communication and information.

Various European programs were introduced during this period, several having similar characteristics to the Magnet program and thus offering obvious points of reference for a comparative study of collaborative GR.[4] In particular, we refer to the Finish collaborative GR program operated by TEKES (The Technology Development Center sponsored by The Ministry of Trade and Industry); the Swedish IT4 program; the British ALVEY program; and to a lesser degree to the European Community framework programs. Like Magnet, these programs are essentially supported by national governments, they are flexible as to the types of consortia they allow, and they typically provide subsidies on the order of 50 percent (Magnet's subsidy level is 66 percent). In contrast, collaborative GR in the U.S. is mostly done in central laboratories adjacent to public technology or research centers. In Japan, on the other hand, government involvement is even more extensive, frequently reaching beyond the program level to the point of defining projects, with subsidy rates sometimes reaching 70 percent (Ouchi 1989).

2. The MAGNET Program

Although Israel was late to initiate programs to promote collaborative GR, applied research and R&D carried out by an individual firm has received massive governmental support since 1967. These applied projects have made significant contributions to the evolution of the high-tech industry. Since the late sixties the Ministry of Industry and Trade has supported industrial R&D by covering 50 percent of most R&D costs. The goal of this support is to promote highly applied and close-to-the-market research. A number of factors led the Ministry of Industry and Trade to initiate the Magnet program, including: the gradual increase of GR throughout the world, the duplication of local research projects, growing pressure by the Treasury and some industrialists, and the insistence of industrial and technology think tanks.

The Magnet program is designed to support projects that fulfill two basic requirements: first, projects must be *generic*, and second, they must involve collaboration between at least two firms and at least one academic institution. Beyond these restrictions, consortia are given considerable programmatic, organizational, and operational flexibility. As is the case in the applied R&D program, MAGNET does not set priorities with regard to particular industries,

[4]In addition to programmatic similarities, another reason to base our comparative analysis on programs in small European countries is that these countries are much closer in size to Israel than the U.S. or Japan.

technology areas, or product class. With regard to organization, consortia can be structured along either horizontal (collaboration between competitors) or vertical (collaboration between suppliers and customers) lines. And operationally, consortia can choose between conducting research in central laboratories or dividing the project into subprojects carried out by the different partners.

During Magnet's first two years of operation seven R&D projects and two dissemination programs were initiated; three new R&D projects and one other dissemination program are expected to be approved during 1995 (see Table 1).

Table 1: MAGNET R&D Projects and Dissemination Programs

Project Name	Active since	Total Budget (Million $US)
R&D Projects		
Ground Stations for Satellite Communications	1993	$27
Wireless Communications/Digital Receiver	1994	$12
Broadband ISDN and Multimedia Communications	1994	$40
MMIC-GaSa Devices	1993	$18
F1 Hybrid seeds and Flowering Control	1993	$1
Optimization of Algae Production and Extraction of Fine Chemicals from Algae	1994	$6
SSR DNA Markets	1994	$2.5
Network Management System (NMS)	1995*	$16
Manufacturing and Inspection of 0.25 Micron Devices on 12" Wafers	1995*	$30 (est.)
Applying Focused Solar Energy to the Generation of Electricity and Other Industrial Uses	1995*	$14
Dissemination Programs		
Surface Mounting Technology (SMT)	1993	$0.3
Very High Design Language (VHDL)	1993	$0.2
NMS Standards and Technology Assessment	1995*	$0.4 (est.)

Source: Ministry of Industry and Trade: Magnet Program, 1995.

* Pending approval.

These programs represent a total budget commitment of approximately US$160 million over the first five years. The Magnet program funds 66% of the approved projects' budgets by way of direct grants (royalty payments in case of success are not required). Academic research institutes participating in R&D consortia are entitled to an 80% grant of their approved budget.

It is important to note that these are relatively large and costly projects, in some cases equivalent in funding to that received by an entire European program or subprogram, and far in excess of the funding typically received by a single project. Nonetheless, in the first years of operation, only a small part of the Magnet approved budget has been used, a fact that might suggest that major market failures are associated with cooperative activities or with GR. It would appear that during the initial stage of new policy programs, financial support by itself is not sufficient to alleviate these market failures, as has been also observed in the case of applied research (Teubal 1983; Toren 1990). Apparently, other nonpecuniary means of intervention are also needed.

3. Stimulating Collaborative GR: A Dynamic Perspective

The Magnet experience suggests that massive financial support is insufficient to alleviate the above-mentioned market failures. Despite substantial financial support in the form of a 66 percent subsidy, only seven projects have been implemented under Magnet sponsorship. In contrast, European programs supporting collaborative GR, during their first years of operation, have been much more successful in generating projects, even though their subsidy levels are often lower than those offered by Magnet. For example, The Alvey program supported 198 projects over its five years of operation (Quintas and Guy 1995). The Swedish IT4 program supported about 46 projects over its first 2 years (Arnold and Guy 1989). This indicates that factors other than financial support play important roles in stimulating the formation of GR consortia. To identify these factors we employ Teubal's (1995) analysis of policies that alleviate market failures.

Teubal stresses the importance of analyzing market failures within a dynamic context because the static neoclassical view fails to consider the evolutionary nature of markets. In contrast, Teubal emphasizes that institutions can and often do change over time. Thus, an analysis of the market failures associated with GR should take into account the dynamics of institutional frameworks and the learning capability of the market. For example, we earlier noted that one reason why pure market forces might fail to initiate sufficient private cooperative GR projects is the existence of antitrust laws which specifically forbid – or are perceived to forbid – such collaborations. But in 1984 both Europe and the U.S. adjusted their antitrust laws to allow cooperative GR. Similarly, the extensive learning associated with most GR cannot be properly treated within a static approach. For example, over time firms learn to identify partners, optimize the process of reaching a contract, change organizational routines, and disseminate and exploit results. In this regard the Israeli experience can teach us an important lesson: whereas reaching a contract took more then one year for the first consortia, this process is continuously shortened by accumulated experience and takes no longer than six months today. An interesting phenomenon which demonstrates the existence of learning effects is evidenced by the fact that firms that already participate in one consortium tend to participate more easily in new consortia. Most of the projects anticipated for 1995 involve partners who are already participating in projects initiated during 1994. Taking into account the relatively small portion of industry which participates in the Magnet program, this is a significant point.

The static view overemphasizes financial incentives and pays too little attention to institutional change, learning and capability creation, and policy design; it is these latter which necessitate a constant updating of policies. In the following sections we characterize and contrast GR programs and related policies in order to identify the technological infrastructures important to promote collaborative GR.

4. The Role of Industrial Support Centers

The initiation of collaborative GR projects is not uniquely a bottom-up process; usually, industry by itself cannot create a critical mass of projects. This stems from the difficulties firms face in identifying potential projects and searching for appropriate partners. Finding partners is an especially difficult task for the individual firm because of the risk and fear of losing competitive knowledge. On the other hand, government officials are in an even more disadvantageous position for forming consortia because they generally have limited information about markets and technological opportunities. From the European experience we learn that *Industrial Support Centers* (ISC) can effectively overcome these difficulties.

Industrial support centers often take the form of technological centers or national laboratories. However, ISCs also include a large variety of institutes like public and private research centers, regional centers, industry information centers, and national sectoral committees. ISCs represent institutional interfaces that specialize in collecting and using industry-specific knowledge, making them important sources for anticipating and projecting expected developments, and for keep abreast of relevant technological change. As such, ISCs are capable of identifying industrial needs and market opportunities, proposing and designing good GR projects, and matching potentially suitable partners. Their neutral position and credibility is conducive to convincing industrialists about the merits of potential projects. In many cases ISCs may be able to coordinate projects or even to carry them out.

The centrality of ISCs in promoting consortia has led many governments to explicitly direct existing ISCs to undertake support of consortia as one of their activities; in other cases temporary ISCs have been established specifically for this purpose. In Finland, the Finnish national laboratories (the VTT) are strongly encouraged to undertake these activities, as evidenced by the fact that conducting collaborative GR is an important item in their budgets (Finnish Academy of Technology 1993). Throughout Europe peripheral technology centers often take an active part in bringing GR programs closer to industry by assisting in the project identification process, partner selection, the preparation of proposals, the provision of legal counsel, and in some cases even the coordination of the research. An example of such centers are the VALUE Relay Centers which are located in most EC countries and provide all of the above-mentioned services (excluding project coordination).[5]

In contrast to these European examples, Israel has no explicit policy towards ISCs; however, there is an ongoing debate about the need for ISCs and about how to operate and support them. As a result, there are few ISCs and their involvement with local firms is often limited (Betzaleli-Engel et al. 1994). This hinders activities aimed at either coordinating GR consortia or carrying out the research. The Magnet committee has recently begun to recognize the importance of such entities. The fact that almost half (three of seven) of the projects supported by Magnet are in the field of biotechnology was attributed to initiatives taken by the National Committee for Biotechnology, a national sectoral committee launched several years before Magnet. Consequently, the Magnet secretariat decided to form several additional sectoral

[5]A presentation on the role of Value Relay Centers was given by Dr. A. LeCorre-Frisch, Value Relay Center, Stuttgart, Germany.

committees, each specializing in a field of technology or industry: optoelectronics, microelectronics, alternative energy sources, information technologies, materials, and food technologies. However, because these committees are still in their infancy and enjoy very limited budgets, it is doubtful whether they will soon have a significant impact on the process of project creation.

We think that the absence of well-functioning ISCs in Israel is one explanation for the small number of consortia formed in Israel thus far.

5. Setting Program Priorities

Another difference between Magnet and most OECD programs lies in the setting of priorities. While Magnet implements a neutral policy, European, Japanese, and U.S. programs target certain research fields for support. There is an ongoing debate about the relative merits of *neutral* versus *selective* policies. Teubal (1995) advocates a mixture of neutrality and selectivity, with the balance changing over time. In general, at the early stage of a new program the emphasis should be on neutral support which allows for extensive learning, rather than on selecting specific technologies or sectors. The goals at this early stage should be the formulation of criteria to assess projects, and the development of complementary infrastructure required for implementing the new policy. At more advanced stages when government officials and firms have acquired the necessary new capabilities, selectivity should increase to ensure a critical level of support for industry. But a case can also be made that a clear and effective set of priorities are important at the early stages of a GR program in order to generate a substantial number of projects.

The absence of a clear set of priorities can partly explain why the neutral Magnet program in Israel has been less successful than Europe's selective GR programs in generating projects and attracting participating firms. Selectivity in early stages of GR programs may encourage the formation of consortia through both psychological and practical effects. By targeting GR the government provides a signal to firms and research organizations that their research activities are of national importance, and that the likelihood of gaining additional public support is high. Further, once priorities are set, firms do not have to continually justify the importance of public support for their research field; at the practical level, this can reduce transactions costs associated with initiating a project and especially reduce proposal preparation costs. Compared to a neutral approach, the setting of priorities can create stronger incentives for firms – and ISCs – to invest in proposals for collaborative GR.

We think that selectivity during the early stages of a GR program is best formulated by programmatic statements of areas of interest with a flexible interpretation of what these areas include. Setting a moderate level of selectivity signals firms of the readiness of the program to fund projects in their technological field, thereby encouraging them to formulate collaborative projects. A large number of projects is crucial in order to bring about extensive learning. As the program matures, policy makers can weigh the relative advantages of increasingly defining priorities and adopting clear criteria for project inclusion and exclusion. However, at this mature stage it is crucial to assure project coherence within each priority area in order to:

- avoid gaps in critical technological capabilities,
- maximize interproject technology transfer, and
- assure a large number of commercially relevant technological capabilities.

Assuring the complementarity of projects within a program which supports a particular technological field is crucial in order to guarantee that the entire spectrum of capabilities in this field is developed and in order to maximize commercialization and exploitation of the program's results.

Magnet's neutral policy reduces the potential for creating the critical mass of capabilities necessary for developing national competitiveness in certain fields. While it might be argued that this is of lesser importance in early stages, when learning is the main goal, the fact that the Magnet program serves as a framework for uncoordinated individual projects also diminishes the likelihood for complementarity in the future.

6. The Institutional Framework

Cooperation among selfish actors does not necessarily yield an equilibrium in the neoclassical market, as has been demonstrated by many researchers (Axelrod 1981; Myerson 1991; Friedman 1991). Collaboration itself can only be sustained when certain conditions are fulfilled. According to Axelrod the most important conditions are:

1. frequent interactions between collaborators in order to create mutual dependence;
2. potential for significant future profits in order to reduce the incentive for short-term opportunism; and

3. the ability to recognize partners and remember previous behavior in order to avoid collaborating with disreputable partners.

In the neoclassical market these conditions are often not satisfied and therefore appropriate policy tools have to be created (Schotter 1981; North 1990). Indeed, most GR programs in OECD countries were careful to create separate program secretariats that provided these services, rather than rely on existing mechanisms that were established and designed to support other types of research.

In the case of collaborative GR, four sets of tasks must be undertaken in order to assure that the conditions for sustainable collaboration are met. The policy mechanisms (like secretariats) that are created must be capable of the following:

- widespread dissemination of program opportunities to expand the pool of potential participants and to increase interaction;
- providing legal support to assist firms with contract formulation and enforcement in order to create a milieu of trust and increase the expectations of future profits (Axelrod's second point);
- assisting participants to identify potential partners and track their records (Axelrod's second point); and
- assuring the effective operation of the consortium by creating work procedures that facilitate frequent and effective interaction among participants (Axelrod's third point).

The following subsections examine each task in turn.

6.1 Disseminating Program Information

Because of its distinct and special nature, GR requires explicit and particular dissemination efforts in order to increase industry awareness. The Israeli experience teaches us that providing general information about the program is not sufficient to generate substantial and active firm interest in such a novel activity. Rather, the dissemination of new programs requires some marketing elements since both the concepts of *collaboration* and of *generic research* are too far from firms' managerial and organizational configuration.

In implementing the Magnet program, the prevailing view was that the high rate of financial subsidy meant that simply making program information available would be sufficient to attract substantial participation in the program. However, the relatively small number of Magnet projects contradicts this assumption. Dissemination of such a program requires extensive industry involvement by the Magnet secretariat by means of seminars, publications in professional journals, e-mail, and brokers. The absence of well-functioning ISCs hindered the ability of Magnet to be more involved at the industry level. It is expected that the various national technology committees will succeed doing so.

6.2 Legal Support

At early stages of GR collaboration one of the central problems confronting firms is the lack experience with the legal issues that emerge during negotiations of contracts. This inexperience often results in the drafting of cumbersome contracts which address an excessive number of often-hypothetical situations that may arise. Negotiating such complex contracts adversely affects the formation of collaborations. Moreover, the process of negotiating the contract might be negative by itself. Quintas and Guy (1991) found that this process negatively affected 48 percent of projects within the Alvey program. In many cases the agreements took over a year to be completed and caused projects to have staggered starts. Raising the possibility of potential controversy can also exacerbate existing initial doubts about cooperation. Even when firms are not dissuaded by these misgivings, contract complexity substantially increases transactions costs and may delay the initiation of the collaboration.

Many secretariats have designed model contracts to help firms deal with the complex legal issues pertaining to collaborative GR. Some programs adopt an even more rigid approach and impose standard contracts, as is the case with European Community GR programs. In addition, program secretariats often subsidize legal costs or even provide legal advice themselves. Over time as participants and lawyers become more familiar with contract issues specific to collaborative GR, the need for legal support diminishes. As firms acquire experience negotiating and implementing collaboration contracts, many concerns are reduced and the process of reaching contractual agreements becomes easier.

The Magnet program encountered a number of problems involving contract negotiations. Initially, the secretariat was not aware of the importance of legal support for the creation of consortia. Only when it became apparent that the formulation of contracts was a major obstacle which was seriously impeding the formation of consortia, did the secretariat prepare a model contract. Another problem was the absence of legal advisors who were acquainted with the special aspects of formulating GR contracts. Naturally this weakness diminished over time as specialists gained experience and developed expertise.

The heterogeneous nature of participants, in terms of the type of firms, their characteristics, and their managerial patterns, also proved to contribute substantially to the difficulties encountered in negotiating collaboration agreements. Particular problems were found in consortia that involved combinations of civilian and military firms, public- and private-sector organizations, or large firms and SMEs.

The actual process of negotiating the collaboration contract presented yet another complication. When representatives to the negotiations were unable to make final decisions, the process was prolonged as each decision had to be reported to and ultimately confirmed by management. Besides being time consuming and complicated, such a procedure tends to increase the rigidity of negotiating positions. This is especially harmful since flexibility is an essential quality for concluding workable collaborative agreements. Fortunately, over time some of the difficulties involved in contract negotiations were overcome through the experience of participants. This is a good example of how capabilities can be developed to facilitate the new program.

6.3 Partner Selection

All researchers point to the importance of partner selection in determining the success of any cooperative activity. Three elements of partner selection are particularly critical. A primary condition is to find partners who can provide the firm with *complementary* assets for undertaking the research program, such as physical assets, human capital, and markets, as well as future potential capabilities. Another factor is the *compatibility* of these assets regarding, for example, standardization, sophistication, human capital, and computerization. The third consideration concerns the *strategic* aspect of partner selection. This involves choosing partners with whom the firm can expect to develop a set of future relationships that expand beyond the confines of the current collaboration. In a survey of 445 consortia in the EC, Linne (1991) asked partners to rate the importance of each of these characteristics to the success of a collaborative project; he found that over 70% of all partners rated the complementarity of resources as high or very high, 65% indicated the same for compatibility, and 53% judged strategic aspects to fall in those categories. Similarly, an evaluation of the European Eureka program

(1993) found that 74% of participants mentioned complementarity to be the main reason for participants to join their Eureka projects.

The abilities of program secretariats to assist firms with the process of partner selection are limited but still can be of importance. Some European program secretariats facilitate collaborations by building data bases with information on project participants; examples include the EC Framework and Eureka Programs. These data bases contain information on firms looking for partners to join them in prospective projects. In many cases secretariats provide informal information about firms searching for participants, potential projects, and firm reputation. This kind of assistance is especially important in the formation of international projects. Most secretariats try to make partner information widely available, even to the point of making the data bases accessible by e-mail (e.g., the CORDIS service of the EC).

The Magnet program did not provide any systematic support in partner selection. To a large extent, however, the small size of Israeli industry diminished the importance of such support. Most Israeli firms can easily identify potential partners and gather information on their reputation using existing contacts and the informal networks which the Magnet secretariat officials provide.

One disadvantage of the Israeli approach is that firms tend to be informed about projects only after being approached by other companies. The absence of a systematic scanning mechanism diminishes the involvement of entire industry segments which might otherwise benefit from GR collaboration. This is especially true for small and even medium-size firms which tend to be ignorant of GR activities. Unfortunately, this ignorance reduces the overall interests of industry in GR activities and contributes to its marginalization.

It is clear, as already discussed in section 4, that ISCs could play an important role in this process. In many cases their acquaintance with the industry enables them to identify suitable partners for specific projects. Also, their developed paths of communication with industry enable them to disseminate information about potential projects to prospective partners.

6.4 Interaction, Routines, and Monitoring

To ensure effective and continuous cooperation, research has to be organized so as to facilitate and compel frequent interaction among participants. Frequent interaction is important to maintain the continued complementarily of the different research components, to facilitate dissemination of knowledge among participants, and to help assure the timely achievement of project objectives. Interaction among project partners is what distinguishes collaborative research from standard in-house research. The quality of this interaction is critical to the success of GR projects.

At the proposal stage, every project is designed for – and participants commit themselves to – frequent interaction and the development of strong communication paths. At later stages when a project takes off, however, other forces may mitigate against fulfillment of these obligations. It is at this point that two major components of collaboration-design can help sustain appropriate levels of interaction: rooted routines and external monitors.

The creation of routines is commonly based on accumulated experience gained in previous projects. This task is the responsibility of the program management. Routines mainly involve the establishment of communication channels that allow for frequent interaction, and even determination of how these channels are to be used. Routines may take the form of e-mail communications for questioning day-to-day research issues and transferring reports; regular dissemination of clear and concise research reports; scheduling frequent formal and informal meetings among researchers; and periodic seminars. However, it sometimes proves difficult to sustain routines in the absence of external monitoring. When routines are well developed, the work of monitors is clearly defined and easily carried out. But the absence of collaborative routines requires monitors to define by themselves how to assure satisfactory levels of interaction among partners. In such cases it is more likely that tension will develop between the external monitor and the project leader.

Most European programs employ monitors to make sure firms interact effectively and that they meet project objectives. The specific degree of involvement by monitors varies from program to program. In some cases monitors oversee most cooperative procedures of the project, organizing seminars, scheduling meetings, and requiring periodic reports. In other cases their role may expand to include evaluating whether milestones have been reached or reviewing periodic reports.

Often, an inverse relation exists between the freedom firms have to formulate contracts and determine work procedures, and the level of involvement of monitors. For example, the Finish GR programs offer only general advice on contract formulation and research organization, but they employ monitors who constantly scrutinize collaborative work, read reports, and make sure that regular meetings and seminars are attended. In contrast, the European Community GR program demands that firms adhere to a rigidly formulated standard contract, but imposes minimal monitoring requirements; in this case program monitors are content to receive regular progress reports.

In the Magnet program the role of monitors is still rather vague and monitors themselves have to determine their responsibilities and level of involvement. In some cases this leads them to become involved in all aspects of the collaboration. Monitors may take an active role in the overall operation of the consortium in addition to being involved in the daily details of the collaboration. In many cases the boundary lines between the responsibilities of the monitor and those of the project manager are not clearing defined. In other cases monitors are much less involved, limiting themselves to reading and commenting on periodic reports.

This situation is best explained by the fact that the Magnet program is still in its infancy and has yet to develop established routines. Indeed, as the program progressed the role of monitors in Magnet became more defined; their function is now focused on assuring constant transfer of knowledge among project participants (through meetings and reports, for example), coordinating academic and industrial partners, and assuring attainment of milestones and budget management subject to research progress.

7. Conclusions and Policy Implications

This chapter has analyzed the Israeli experience with the Magnet program. By contrasting Magnet with other programs, we have been able to shed light on specific program elements that are crucial for implementing GR programs. The main conclusion arising out of the comparison of Magnet and several European programs is that effective promotion of GR collaboration requires a variety of complementary policies in addition to financial support.

The contention that collaborative GR suffers from market failures is supported by the observation that all the participants in the various Magnet-sponsored projects stressed that the GR projects would not have been undertaken in the absence of the program. And yet, the small number of projects supported by Magnet indicates that financial support alone is insufficient for introducing this new kind of research into industry. This suggests that the market failures associated with GR should be analyzed within a dynamic context. In the early phase, there is a need to identify critical elements of technological infrastructure which are required to support the desired policy. At this stage efforts should be directed toward the creation of these elements. For example, this study points to the importance of industrial support centers (ISCs) for the implementation of GR policies. Another problem at this stage, similar to that experienced in both the U.S. and Europe, is the need to adjust or build appropriate institutions, such as effective intellectual property rights protection and antitrust laws. In a later stage, when the required new routines, capabilities, and institutions have been established, market failures like those inherent to large indivisible costs need to be addressed.

Our analysis suggests one vital element for promoting the concept of GR is the industrial support center (ISC). The importance of these centers are fourfold:

1. They form a bridge between government and industry which is useful for the dissemination of the program.
2. Their familiarity with industrial needs helps them to identify good projects and locate suitable partners.
3. Their impartiality places them in a good position to coordinate consortia.
4. In those cases where ISCs have in-house research facilities, they can conduct the research and serve as a central laboratory; this is especially important for the involvement of small firms whose own research capabilities are often limited.

The Israeli case indicates that the absence of well-functioning ISCs is a major obstacle to the introduction of collaborative GR. The experience of the biotechnology committee (whose existence predates Magnet) which was responsible for the creation of almost half of the projects, encouraged the Magnet secretariat to support similar committees in various fields of technology and industry such as: microelectronics, optoelectronics, food, alternative energy sources, information technology, and materials.

A competent program secretariat is another critical institutional component that we identified. The secretariat needs to provide the environment in which collaborative projects can thrive. Their major responsibilities include program dissemination, assistance in finding suitable partners, provision of legal assistance, and the establishment of research and monitoring procedures. Initially, the Israeli Magnet secretariat did not address these issues systematically. However, as Magnet officials came to recognize the importance of such support, they increasingly provided these services.

In the area of legal services Magnet developed a contract which is a mix between a model contract and a formal contract. Although the model includes many obligatory sections concerning mainly issues of property rights, it leaves partners free to adjust the contract to their specific needs.

In the area of partner selection Magnet involvement has been relatively limited. Most assistance provided to firms is on an informal basis. Because no systematic mechanism for searching for partners exists, many firms, especially small and medium-size enterprises, do not get to know about existing projects.

Another issue not yet satisfactorily addressed by the Magnet program is the dissemination of program information. Our research results indicate that many firms were not acquainted with the program, or that they only learned of it incidentally.

The role of selectivity in collaborative GR support programs also proved to be an important issue. Most programs in OECD countries define broad fields of priority (e.g., information technology and biotechnology). In contrast, the Magnet program is a horizontal policy with neutral incentives. We found reason to support a somewhat balanced view. Despite Teubal's arguments supporting neutrality during the early stages of a new policy program, we think that in early stages of complicated and often-indefinite activities such as those that characterize GR programs, priority setting may be helpful in directing firms to take active part in the new activities and in raising the overall interest in the new program. However, selectivity during these early stages is best formulated by programmatic statements of interest with a flexible interpretation of what is and what is not included in priority technological areas. Such a policy sends signals to specific industries about opportunities that are embodied in the new program, thereby increasing the number of proposals. Generating a critical number of proposals and approved projects is essential in the early stages of any new program in order to amplify learning. As the program matures, policy makers can weigh the relative advantages of adopting increasing levels of selectivity and create tools for selecting appropriate research fields.

The implications of the Magnet experience for policy are rather straightforward: before implementing a collaborative GR program, governments must identify the required complementary elements (such as ISCs) and ensure their existence. The Israeli experience shows that if parts of the necessary infrastructure or other complementary elements are absent, government initiatives may be ineffective unless adjustments in policy are made. The problems arising out of Israel's lack of effective ISCs is a case in point. Fortunately, after more then one year of operation, Magnet program officials came to realize the crucial role played by ISCs in the promotion of GR and began to support their development. However, a reactive policy is less than ideal since bad experiences by firms might deter them from participating in future projects and programs, even when the bad experience is the result of infrastructure problems.

Most of Magnet's inefficient functioning can perhaps be attributed to "labor pains". Learning through trial and error is, to some extent, inevitable and even desirable at the infant stage of any technological infrastructure policy. Nonetheless, it is likely that some of the difficulties we have identified could have been avoided had careful studies of existing collaborative GR programs in other countries been undertaken before implementing such a complicated program. Thus, during infant stages of TIP, emphasis should be placed on maximizing and systemizing learning opportunities.

References

Arnold, E. and K. Guy. 1989. *Evaluation of the IT4 Program Phase I.* Stockholm: Booz, Allen & Hamilton.

Arrow, K. 1962. Economic welfare and the allocation of resources for invention. In *The Rate and Direction of Inventive Activity*, edited by R. Nelson. Princeton, N.J.: Princeton University Press.

Axelrod. 1981. The emergence of cooperation among egoists. *American Political Science Review* 75: 306-318.

Betzaleli-Engel, G., L. Kravitz, and M. Teubal. 1994. *The Role and Activities of Industrial Centers in Israel.* The Jerusalem Institute for Israel Studies.

Chesnais, F. 1988. Technical cooperation agreements between firms. *STI Review* 4: 51-119.

Commission of the European Communities, Directorate General. 1992. *EC Research Funding*. Brussels.

Dekker, K. 1992. *The Finnish Technology Programme: Industrial Building Technology*. Technology Development Center (TEKES). Helsinki.

Dove, G.A. 1989. Cooperative research at MCC: A focus on semiconductor-related efforts. *Proceedings of the IEEE* 77 (9): 1364-1375.

Evaluation of the Eureka Industrial and Economic Effects. 1993.

The Finnish Academy of Technology. 1993. *High Technology in Finland.*

Friedman, J.W. 1991. *Game Theory with Applications to Economics*. Oxford, USA: Oxford University Press.

Jorde, T.M. and D.J. Teece. 1990. Innovation and cooperation: implications for competition and antitrust. *Journal of Economic Perspectives* 4: 75-96.

Katz, M.L. and J.A. Ordover. 1990. R&D cooperation and competition. *Brookings Papers on Economic Activity, Microeconomics*: 137-191.

Linne, H., R. Magnaval, and J. Removille. 1991. *Key Factors for Industrial Partnership in EC Programs.* Commission of the European Communities.

Myerson, R.B. 1991. *Game Theory: Analysis of Conflict.* Cambridge, USA: Harvard University Press.

North, D. 1990. *Institutions, Institutional Change, and Economic Performance.* Cambridge: Cambridge University Press.

Ouchi, W.G. 1989. The new joint R&D. *Proceeding of the IEEE* 77 (9): 1318-1327.

Quintas, P. and K. Guy. 1991. *Collaborative, Pre-Competitive R&D and the Firm: Some Lessons from the UK Alvey Programme.* Science Policy Research Unit University of Sussex. Brighton, UK.

—. 1995. Collaborative, pre-competitive R&D and the firm. *Research Policy* 24: 325-348.

Schotter, A. 1981. *The Economic Theory of Social Institutions.* Cambridge, London: Cambridge University Press.

Soete, L. 1991. *Technology in a Changing World.* OECD Technology Economy Program. Policy Synthesis.

Tassey, G. 1991. The function of technology infrastructure in a competitive economy. *Research Policy* 20 (4): 345-361.

Teubal, M. 1983. Neutrality in science policy: The promotion of sophisticated industrial technology in Israel. Chapter 8 of *Innovation, Performance, Learning and Government Policy*, edited by M. Teubal. The University of Wisconsin Press. 1987.

—. 1995. *A Catalytic and Evolutionary Approach to Horizontal Technology Policy (HTP).* Industrial Development Policy Group (IDPG), The Jerusalem Institute for Israel Studies, Jerusalem, and STEP Group Storgaten 1, N-0155 Oslo, Norway.

Toren, B. 1990. Research and development in industry. In *Industrial-Technological Policy for Israel*, edited by D. Brodet, M. Justman and M. Teubal. The Jerusalem Institute for Israel Studies. Jerusalem, Israel.

III. TIP for SMEs in Established Industries

Transfer of Technology to Small and Medium Enterprises (SMEs): Conceptual Changes and Lessons from the Two Banks of the Rhine

Patrick Cohendet*

Technology transfer is now recognized as a decisive element in business firm strategy. A fuller recognition of the significant role played by the diffusion of innovations in the competitive process has brought about an increase in the scope and diversity of technology transfer in industry. Increasingly, the study of technology transfer (which used to refer only to problems of international north-south transfer) has come to focus on this phenomenon within the firm and in its industrial environment. Using the term in this sense, public authorities in France and other countries have come to recognize the strategic role played by technology transfer, and have established over the last few years a large number of services to promote technology transfer. The Regional centers for Innovation and Technology Transfer (CRITTs) are a leading example of technology transfer services that have been set up in France over the last ten years; others include: services for funding innovation, help for patenting, promotion of innovation (ANVAR – Agence Nationale pour la Valorization de la Recherche), training, and auditing. This often complex set of services – firms find it difficult to navigate through the maze – implies that new service relationships between client firms and the providers of services are constantly being established.

This chapter offers some thoughts on the modes of implementation of technology transfer in France, organized around two main working hypotheses:

1. The policy of technology transfer is closely connected to the conceptualization of technology transfer within the process of innovation diffusion. We will therefore try to show that the type of policy in place depends on the theoretical view of technology transfer. The initial development of services in technology transfer was based on a restrictive interpretation of the notion of technology (as a form of codifiable information) and a linear representation of the diffusion process. More

*I am grateful to Monique Flasaquier for the translation.

M. Teubal et al. (eds.), Technological Infrastructure Policy, 271–283.

recently, as the interactive theoretical view of the process of innovation diffusion became established, a more complex – but more appropriate – type of technology transfer policy emerged, based on a concept of technology as tacit, firm-specific knowledge.

2. The service relationship in technology transfer is closely dependent on the concrete context of the local innovation system within which it develops. The nature of this local innovation system (i.e., the habits, routines, institutions, and rules specific to a given milieu) must shape the development of technology transfer services.

These two hypotheses explain how one moves from mediating services characterized mainly by commercial relations to new forms of services characterized by relations which presuppose greater interaction between the transferring agents and their clients, among whom SMEs figure prominently.

1. Technology Transfer and Theoretical Representation of the Innovation Process

1.1 The Traditional Vision

The initial form of technology transfer services appeared in the early 60s, based on the *traditional* view of innovation. According to this view, technology can be considered as information of a strongly generic character, which means that it appears *a priori* in the same way to everyone and that it exists independently of its implementation in production processes. Traditional theory acknowledged the possibility of suboptimal allocation mechanisms; the case of underinvestment in research activities was attributed to three classical causes of market imperfection: indivisibility, uncertainty, and above all, inappropriability. These imperfections may also account for unequal distribution in the acquisition of technologies, some firms being more advanced than others in the use of a given technology. The existence of a technological gap between firms can take on two main forms: within a given sector, some firms may appear more advanced than others; and some high-tech sectors may be considered more advanced than others. Technology transfer policy in this early form dealt mainly with the latter case.[1]

[1]Of course the problem of technology transfer in this view can also bear on the transfer of information between fundamental research and industry, in particular transfer from university laboratories to industry.

In this representation, the establishment of technology transfer services is structured around three types of agents. Transfer starts from a *sending* firm, where the technology originates and ends with a *receiving* firm which applies the technology. In between is the third agent, a mediating organization. The major role played by this service firm, according to the prevailing theoretical representation, is to promote the circulation of information between the sending and the receiving firm, with the underlying idea that the latter lags behind the former in its technological knowledge. The establishment of firms specializing in technology transfer around technologically advanced centers such as space, aeronautics, nuclear energy, and electronics (i.e., firms aiming at promoting the technnologies resulting from these sectors) was therefore one of the main characteristics of technology transfer in the 1970s. The objective in setting up such firms was to transfer know-how coming from a highly technological milieu to sectors with a lower level of complexity (Bach, Lambert, and Shachar 1989).

The type of service relationship corresponding to this model is specific: the service firm plays the role of mediator in the context of commercial relations. In abstract terms, one can say that this firm "buys" technology from the sending firm in order to "sell" it to the receiving firm. This representation most often posits an intermediary stage in which the service firm assumes temporary ownership of the information which embodies the technology; thus the transfer of technology involves transfer of ownership.[2] Less frequently, the service organization is limited to helping graft a technology coming from the sending firm onto a receiving firm, without the intermediate step of ownership by the service firm.

The relationship of this mediating agent to potential receivers assumes that the client can easily integrate the new technology, implying that no attention is devoted to any training process; the main obstacle to the diffusion of new technologies consists of the transactions costs resulting from the search for clients interested in acquiring the technology. The firm engaged in technology transfer has therefore to "seduce" potential receivers using appealing catalogues or brochures, colloquia, demonstrations, and other marketing activities.

The typical model of the service firm corresponding to this commercial view of technology transfer is therefore that of a mediating unit coordinating

[2]In this case, the basic scheme of the service relation proposed by Jean Gadrey based on the definition provided by Peter Hill (cf. Gadrey 1992, 18) becomes more complex so as to take into account the double transfer of ownership.

the bilateral relationships which rule the transfer of ownership rights to technology. On one side stands a sending firm whose main transactions costs involve coding the technology, and on the other stands the receiving firm for which the main transactions costs are *search* costs.

Many of the service firms created in the 1970s corresponded to this model, facilitating technology transfer from fundamental university research to the market, from advanced to basic industries, or from large firms to smaller units. All large universities in France established technology transfer departments and specific efforts were initiated to transfer technologies from the advanced sectors of space and nuclear energy to other industrial sectors. Organizations like NASA developed this type of service firm at great cost, with the aim of "selling" space technology and thus better justifying the catalytic role played by space projects in the creation and diffusion of new ideas. This policy met with some initial success but quickly petered out. Some of these service firms continue to exist though they never really reached a mature stage, and some have evolved into a different type of agent of change (e.g., Bertin in France), calling into question the linear model of innovation diffusion.

The studies carried out by BETA (Bureau d'Economie Théorique et Appliqué) on the indirect effects of space projects (Bach et al. 1992) confirm that the obstacles to the linear transfer of innovation are substantial. These studies showed that for the contractors of European space projects, only a small share (15%) of technology transfers were external to the firms, i.e., transfers between firms. The remaining 85% were transfers within firms. The managers who were interviewed mentioned precisely the risks linked to commercial relationships (for instance, the risk of not being certain of the appropriability of the information before it is sold) to explain the weakness of transfers out of the contracting firms.

All these transfer structures rely on an explicitly linear view of the process of innovation diffusion: innovation diffuses in a single direction from fundamental research towards the market, or from advanced industries toward basic industries. The interactive view of the innovation process goes beyond these limitations of the traditional linear model. In this context the notion of technology transfer takes on a totally different significance requiring different modes of implementation.

1.2 Technology Transfer and the Interactive Concept of the Innnovation Process

As soon as one accepts the idea that technology is not the same as information but relies essentially on tacit knowledge which is specific to the firms that have learned to use it, and that because of this tacit character technologies are difficult to transfer and to reproduce, the issue of technology transfer must be totally reformulated (Pavitt 1986). This concept of technology leads to a profoundly interactive view of the process of innovation diffusion, which, rather than being limited to a linear scheme, relies on a dense network of reciprocal relations between the different components of the process, and thus leads to a totally different notion of technology transfer (Kline and Rosenberg 1986).

Under this reformulation the technology transfer service firm no longer fills the simple task of mediating information emanating from a sending unit to a receiving one, but assumes the more complex role of ensuring a better diffusion of innovation within an interactive milieu. In this case, client firms are not ambiguous producers or consumers of well-defined technology products; instead each component may be either sender or receiver at different stages of the process, and technology is strongly dependent on tacit knowledge and specific learning processes. This requires the mediating agent to intervene in the complex individual innovation processes of the client firms, each with its own specific learning procedures.

In this context, firms need not worry about uncontrolled transfers because transactions costs are by definition too high (Cohendet, Heraud, and Zuscovitch 1992). On the contrary, these costs promote explicit cooperation, leading to the application of a reciprocity principle that ensures compatible incentives between agents. The rapid development of cooperation mechanisms results in a network which is superimposed on the traditional industrial structure. Where production systems are not standardized and where there exist a large number of micromarkets for specific applications, it is necessary to master a growing body of knowledge, and the only way one can have access to such diversity is by joining forces with other suppliers and users. Thus, the new model of innovation incorporates a new mechanism of appropriation. This mechanism is connected with the contextual self-protection of the firm, resulting from its original organization and control of interfirm diffusion. This spontaneously higher appropriability contributes to accelerating the general rate of progress. Not only does the industrial structure tend toward a network organization, but the firm itself becomes a network of scientific and technical qualifications able to adopt new technologies and to select its own development trajectories among the various opportunities; the inter- and intrafirm levels are not independent. In this context, the ability to have access to the know-how of firms active in neighbouring fields enhances the human resources of the firm. This can only reinforce the incentive to cooperate. This is a fundamentally different perspective on the role of technology transfer from the traditional view. The point is no longer to transfer technology from a sending firm to a receiving firm, but to meet the technological needs of firms, and, in particular, to develop their creative capabilities and their ability to assimilate and learn new ideas.

The concrete forms taken by newer firms delivering innovation services have been developing heavily in the early 1980s. There is, in fact, a proliferation of companies that intervene in quite varied fields: financial consulting and assistance, commercial consulting, patenting, technical assistance, assistance for normalization and certification, technological auditing, assistance for implementing new organizational structures, specific training, and legal expertise. This intense proliferation arising spontaneously in some cases, initiated by the public sector in others, corresponds to a type of progressive division (into distinct services each based on its own know-how) of the essentially tacit quantity of know-how necessary to implement the process of innovation creation and diffusion. Even though it is not possible, by definition, to identify all the knowledge which constitutes a given firm's tacit know-how, in order to provide the firm with a globally integrated service to support its development efforts one has to try and *recreate* the knowledge bases on which the tacit knowledge of firms rests. This is accomplished by a multitude of service providers, each specializing in its own narrow field. Thus all ingredients necessary for the evolution of the firm's technology are put together in specialized units which play the role of transfer-service companies. Major modifications, compared to the original concept of technology transfer, go along with this move; the notion of technology transfer can only be conceived through the networking of the various agents involved (research laboratories, firms, service companies, and public organizations), involving commercial relations, hierarchical relations, and informal relations (confidence, fame, etc.).

Theoreticians and decision makers increasingly focus attention on the significance of these informal relations in the circulation of tacit knowledge. Among the main determinants of the quality of transfers, one can cite reciprocity, obligation, and confidence aspects based on the reputation and repeated interaction between agents. "When there exists a strong probability of future association, individuals not only wish to collaborate but they tend also to punish both opportunism and the refusal to cooperate. With the development of repeated exchanges, the quality of relations becomes more important than their quantity. The reputation of the participants constitutes the most visible sign of their reliability and plays also a major role in the establishment of numerous relations of the network type" (OCDE 1992).

In this context the service company should be considered as a learning place which interacts continuously with the other agents of the network. And the quality of the service is largely dependent on the nature of the interaction (particularly on the informal elements) between the service company and other agents, be they other service companies of the network (whether competing or collaborating), research laboratories, or client firms. The service relationship cannot be confined in this case to a simple bilateral relationship expressing a transfer of ownership. What determines the service relationship is the ability of the service company to coordinate different interactive learning processes and to accumulate specific know-how. This ability is all-the-more important when the network of which the service company is part is dense, varied, and interactive. There is a *collective* dimension in the transfer service which should not be neglected. Of course, the quality of the service relationship depends also on the ability of the client firm to express its needs in terms of technology, to assimilate new ideas, and, if need be, to modify its organization in consequence. In particular, the *joint production* of the service will be possible only if there is a critical mass of expertise within the client firm which allows it to interact efficiently with the service company.

This aspect underlines the specific weakness of SMEs with regard to technology transfer: they usually do not possess sufficient knowledge to efficiently sustain an exchange of knowledge. The establishment of specific mechanisms, institutions, and transfer centers to facilitate the transfer of technological information for the benefit of SMEs has therefore become more intensive in all countries; local technology institutes in Japan, a variety of state programs in the United States, and CRITTs in France represent recent examples of the specific efforts made with regards to SMEs.[3]

[3]However one must always pay attention to the accepted definition of SMEs – 50% of them are either subsidiaries of larger companies, or belong to a stable network of subcontractors of a holding. In these cases the technology transfer comes mostly from the larger company to which the SME is related.

2. Technology Transfer, Service Relationships, and Local Systems of Innovation

The second hypothesis which explains how service relationships arise and evolve in the field of technology transfer, is that the mode of the relationship which develops between service suppliers and client firms is strongly influenced by the type of local system of innovation of which they are a part. By a local system of innovation we mean, following the work of Lundvall (1988) and others, a coherent set of institutions, rules, norms, habits, and routines which impart a specific local character to the phenomenon of innovation creation and diffusion. A local system of innovation presupposes a specific mode of usage and diffusion of technology that is highly dependent on local characteristics, such as, the degree of interaction between firms, the type of institutions that exist, the level of decentralization of political power, and the cultural characteristics of consumption.

The local character – which depends on the context and on past history - of the process of technological learning limits even further the transferability of technology, either because it requires a specific division of labor, or because the specification of the technological solution does not permit its easy application to other problems. In such a context, technological knowledge is so difficult to transfer (at least in codified form) that strong protection of the technique by the firm is no longer necessary. Nor is it desirable, as it would hamper innovation efforts at the level of the industrial sector, undermining the efficiency of its cumulative progress. Viewed in this light, the problem of transferring technology from one firm to another hardly makes sense. It is the quality of the local context which is important, as some local systems of innovation are more efficient than others in stimulating the circulation and assimilation of new ideas or the creative abilities of firms which are part of the local system. The place, the role, and the mode of technology depend, in large part, on local context. What is more, so does the nature of the mediating function.

To give a concrete example of this property, consider two cases of local systems of innovation: those of Baden-Wurttemberg and of Alsace. They exhibit strong differences of character, and clear distinctions can be observed in the types of mediating agents, and in the way they relate to their clients (Cohendet and Ledoux 1990).

The innovation system in Baden-Wurttemberg relies on a dense net of traditional relations between the SMEs that comprise much of the industry in the region and academic institutions – the Technische Universitaten and Technische Hochschulen, alongside which vocational schools have developed, Fachhochschulen – from which the SMEs draw their managers. Other important elements of this local system is a very close proximity between its banking and industrial communities. In many regards, Baden-Wurttemberg is one of the most convincing examples of the interactive innovation model (see Box).

Baden-Wurttemberg was the first Land in Germany to implement programs to support technology; in 1976 it devised a comprehensive programme which focused on SMEs. But it was really in 1984, with the second programme, that innovation consulting and technology transfer were accorded leading priority in the Land's technology promotion effort. In 1983, the position of Government Director for Technology Transfer was created. The Director is, at the same time, Chairman of the Steinbeis Foundation, which was set up in 1971 to stimulate the development of SMEs in the Land.

The technological infrastructure of the Land is the following:

- 3 big national research centers, e.g., the Karlsruhe nuclear research center, and the German research and training office for aerial and space navigation;
- 14 institutes of the Max Planck Society which concentrates on fundamental research but is turning increasingly to applied research;
- 13 institutes of the Fraunhofer Society which is the most important German organization for applied research.

These three types of structures receive, for the Land, one-fourth of the total funding granted by the Bund to these organizations in the whole country.

- The Steinbeis Foundation mainly stimulates and coordinates the technological network, at the center of which one finds the Steinbeis technology transfer centers. It is composed of two types of organizations:
  - the technical consulting departments, all of which are connected with a Fachhochschule;
  - transfer centers, most of which are connected with a Fachhochschule.
- 11 institutes of collective research, joint industrial research centers, belonging to the National Association for Industrial Research (AIF),
- 28 innovation-consulting offices attached to chambers of trade, industry associations, vocational organizations, and Fachhochschulen;
- 10 technology parks or centers, near the research institutions.

Most exchanges (between firms and universities, firms and banks, and among firms) are direct; they do not require the mediation of specialized service firms. This derives from the long-existing maturity of the partners in their interaction with each other, and even more, from their critical know-how, which limits their need for a mediator. However, the interests of all partners are managed in the framework of a very powerful specialized foundation (the Steinbeis Foundation) which gathers all information and plays the role of a common platform for technical advice and assistance in the field of transfer, as well as the indispensable role of a collective memory of transfer activities. It is clear that such mediating agents progressively accumulate the experiences they have acquired by working for clients. These experiences build up a kind of collective memory that stresses the interactive character of the service relation. Thus, the mediating agents find a very specific role for themselves within the network of transfer activities, with their collective memory of experiences becoming a prerequisite for the evolution and dynamic efficiency of the network.

The case of Alsace is fundamentally different. The regional innovation system of Alsace is characterized by the parallel workings of two distinct innovation subsystems – one based on *science-push* and the other on *demand-pull* – which hardly interact and which do not manage to enrich one another. On the one hand, *fundamental* public research of high quality gives birth to high-tech firms in the traditional fields of the university (biology, chemistry, materials, optoelectronics); and on the other hand, a set of dynamic SMEs further the innovation process downstream, in response to demand signals, but in sectors that do not draw on the know-how of public research, such as mechanics and textiles. In such a context, technology transfer is characterized by a proliferation of mediating firms, each specialized in a particular field and responding to the specific context in which it functions. The CRITTs, for instance, were originally conceived as totally autonomous from public research in order to distance them from fundamental research and draw them closer to SMEs.

In the same way, since mediating firms lacked competent partners for their purposes in client firms, mediation was limited at first to simple service functions. However, the service relationships evolved progressively towards long-term cofinanced studies, on the basis of longer-term agreements. This was achieved by implementing training programs and through support for employment of researchers in firms.

The differences in transfer services between the Alsatian and Baden-Wurttemberg local systems of innovation appear clearly. In Baden-Wurttemberg the various production units have acquired the ability to express their needs in terms of technology, to assimilate new ideas, and to compare their experiences through direct interaction. The role of the mediating firm (Steinbeis) is to stimulate the network, to coordinate various local transfer efforts between research centers or firms, and to ensure that experiences contribute to local memory. In the Alsatian system the firms – mostly SMEs – have not yet reached a sufficient critical mass of internal know-how to have interactive relationships within a network. The creation of resource centers such as the CRITTs, together with the efforts of such organizations as ANVAR or the Alsace Region to increase the number of researchers in firms aim at making up for this lack of interactivity. But this type of mechanism entails the risk that the (public) resource center may monopolize the transfer function with the creation of a star-shaped subnetwork around the center. This polarization of relations may be initially desirable to improve the efficiency of transfer; but there is a risk that it might undermine the development of interrelations between production units by imposing the mediating organization as a necessary go-between.

These observations demonstrate the importance of the evolution over time of the missions of the mediating organizations, and in particular, the CRITTs which are supported by public authorities. In this context, the mediating relations appear to be dependent on the nature of the local system of innovation rather than exogenous to the system as in the traditional view of commercial mediation. From the above remarks one might conclude that systems with limited interactions further the creation of narrowing specialized mediating organizations which help firms in very specific areas. On the contrary, in a very interactive local system the need for specialized service firms is lower, because bilateral relations form naturally between the various agents. On the other hand, the need for mediating services remains high, if only to ensure that experiences become part of the "collective memory", and that information is centralized.

3. Conclusion

This chapter described the evolution of the mediating function in technology transfer, characterized initially by a strong commercial component, but then progressing to a form which stresses assistance and coproduction between the mediating organization and its client firms. Thus, the whole approach to technology transfer changed following the shift to an interactive model of innovation based on the hypothesis that firms' know-how is essentially tacit.

These conclusions indicate that firms acting as agents of technology transfer based on commercial mediation which only compensates for absent market mechanisms are, by their nature, hardly viable. The difficulties encountered by mediating agents located within American universities to commercialize academic knowledge, and the numerous failures of other intermediaries created for the purpose of applying the results of high-tech activities to other sectors can be explained by their excessively restrictive vision of the process of innovation creation and diffusion.

The development of mediating organizations within a framework of interactive diffusion and innovation, tends to supplant the earlier type of mediation service, emphasizing interactive relationships within a network which supposes a common training of the client firms and mediating agent. However, even if it were possible to show the richness and the development potential of this type of mediating agent, within the framework of an interactive network model, it can take a broad spectrum of concrete forms. Local innovation systems favor particular types of relationships which meet their unique needs. The density of the network seems to be decisive in this regard.

Public authorities now largely accept this concept of technology transfer based on interaction and partnership, and realize its strategic importance for competitiveness. Many recent public initiatives aimed at supporting technology transfer (e.g., ANVAR) openly favor actions which promote partnership and a denser network of relations between the participating economic agents, thus stressing the decisive importance of the intreractive context in mediating technology transfer, and highlighting the possibilities of creating an environment which favors positive externalities for the process of creation and diffusion of innovations.

References

Bach, L., P. Cohendet, G. Lambert, and M.J. Ledoux. 1992. Measuring and managing spin-offs. In *Space Economics*, edited by J.S. Greenberg and H. Hertzfeld. Washington: AIAA.

Bach, L., G. Lambert and J. Schachar. 1989. Les transferts de technologies spatiales en Europe. In *ESA Promotion of European Space Technology Transfer*. Paris: ESA Report, ISBN 9290920602.

Cohendet, P., J.A. Heraud, and E. Zuscovitch. 1992. Apprentissage technologique, reseaux economiques et appropriabilite des innovations. In *Technologie et Richesse des Nations*, edited by D. Foray and C. Freeman. Paris: Economica.

Cohendet, P. and M.J. Ledoux. 1990. La recherche industrielle en Alsace. *Annales des Mines* (June).

Dasgupta, P. and P. Stoneman. 1987. *Policy and Technological Performance*. Cambridge, England: Cambridge University Press.

Gadrey, J. 1992. *L'economie des services*, Reperes. Paris: Editions La Decouverte.

Gruber, W. and D. Marquis. 1969. *Factors in the Transfer of Technology*, Cambridge, Massachusetts: MIT Press.

Hill, P. 1977. On goods and services. *The Review of Income and Wealth*: 315-338.

Kline, S. and N. Rosenberg. 1986. An overview of innovation. In *The Positive Sum Strategy*, edited by R. Landau and N. Rosenberg. New York: National Academy Press.

Lundvall, B. 1988. Innovation as an interactive process, from user-producer interaction to the national system of innovation. In *Technical Change and Economic Theory*, edited by G. Dosi et al. London: Pinter Publisher.

OCDE. 1992. La technologie et l'economie. Paris.

Pavitt, K. 1986. Technology, innovation and strategic management. In *Strategic Management and Research: A European Perspective*, edited by J. MacGee and H. Thomas. New York: Wiley.

Modernizing Small Manufacturers in the United States and Japan: Public Technological Infrastructures and Strategies

Philip Shapira*

Although the United States and Japan differ in many aspects of industrial and technology policy, in recent years they have shared an increased concern about the modernization and technological upgrading of small and mid-sized manufacturing enterprises (SMEs).[1] American policymakers have been worried about the slowness of smaller firms to adopt and fully utilize new manufacturing technologies and techniques, and the resulting adverse effects on industrial competitiveness, domestic supply chains, regional economies, and the stability of high-wage manufacturing jobs. To assist American SMEs, a series of technological infrastructure initiatives and programs have been put into place by federal and state governments, academic and industry organizations, and other groups. These efforts include new legislation and policies to promote industrial modernization and technology transfer, the expansion of industry assistance centers, the stimulation of industrial networking, and support for the conversion of defense suppliers to civilian technologies and markets.

Interest in industrial modernization has grown in Japan too, adding a new element to that country's long-held "dual view" of smaller enterprises. SMEs have often been seen as technologically backward and distinct from Japan's successful large companies, with the latter traditionally receiving greater policy priority. Simultaneously, the strong vertical linkage between Japanese SMEs and their larger customers has been identified as an important

* An earlier version of this chapter was presented as a paper at the Conference on the Future of Research and Technology Organizations in Europe, SPRINT, Commission of the European Communities, Brussels, November 16-17, 1993. The author has drawn upon research in the United States and Japan supported by the National Institute of Standards and Technology of the U.S. Department of Commerce, the Japan Institute of Labor, and the Georgia Tech Foundation. The views expressed in the chapter are the author's, and not necessarily those of these research sponsors.

[1]A "small and mid-sized" manufacturing enterprise is usually defined as one with fewer than 500 employees in the United States and fewer than 300 employees in Japan.

M. Teubal et al. (eds.), Technological Infrastructure Policy, 285–334.

source of technological strength. Both perspectives are now changing. Japanese policymakers want SMEs to assume a more prominent role in national and regional economic and technology development – induced, in part, by the internationalization of the Japanese economy, which has largely taken big firms out of the ambit of national policy control. Meanwhile, many SMEs themselves want to reduce traditional vertical dependencies and build stronger horizontal and lateral ties with a wider variety of other enterprises and with research centers to secure business survival and greater technological autonomy. To support this direction, new regional technology initiatives focused at SMEs are being added to Japan's long-established programs for small-firm modernization.

This chapter compares the problems of small-firm industrial modernization and the evolving public infrastructures serving SMEs in the U.S. and Japan. It begins with a discussion of the comparative environments for technological infrastructure policy in each country. This is followed by a detailed examination of the role of SMEs in industrial modernization in each country, a survey of their specific modernization policies, and a critical assessment of each country's approach. The chapter concludes with discussion and analysis of the similarities, contrasts, and insights from each country's efforts.

1. Technological Infrastructure Policy Environment in the U.S. And Japan

In both the U.S. and Japan, the underlying rationale for bolstering the public technological infrastructure for SMEs has combined considerations of industrial and technological competitiveness with those of employment and local economic development. This mix of objectives is understandable given the large number of SMEs in each country and the number of jobs associated with smaller industrial firms (Table 1). Japan is particularly remarkable here, with almost three-quarters of its manufacturing employment in SMEs with 300 or fewer employees, and in absolute terms, more than twice the number of small manufacturers than in the United States. The pressures on American and Japanese SMEs have grown as their respective economies have become more internationalized, as large customers have changed supply policies and sources (becoming "leaner" and "meaner" according to one recent account – Harrison 1994), and as the pace of technological change has increased. Many U.S. SMEs have been affected also by America's defense cutbacks. For their part, Japanese small manufacturers have been hard hit by the post-1991 collapse of the "bubble-economy" and the rising international value of the yen, making exporting more difficult, offshore sourcing more favorable, and importing more attractive even in Japan's difficult-to-enter markets.

Table 1: Small and Mid-Sized Enterprises, United States and Japan

	Enterprises		Employment	
	U.S. 1990	**Japan 1991**	**U.S. 1990**	**Japan 1991**
	Thousands		**Thousands**	
Small and Mid-Sized Enterprises	415.4	840.3	7,666	8,888
of which:				
under 20 employees	324.1	749.5	1,800	5,371
20-99 employees	74.0	90.9	2,853	3,517
Large Enterprises	4.2	3.9	13,640	3,168
Total	419.6	856.9	21,306	14,087
	Percent		**Percent**	
Small and Mid-Sized Enterprises	99.0	98.1	36.0	63.1
of which:				
under 20 employees	77.2	87.5	8.4	38.1
20-99 employees	17.6	10.6	13.4	25.0
Large Enterprises	1.0	0.5	64.0	22.5
Total	100.0	100.0	100.0	100.0

Note: Small and mid-sized enterprise: United States = less than 500 employees; Japan= less than 300 employees.

Sources: Small Business Data Base, USEEM file, version 8, 1991, reported in U.S. Small Business Administration, *The State of Small Business: A Report to the President*, USGPO, Washington, D.C., 1992 (Tables A.7 and A.22); Small and Medium Enterprise Agency, *White Paper on Small and Medium Enterprises in Japan 1993*, Ministry of International Trade and Industry, Tokyo, 1993 (Appendix, Tables 1 and 2).

These developments have intensified demands in each country for increased policy and programmatic assistance to upgrade SME capabilities. In Japan, there has been renewed debate about what was once viewed as mainly an American problem – the "hollowing-out" of the national manufacturing base, with the collapse of many small manufacturers and the weakening of local economies dependent on industrial production (Sumiya 1994). The technological upgrading of Japanese SMEs is seen as an important national and local strategy to address this situation. Even in the United States, far more ideologically disposed to free-market policies than Japan, the practical reality is that there are simply too many businesses and jobs involved in SMEs to defer to free-market outcomes without any government intervention. This has been especially apparent to policymakers at the state and local level who have formulated an array of programs to promote small firms as part of their overall economic stabilization and development strategies (Eisinger 1988). In the 1980s many state governments added SME-targeted industrial modernization and technology deployment programs to their portfolio of business assistance measures. Infrastructure initiatives to aid industrial modernization in the U.S. were further expanded in the first part of the 1990s, stimulated by large increases in federal funding and programmatic support.

Despite the broad similarities in American and Japanese policy and program goals for SME modernization, several fundamental disparities remain. There are important differences in the overall industrial, labor market, macroeconomic, technological, and regulatory contexts – otherwise known as *national innovation systems* (Freeman 1987) – within which U.S. and Japanese SMEs operate. For example, the tight relationships found in Japan between SMEs and their larger customers are still much less common in the United States. Similarly, while perhaps frayed at the edge during its deep post-1991 recession, Japan has maintained a policy commitment to employment stability – at least for male prime-age workers, while the U.S. labor market is far more flexible. Such variations between the countries have significant influences on respective SME approaches to training, capital investment, and work organization. There are also obvious contrasts in policy leadership and program implementation. Japanese policymaking is centralized, with few local variations in program operations. U.S. policymaking occurs in a federal-state system, where subnational governmental units often assume leadership, resulting in considerable program diversity and experimentation across the country. Additionally, while the Japanese central government has been a steadfast patron of public industrial modernization policies, U.S. federal backing has oscillated with political changes in executive and legislative administration.

The issue of small-firm industrial modernization in the U.S. and Japan thus presents several points of comparison and contrast. Each country is extending its infrastructure of policies, programs, and measures to promote SME modernization. But these infrastructure initiatives are being carried out in diverse industrial systems with different policy and programmatic styles. When examining details, the broad correspondence between the modernization goals of the two countries gives way to rather significant variations in program operation, stability, funding, and management approach. However, this is a dynamic situation and there is considerable cross-learning between these two huge economies. For instance, many American manufacturers are trying hard to apply "Japanese" industrial methods such as just-in-time delivery, in-process quality control, and team work. U.S. industrial modernization efforts seek to accelerate the deployment of such techniques. On the other hand, Japanese SMEs and policymakers are searching for ways to replicate the vitality of innovation seen in the best U.S. technology start-up companies. A new thrust of Japanese modernization policy is to stimulate a higher level of technological innovation and fusion among SMEs. Interestingly, U.S. analysts have pointed to Japan's large resource commitment toward industrial modernization in arguing (successfully) for increased resources for new American programs. Equally, Japanese policymakers are now contemplating a major overhaul of their modernization system to match emerging U.S. efforts.[2]

[2]Cross-learning between U.S. and Japanese industrial modernizers has been boosted by exchanges between the two countries. Japanese policy officials have visited U.S. manufacturing technology centers and they monitor developments in U.S. federal and state policies. Similarly, U.S. modernization program managers have inspected Japanese programs and firms. The U.S. Department of Commerce's Japan Technology Program, the Tokyo office of the National Science Foundation, and many other U.S. organizations track developments in Japanese industrial and technological policymaking.

2. SMEs and Industrial Modernization in the U.S.[3]

The backdrop to the emergence of U.S. industrial modernization policies and programs has been the growing concern about the overall performance of American industry and the loss of employment in the manufacturing sector.[4] Between 1979 and 1992, manufacturing productivity growth – averaging 2.4 percent annually – increased at only two-thirds of the rate in Japan, the yearly merchandise trade deficit expanded to $96 billion, and total manufacturing employment declined by three million jobs.[5] America's poor manufacturing performance has been associated not only with declining employment, but also with a shift in the structure of firms comprising the U.S. industrial base. While many large manufacturing firms have rationalized or closed U.S. production facilities, the number of smaller manufacturers has increased. Manufacturing establishments with fewer than 250 employees rose by 19 percent, from 303,500 in 1980 to 359,900 in 1990. Over the same period, very large (1,000 or more employees) establishments fell by almost one-quarter, from 2,440 to 1,880.[6]

Different explanations have been put forward for this tilt toward smaller manufacturing units. One interpretation suggests that standardized, large-scale, mass production systems of the post-World War II era (sometimes typified by the term *Fordism*) are now being edged out by more specialized and flexible forms of manufacturing. In this new phase, competitive advantage is accruing not to the old industrial giants, but to networks of small, innovative, flexible, and specialized production units (Piore and Sabel 1984; Scott 1988). These systems of flexible specialization are often associated with geographical concentration and highly collaborative linkages among firms, and with business-support infrastructures. California's Silicon Valley has been submitted as an example of the leading edge of this new production system, although

[3]The following discussion of modernization issues and barriers, and the subsequent review of U.S. modernization policies draws, in part, on Shapira, Roessner, and Barke 1995.

[4]See, for example: Dertouzas et al. 1989; U.S. Congress, Office of Technology Assessment 1990a.

[5]Manufacturing employment change calculated from establishment data reported in U.S. Department of Labor 1993, Table B-1; productivity comparison is measured by output-per-hour in manufacturing, from U.S. Department of Labor, Bureau of Labor Statistics 1993; merchandise trade deficit reported in *Economic Report of the President* 1994, Table B-105.

[6]Data from U.S. Department of Commerce, Bureau of the Census, *County Business Patterns*, various years.

other locations in the U.S. (and Western Europe) have been highlighted too (Saxenian 1994; Scott 1988). However, there is another explanation which identifies the cause of the shift toward smaller units, not in any intrinsic new advantage to small and mid-sized firms, but in the changing strategies of large corporations. Here the argument is that large firms – responding to global competition – are concentrating on core-business elements, using lean production techniques. Other operations such as parts production, subassembly, transportation, or maintenance are subcontracted out to smaller, less-costly suppliers (Harrison 1994). The result – the downsizing of big production units and a growing number of smaller ones – still leaves unscathed the dominance of larger parent corporations.

Significantly, whichever explanation turns out to be more accurate (and it may be that both processes are occurring simultaneously), the prospects for U.S. SMEs are not necessarily favorable. There is little disagreement that in the past, U.S. manufacturing has emphasized large-batch production, with slow change in product lines, and subordinate supplier relationships. With some exceptions (for example, in very high technology sectors), this approach has promoted a small-firm manufacturing sector that often lacks dynamism and strong internal technological capabilities. Indeed, despite increased employment in small plants and the trend toward smaller manufacturing enterprises, the U.S. still has a lower share of jobs and value-added in enterprises with less than 500 employees than in Japan and some European countries (Sengenberger, Lovemen, and Piore 1990). Thus, if the balance of industrial advantage is shifting toward flexible small-scale units, the U.S. is not especially well placed at present. On the other hand, if large corporations remain dominant and aggressively pursue lean production methods, small U.S. manufacturers will face tremendous pressures, both to reduce unit costs and to improve other aspects of manufacturing performance. U.S.-based large firms will no doubt continue to seek sources of supply on a global basis, putting U.S. suppliers under the strain of matching foreign competitors or losing business, especially for more routine types of production. Moreover, where larger customers place orders in the U.S., they will increasingly require contractors to pay greater attention to quality and on-time delivery and, in some cases, take considerable responsibility for design and subassembly. As U.S. SMEs face these increasingly demanding commercial markets, many will also need to deal with the loss of business from reduced U.S. military procurement over the coming years.

Many SMEs have responded to these structural changes in manufacturing systems and corporate strategies by bidding down wages and working conditions, also known as the "low road" strategy of industrial competitiveness. Among small firms in sectors like apparel, electronics, and metalworking, low-wage employment has proliferated.[7] But another approach is for smaller manufacturers to upgrade their production systems, improve products, enhance design capabilities, invest in workforce skills, and develop new customers and markets in the U.S. and in foreign countries. This "high road" strategy is more likely to maintain high-wage manufacturing sectors (although the distribution of jobs is sure to change, in favor of off-the-floor technical work rather than traditional manual production employment) and strengthen regional and national technological capabilities.

While numerous U.S. SMEs are already pursuing a high-road approach, far more are slow to modernize their manufacturing technologies, methods, and relationships. This has been confirmed in several studies conducted over the last few years.[8] For example, U.S. small and mid-sized manufacturers not only lag behind larger U.S. units in adopting new manufacturing technology, but also trail the technology adoption rates of small and mid-sized firms in Japan and Germany. Moreover, the gap in new-technology use between large and small units is wider in the U.S. than in Japan (Table 2). Perhaps of even greater concern, small and mid-sized U.S. firms are not only slow to adopt new "hard" manufacturing technologies, but also lag in using "soft" people-based technologies and techniques. Such methods as statistical process control, just-in-time manufacturing, cell-focused manufacturing, or greater attention to manufacturability in design can lead to significant improvements in productivity without large capital expenditures. Similarly, the use of simple cost-estimation spreadsheets and planning techniques can save managers considerable time in bidding for jobs. But smaller firms do not often use these methods and techniques.[9]

[7]Examples and explanations for new low-wage small-firm complexes in the U.S. are discussed in: Sassen 1989; Teitz and Shapira 1989; Harrison 1994, especially chapter 9.

[8]For examples of studies, see: Industrial Technology Institute 1987; Kelley and Brooks 1988; U.S. Department of Commerce, Bureau of the Census 1989; Young, Francis, and Young 1993; Rephann and Shapira 1994.

[9]Evidence on the low adoption rates of "soft" technologies and techniques by small and mid-sized firms is provided in: Industrial Technology Institute 1987; Shapira and Geiger 1990.

Table 2: Use of New Technology in Manufacturing, U.S. and Japan, 1988

Types of New Manufacturing Technology Japanese definition (closest U.S. definition in brackets)	Technology Users By Employment Size				Technology Use: Comparative Ratios			
	Japan		U.S.		Large/SME Ratio		Japan/U.S. Ratio	
	SME [a] Enterprise Employment Under 300 %	Large [b] 300+ %	SME [c] Establishment Employment 50-499 %	Large [d] 500+ %	Japan [e]	US [f]	SME [g]	Large [h]
NC/CNC Machine Tools (NC/CNC Machine Tools)	57.4	79.4	39.6	69.8	1.4	1.8	1.4	1.1
Machining Centers (FMS Cells or Systems)	39.4	67.4	9.1	35.9	1.7	3.9	4.3	1.9
Computer-Aided Design (and Computer-Aided Engineering)	39.1	75.2	36.3	82.6	1.9	2.3	1.1	0.9
Automatic Transport Equipment	34.9	68.3			2.0			
(Automated Guided Vehicles)			0.8	13.1		16.4		
Automatic Inspection/Measuring (Automatic Inspection: Final Product)	30.1	66.7	10.5	44.3	2.2	4.2	2.9	1.5
Automatic Warehouse Equipment (Automatic Storage and Retrieval)	22.6	62.2	5.5	43.3	2.8	7.9	4.1	1.4
Handling Robots (Pick and Place Robots)	10.9	44.9	1.9	24.4	4.1	12.8	5.7	1.8
Assembly Robots (Other Robots)	8.3	41.4	3.9	35.0	5.0	9.0	2.1	1.2
Welding/Painting Robots	16.0	42.7			2.7			

Notes: Japanese technology definition is given (with closest matching U.S. technology definition in brackets). The comparisons between Japan and the U.S. are approximate since there are differences in technology definitions and employment size categories. Additionally, the Japanese data is enterprise based, while the U.S. data is establishment based. SME = Small and medium enterprise.

Sources: [a], [b] "Current Survey on the Manufacturing Industries 1988," reported in Ministry of International Trade and Industry, Small and Medium Enterprise Agency, *Small Business in Japan 1989: White Paper on Small and Medium Enterprises in Japan*, Tokyo, 1988.

[c], [d] U.S. Department of Commerce, Bureau of the Census, *Manufacturing Technology 1988*, Current Industrial Reports SMT(88)-1, Washington, D.C., May 1989.

[e] = [b]/[a], [f] = [d]/[c], [g] = [a]/[c], [h] = [b]/[d].

In an aggregate sense, U.S. SMEs are thus caught between two competing systems, being neither fully lean (particularly compared with Japanese counterparts) nor fully flexibly specialized (outside of Silicon Valley and some other unique cases). Several factors have contributed to this situation. At the firm level, small and mid-sized manufacturers face a series of barriers, including the lack of technical expertise, access to information, issues of finance and cash flow, and limited strategic planning. These barriers are perhaps universal to all SMEs, being a function of small size. U.S. SMEs may have more difficulties in addressing some of these barriers due to internal differences (e.g., fewer engineers in SMEs) and broader U.S. business practices and policies. It has been suggested that the short-term character of supplier-customer links and the weakness of small-firm, technology-focused networks puts U.S. small and mid-sized firms at a disadvantage in pursuing modernization when compared with small and mid-sized firms in Japan, Germany, Italy, and other European countries.[10] U.S. smaller firms operating in end-use markets do so in an atmosphere of great uncertainty, generally squeezed for investment capital and typically in a weak position to bargain with customers or vendors. Trade associations could help, but most American trade associations tend to be reactive, responding to government actions or dealing with business regulation issues. Only a few associations actively focus on helping their member firms improve technology.

The public infrastructure for modernization in the U.S., including education and training systems, technology transfer programs, and other public services, is also an area of concern. At the local and regional level, where the workings of this infrastructure are most relevant for small firms, many gaps and problems are apparent. For example, universities typically stress research and advanced technology missions, and reward their faculty for excellence in these areas. Technology deployment for small and mid-size firms is generally not high on the university institutional priority agenda. Equally, state and local economic development programs are felt to offer little in the way of appropriate technological expertise and infrequently focus on the needs of existing small and mid-sized manufacturing enterprises; traditionally, most economic development efforts have concentrated on recruiting large firms, aiding new startup businesses, or building physical infrastructure. Until recently, few states had active industrial modernization programs. The educational system, including vocational education, is frequently criticized for producing neither

[10]For international comparisons, see: Dore 1987; Best 1990; Pyke, Becattini, and Sengenberger 1990.

well-rounded competent students nor well-trained technical specialists. Smaller firms often employ workers who have less education and experience higher turnover. Such firms generally do not provide formal training and tend not to participate in public training programs, which are frequently not suitable for small firm needs, especially for upgrading existing employees. In a mutually reinforcing downward cycle, the general weakness of training makes many small-firms reluctant to invest in training themselves because they fear workers will leave to obtain higher wages elsewhere (U.S. Congress 1990b; National Coalition for Advanced Manufacturing 1993). Inadequate training means that smaller manufacturers are then unable to develop and retain skilled labor, constraining their abilities to pursue modernized manufacturing methods and upgrade pay and working conditions.

U.S. policymakers, in designing a series of fresh federal and state initiatives to promote industrial modernization, hope to address these barriers and stimulate the upgrading of more SMEs to be competitive in high-road terms. Yet success is not guaranteed, even if the new initiatives are well implemented. For example, much depends on the stance of larger U.S. firms. If they singularly pursue short-term, low-price relationships with their smaller suppliers, those suppliers will continue to find it difficult to upgrade, resulting in an industrial variation of the "tragedy of the commons" where both large and small firms will be adversely affected in the long run. Equally, if basic regional social structures (such as education and training systems) remain below par, U.S. SMEs will be disadvantaged vis-a-vis their counterparts operating within stronger regional systems elsewhere. If they are to be effective, U.S. SME modernization strategies need to incorporate a two-part strategy involving improvements in broader systems of relationships, such as customer-supplier links and public and private support services, as well as focusing on the upgrading of individual firms.

3. U.S. Modernization Policies

Over the past decade, there has been an expansion of public and public-private technological infrastructure initiatives and programs to help small firms with industrial modernization and technology deployment in the United States. Until quite recently, most of these industrial modernization activities were initiated at the subnational level by state governments, universities, colleges, and other local and nonprofit organizations. During this period (covering much of the 1980s) the federal government provided only limited and uncoordinated support for state and local modernization efforts, preferring to focus the bulk of federal technology resources toward basic R&D and defense.

The federal government's industrial modernization role began to change with the passage of the 1988 Omnibus Trade and Competitiveness Act. This Act recognized that America had a problem of industrial competitiveness and mandated the National Institute of Standards and Technology (NIST), within the U.S. Department of Commerce, to promote manufacturing technology deployment.[11] NIST started programs to establish regional manufacturing technology centers and promote state technology extension programs, marking a first round (1988–1992) of increased federal support for U.S. modernization efforts.

This first period of federal involvement was motivated largely by the U.S. Congress – under Democratic party control, but with a bipartisan consensus – stimulating President Bush's Republican administration into action. In the subsequent much bigger phase, the impetus came from the executive side. After taking office in 1993, President Clinton – reinforcing statements he made as the Democratic party candidate during the 1992 election campaign – pledged to build a national system of manufacturing centers to help small and medium-sized manufacturers adopt new technology, production techniques, and business practices (Clinton and Gore 1992; 1993). This was one part of the administration's plan to promote U.S. civilian technology and strengthen the country's technological infrastructure.[12] Initially, Clinton proposed 170 U.S. manufacturing centers, matching the number of equivalent technology centers in Japan.[13] Subsequently, the administration pared back its goal to establishing "100 manufacturing centers nationwide by 1997." Nonetheless, this remains a significant commitment and the Clinton administration has overseen a dramatic enlargement of manufacturing technology assistance. NIST's manufacturing extension budget has been greatly increased through the administration's defense conversion initiative (known as the Technology

[11]NIST was previously known as the National Bureau of Standards.

[12]For a discussion of the shift in U.S. technology policies in the early 1990s and a contrast with the 1980s, see Bransomb 1993.

[13]This was not a coincidence – Clinton's advisers were aware of the operation of the Japanese program. I should declare an interest (and some small responsibility) here, since I prepared a briefing paper and other materials for the Clinton campaign and transition team, in which the scale of Japan's industrial modernization effort was noted. But other documents, most notably OTA's 1990 ("Making Things Better") study, had also discussed the Japanese program.

Reinvestment Project – TRP), supplemented by additional funds to the agency through the regular Department of Commerce appropriation. These resources are being used to expand already-existing state programs as well as to establish numerous new programs at national and state levels.

However, the expansionary post-1992 stage of federal industrial modernization involvement could end in 1995. The Republican majority that took control of Congress after the 1994 elections promises to stymie further efforts to extend the federal role in industrial modernization.[14] A third phase may thus be beginning during which leadership shifts back to the states to continue the development of their own modernization programs with more-limited federal support.[15]

The following sections describe in detail the development and operation of recent state and federal industrial modernization initiatives.

3.1 State Industrial Modernization Programs

In America's decentralized governmental framework, states have frequently led the federal government in promoting new developmental and public infrastructure policies (Osborne 1988; Eisinger 1988). This has been the case for industrial modernization where several states have run their own programs for many years. For example, in the U.S. South industrial extension programs were started by North Carolina in 1955 and Georgia in 1960 (Combes 1992). These efforts – modeled after the long-established cooperative extension service in agriculture – used locally based professional engineers to help design

[14]Incoming Republican Congressional leaders have stated that the federal government should focus on basic research and back away from applied technology policies (National Coalition for Advanced Manufacturing 1994). Deregulation and a reduction in capital gains and other taxes are preferred ways of stimulating technological innovation, rather than direct government-sponsored support (Andrews 1995). In this context, the Clinton administration will likely lack the Congressional endorsement to fully fund its technology investment plan, including the plan to increase NIST's budget to $1.4 billion by 1997 (up from $381 million in FY 1993).

[15]By the end of 1994, there were Manufacturing Extension Partnership centers in 32 states – 18 with Democratic governors, 13 with Republican governors, and one state with an independent governor (The Modernization Forum 1994). At the state level, industrial and technology policy is generally seen as an augmentation of the long-accepted states' role in economic and business development. States are likely to continue to support a variety of industrial modernization efforts, even if federal assistance wanes (although a decrease in federal funds may lead to a decline in cash match by several states).

technology-oriented, regional industrial development strategies, and to assist local firms to resolve technical problems. In the mid-1960s, state programs such as the Pennsylvania Technical Assistance Program were formulated to diffuse technical information to industry and to solve problems by linking firms with technical specialists.[16] Subsequently, in the late 1970s and throughout the 1980s, new state industrial modernization and technology transfer programs were started in Maryland, Massachusetts, Michigan, New York, Ohio, Pennsylvania, Virginia, and several other states.

By the beginning of the 1990s, 42 industrial modernization programs in 28 states had been established (National Governors' Association 1990; 1991). While this total included a handful of federally supported, state-operated manufacturing technology centers (see next section), these programs were mainly sponsored and funded by the states themselves.[17] There were several reasons for this growth of state interest in industrial modernization. At the broadest level, states were concerned about the real or potential declines of their industrial competitiveness. Industrial modernization was also seen as a new and useful tool for promoting local economic development. In other cases, modernization efforts sought to help small firms hard hit by the restructuring of core industries, or formed part of targeted strategies to promote the implementation of new technologies.

About one-half of the state modernization programs in place by the early 1990s were administered by educational institutions (usually universities, sometimes community colleges), with the others operated by nonprofit institutions, state agencies, or other local organizations. Each program offered its own blend of services ranging from the provision of technical information, seminars, and workshops; to demonstrations, referrals to qualified technical experts, and on-site plant consultations. However, the number of programs with direct and intensive one-on-one field assistance was small. A 1990 NIST study found only 13 state programs using field agents to work on-site with firms to solve technical problems (U.S. Department of Commerce 1990). Despite some further expansion of state efforts, as recently as 1992 most

[16]The Pennsylvania program did receive federal sponsorship through the State Technical Services Act of 1965 (Public Law 89-182) but after federal funding was cut in 1969, the Commonwealth of Pennsylvania continued to support the program.

[17]For its 1991 study, data collected by National Governors Association indicates total spending for surveyed programs at about $79 million, including almost $41 million from state and university sources, $20 million from the federal government, and over $15 million from industry and program income. Most of the federal support went to Manufacturing Technology Centers.

programs still had relatively few resources, services were patchily available, and many states had no programs at all. By 1994, however, this situation had changed dramatically, as increased federal resources leveraged additional state funds to significantly expand modernization program activities through a new federal-state partnership.

3.2 NIST, the MTC Program, and the Manufacturing Extension Partnership

The recent growth of U.S. industrial modernization activity has been greatly aided by the emergence of NIST as a sponsor and champion. For the first time, a federal-level civilian agency is directly concerned with issues of industrial technology and manufacturing performance. The 1988 Trade Act charged NIST with assisting industry to improve technology development, quality, process modernization, product reliability, manufacturability, functionality, cost effectiveness, and commercialization. The agency was authorized to provide technical assistance to state and local industrial extension programs and to serve as a link between these programs and other federal technology services.

While NIST was given a substantial mandate, during the agency's first few years through 1993, its budget for manufacturing technology programs was quite small – under $16-17 million annually. These funds were mostly used to sponsor a group of regional centers for the transfer of manufacturing technology – to provide information and education for local small and mid-sized firms, demonstrate advanced technology, help firms evaluate their needs and implement new technologies, and support workforce training. By 1992, seven of these Manufacturing Technology Centers (MTCs) had been designated through a competitive award process (in Ohio, New York, South Carolina, Michigan, Kansas, Minnesota, and California). The MTC program was initially designed to transfer advanced technologies developed at NIST's Advanced Manufacturing Research Facility in Maryland and at other federal laboratories. Once established, the centers quickly realized that most small firms did not need state-of-the-art technologies which were usually expensive and often untested. For most smaller firms it was recognized that the first job was to improve existing operations, using proven, off-the-shelf technologies, and to strengthen quality, inventory control, design, training, and marketing. MTCs have also found that so-called soft activities in such areas as training, management guidance, information access, referrals, and networking and association are typically more important than hard assistance in helping firms to introduce new machine technologies.

Each MTC has a budget of roughly $6 million a year from federal, state, and industry funds, supplemented by service-fee revenues. Under current legislation, federal funding is provided for six years. For the first three years, NIST funds have to be matched equally by state and other sources. The federal share then declines to zero in year six, with the aim that the MTCs become self-sustaining without direct federal sponsorship. Few now believe this is practical or desirable, and proposals have been made to extend NIST funding to the MTCs beyond the six-year limit (at a level of about one-third of each MTC's budget, with state, private, and other federal sources comprising the balance).[18]

A series of reviews found the original MTCs to be helpful in promoting small firms to modernize and upgrade technology. But these assessments also acknowledged that the small number of MTCs and their limited geographical scope, the relatively low level of funding, and their separation from other federal and state programs limited the total effectiveness of the effort (U.S. General Accounting Office 1991; U.S. Department of Commerce 1992a). To expand the scale of activity and better integrate MTCs with other public technology infrastructures, NIST subsequently established the Manufacturing Extension Partnership (MEP) in 1992 to coordinate and develop a national network of government, industry, and academic resources engaged in industrial and technological modernization. This network aims to coordinate services, facilities, expertise, and technologies to better assist customer firms. The MEP also seeks to strengthen links to other federal agencies and programs, including the Small Business Administration, Labor, Energy and Defense.[19]

[18]Efforts to extend federal funding to MTCs beyond the six-year limit received considerable support in the 103rd Congress in 1994, although legislation was ultimately not enacted.

[19]For an early exposition of the MEP, see: U.S. Department of Commerce 1992b. A more recent elaboration is: National Institute of Standards and Technology, *Manufacturing Extension Partnership*, U.S. Department of Commerce, Gaithersburg, MD, December 1994, electronic document available through NIST's gopher server (telnet gopher.nist.gov) or http://www.nist.gov/item/NIST_Manufacturing_Extension_Partnership.html.

3.3 The Technology Reinvestment Program

With funding from the Technology Reinvestment Program (TRP) – using money diverted from the defense budget – NIST has been able to make a considerable down payment in establishing the Manufacturing Extension Partnership. The TRP combines the missions of maintaining and developing the advanced defense technology base with those of helping the transition of defense suppliers and technologies to civilian markets and stimulating the production of competitive commercial products (U.S. Department of Defense 1993). This program was conceived in Congress before the Clinton presidency, but it has been embraced by the new administration. Three broad areas of activity are defined: technology development, technology deployment, and manufacturing education and training.

Over two fiscal years, beginning in 1993, about $900 million in federal funds was allocated to all TRP activities, with a further $550 million authorized for FY 1995. In FY 1993, as part of its technology deployment effort, TRP allocated $87 million for manufacturing extension and $91 million for defense dual-use extension to assist small businesses, including defense-dependent firms, to improve technology and operations. Support was also provided for technology extension enabling and access services, such as training, electronic linkages, and regional networking. In FY 1995, TRP's allocation for these particular technology deployment activities began to be reduced as NIST's own civilian-side budget for manufacturing extension was increased. However, several other TRP initiatives continued to support related industrial modernization and technology promotion efforts (U.S. Department of Defense 1994).

Awards under the TRP program are based on competitive application and generally require applicants to collaborate and provide matching funds, thereby leveraging the federal resources. The overall lead agency is the Defense Department's Advanced Research Projects Agency (ARPA, formerly known as DARPA) but several other line agencies are charged with program execution. NIST has taken the lion's share of responsibility for carrying out the technology deployment and extension enabling thrusts of TRP, merging these into the MEP program.

The TRP program has stimulated a tremendous response, as state and local governments, educational institutions, industry groups, and companies formed alliances and teams to develop proposals and secure matching funds. In the program's first year, 1993, almost 2,800 TRP proposals were submitted, requesting $8.4 billion. In the technology deployment area, there were 545 proposals, requesting $1.4 billion – almost eight times the available funding (Reilly 1993). Proposals were submitted to expand existing industrial modernization efforts; others sought to establish new programs of service delivery, networking, and enabling assistance. In this process, many new partnership arrangements were established, innovative service approaches devised, and funding commitments secured.

By the end of 1994, more than 45 MTCs or smaller Manufacturing Outreach Centers (MOCs) had been funded through NIST with TRP funds (Map 1). In addition, NIST used TRP funds to establish about 25 pilot programs to expand the infrastructure for manufacturing-modernization service providers. Some twenty states also received smaller State Technology Extension Program (STEP) awards for program development or demonstration projects. State and private contributions typically match federal funds on a 50:50 basis for new centers and at a higher rate for existing programs; considering all sources, the total investment in U.S. public industrial modernization programs going into 1995 will probably exceed $250 million, a substantial increase from two or three years earlier (Table 3).[20] A further round of competitive awards by mid-1995 will bring NIST's portfolio to about 20 MTC-sized programs and 50 smaller MOCs, taking the Clinton administration more than two-thirds of the way toward its goal of establishing 100 centers.

[20]NIST's 1995 (non-TRP) budget for its Manufacturing Extension Partnership program was originally set at $90.6 million. After a reconsideration of national technology programs by the new Republican-led Congress, the budget was cut by $16.3 million. For FY 1996, the Clinton administration sought to expand the MEP program to $146.6 million. However, as Congress reviews this request, 1996 federal funding of between $75 million to $90 million for MEP is the most likely outcome. These federal funds are matched at least equally by state and industry monies. This expanding set of non-MEP technology deployment and transfer centers is supported by other local, state, industry and non-NIST federal funds (Berglund and Coburn 1995). Some, but by no means all, of these centers may be incorporated into the MEP program as it is enlarged.

Table 3: Comparative Measures, U.S. and Japanese Industrial Modernization Programs

	United States		Japan
	Federal and State Programs[1] Annual Rate 1992 - Estimated	**Manufacturing Extension Partnership**[2] Annual Rate 1995 - Projected	**Kohsetsushi Centers**[3] Annual Rate 1993-4
Centers/Programs	30	70+	178
Total Funding	$80 million	$250 million+	$982 million
National Funding	17%	30-50%	10-20%
Staffing	N/A	1,500+ (professional and technical staff)	5,200 (engineers)
Staff Research Time	almost none	almost none	30-50%
Technological Assists[4] (annual cases)	Total: 13,500+ Field service: 6,000	Total: 26,000+ Projects: 13,500+	All contacts: 406,000 Field visits: 11,300 Technological advisors: 9,800
Inspections/Exams	N/A	N/A	684,000
Open Labs/ Demonstrations	1,740	increased	63,500
Diffusion Groups/ Networks[5]	50	150-200+	2,500

Sources and Notes

1. U.S. federal and state programs, 1992. Author's calculations from surveys of state programs.
2. Author's projections. Funding estimate assumes up to 1:1 state and private to federal match for new MEP programs, with a higher ratio for existing centers. Staffing and service levels scaled up from data from a subset of existing MEP programs. Staffing estimate excludes support staff and is for full-time equivalent personnel. "Projects" includes informal engagements, technical assistance projects, and assessments.
3. Data provided by Ministry of International Trade and Industry, Tokyo. Service data is for 1991. Total staff is 6,800 (with support personnel). Currency converted at market exchange rate of ¥100/$US1. This exchange rate does not reflect differences in purchasing power.
4. Definition of technological assists is broader in Japan (includes any type of contact) than for the U.S.
5. In both the U.S. and Japan, the estimate of diffusion groups/networks includes groups sponsored by other organizations. U.S. estimate excludes many quality and user groups sponsored by MEP centers.

N/A = Not available

3.4 Industrial Networking

In a comparatively recent development, several states and regional organizations have started to promote industrial networks or groupings of manufacturing companies to promote shared approaches to modernization. Networking seeks to aid groups of firms, usually within a proximate geographical area, through information sharing, solving common problems such as training, and through cooperation in design, production and marketing. Additionally, networking requires and generates cultural and systemic changes by generating new interfirm linkages and new forms of cooperation. Network advocates also point out the leveraging effects of the approach. By stimulating networks, more firms are expected to modernize by themselves, overcoming the scale problem of the limited public resources available to fund costly one-on-one methods of technology assistance. It is suggested that public resources will go further by leveraging networks, especially if public efforts focus on developing private network brokers who ultimately see it in their interest to form and support networks (Bosworth and Rosenfeld 1993). Practitioners note that networks of similar manufacturers tend not to form autonomously, but they can be facilitated by the provision of skilled assistance and extensive efforts to build trust and cooperation (Giancola 1991; Bosworth 1992; Rosenfeld 1992).

The scale of organized industrial networking in the U.S. has grown rapidly over the past few years. An analysis of 1994 data for 27 states finds around 140 industrial networks involving more than 2,600 firms.[21] Efforts by public agencies, industrial associations, and other hub organizations to further promote networking are now found in more than one-half of U.S. states. Nonetheless, there is a degree of skepticism about the networking approach, especially by some of the more-established programs in the modernization field. Here, the belief in the value of one-on-one professional field service contact is very strong, along with the claim that there is no substitute for good professional staff who can interact with firms on a face-to-face basis. Other practitioners see a potential marriage between networking and individual approaches, with networks providing a way for business firms and technical assistance providers to interact with one another. Firms may not initially form networks with an explicit and major focus on modernization issues; marketing or shared services are more likely initial aims. But once the network is up and running, it is

[21]Collated from data in Shapira and Youtie 1994; Connect, Inc. 1994. For earlier accounts of U.S. networking, see R.E. Friedman 1991; Lichtenstein 1992.

possible that modernization issues can become part of the network's agenda, and at that time they may be more effective than traditional modernization programs. One particularly strong argument in favor of networking is that it more readily allows the underlying modernization issues of interfirm relationships and public-private linkages to be addressed. NIST has recognized this and, partly using TRP funds, has provided additional funding to several local and national networking initiatives, including projects such as USNet – a consortium of fifteen states working together to promote interfirm collaboration.

3.5 Modernization Services and Practices: Two Examples

While there are many differences in the details of organizational design and technological expertise among U.S. modernization programs, in a growing number of centers – particularly those affiliated with the MEP program – it is possible to discern a common core of services and practices. Examples from two centers – one in Georgia, the other in Pennsylvania – illustrate this.[22]

The Georgia Manufacturing Extension Alliance (GMEA) is a partnership between the Economic Development Institute (EDI) at the Georgia Institute of Technology and three other organizations – the University of Georgia's Small Business Development Centers, the state's Quick Start customized training program, and Georgia Power Company's Technology Applications Center. GMEA aims to offer an integrated delivery system linking technology, management, training, and applications assistance to Georgia manufacturers. A central element of this delivery system is a network of 17 regional field offices staffed by full-time personnel with engineering, management, and industrial experience. These skilled professionals can provide reliable and timely in formation to firms, resolve technical problems, and offer guidance to encourage the application of technology and improved business practices. GMEA's services include initial meetings to discuss needs and potential projects with firms, informal engagements through which limited assistance is provided, and in-depth formal assessments leading to specific recommendations. The program then carries out technical assistance projects involving up to five days of staff time to resolve specific problems or to transfer new or existing technology. GMEA also makes referrals to external resources (such as private consultants, university faculty, or federal laboratories) and offers an extensive array of group assistance activities (training, conferences, seminars, user

[22]For sources and further details on these examples, see Shapira 1994b.

groups, ISO 9000 groups, and demonstrations). GMEA receives funding from federal and state sources and provides most of its services without charge to Georgia firms.

The Manufacturers Resource Center (MRC) in Bethlehem, Pennsylvania primarily provides services similar to those of GMEA, despite some differences in organization and funding. MRC is a not-for-profit corporation within Lehigh University and one of eight state-sponsored Industrial Resource Centers. The MRC is funded through state and federal sources (the latter through a recent TRP award) but also generates some revenue through cost sharing and private contributions. MRC's industrially-experienced staff conducts needs assessments of area manufacturing companies, identifying important problems or opportunities and providing some technical assistance. Fees range from zero to a little over $2,000 for this initial service. If a firm requires further assistance, the MRC makes a match with qualified outside consultants, sharing the cost of the external help with the firm in most cases. The MRC also offers a range of information seminars and training programs, supports group activities (such as quality forums and ISO 9000 groups), and coordinates with other area economic development and technology organizations.

Although methods of financing and implementation vary, the core services of each of these two programs include: initial assessment, diagnosis, technical assistance, qualified referral, training, information provision, and group services. Similar arrays of services are found in programs throughout the country. The identification and adoption of such *best practice* program management and service approaches has been encouraged by NIST through conferences, training sessions, research projects, and program reviews.[23] Professional associations have also actively assisted the dissemination of best practices. Innovative or successful services and practices developed in one locality are thus quickly recognized and copied by others. For instance, a technique to systematically identify modernization opportunities by comparing a manufacturing customer to similar firms along a series of critical measures was developed by the NIST/Midwest Manufacturing Technology Center in Michigan (Performance Benchmarking Service 1994). This technique – known as performance benchmarking – is now being adopted (with NIST's support) by other MEP centers. Similarly, the promotion of industrial networks is now attracting broad interest after trial projects in a handful of locations.

[23]The literature describing U.S. modernization best practices includes: Shaw 1987; Technology Management Group 1989; Clifton, Edens, Johnson, and Springfield 1989; Simons 1991; Shapira 1993a.

4. Assessment of U.S. Approaches

We have seen that the concept and practice of industrial modernization for small firms has advanced rapidly in the United States in recent years. Once practiced almost unnoticed and with meager resources by a handful of states, industrial modernization has now gathered considerable political and business support, and has seen the development of new federal and state policies and institutions. Funding has been enlarged and the number of programs and centers has increased across the country, although the coverage is by no means yet complete. Moreover, there has been improved understanding of what constitutes good practice in addressing company, industry, and regional deficiencies, needs and opportunities.

There are still several important challenges to be addressed in further promoting public–private collaboration for industrial modernization in the United States. While it has been possible to distill a series of common services and best practices for industrial modernization, there is still a tendency at the field level to focus on "fixing" individual problems at the firm level, for example, correcting a quality problem or recommending a specific technology. While these *project* services are important, the charge of stimulating deeper systemic change in firms and in relationships with other firms and public institutions is given less attention, probably because such efforts are complex and success is difficult to gauge. But such changes are possible, as illustrated by an effort in Grand Rapids, Michigan, where a modernization program affiliate has effectively brought together customers and suppliers in the office furniture sector to improve mutual linkages and develop common industry directions.[24] The further promotion of systemic activities will involve broadening program mandates, improving staff training, and changes in evaluation systems that shift program horizons more toward longer-term objectives.

Another concern yet to be fully addressed is that of geographical and industrial targeting. The highly competitive award processes used both for MTCs and the TRP program risk overlooking key industries or regions because of poorly drafted proposals or difficulties in obtaining matching funds. Moreover, the case been made that industrial modernization resources, especially from the federal government, should be focused at selected strategic regional industrial clusters, such as the motor vehicle industry complex in the

[24]The reference here is to the Right Place Program, Grand Rapids, MI, which the author visited in December 1993 as part of a NIST review panel.

Midwest (Fogarty and Lee 1991; Luria, Cole, Baum, et al. 1994). However, while perhaps desirable from a pure industry policy perspective, targeting resources to specific locations and industries is difficult in America's federal system, especially since analyses of clusters tend to favor older and more geographically concentrated Northeast and Midwest industrial regions. It is unlikely there would be much support for a targeted approach in the less industrially concentrated South and West, for fear of being left out.

Perhaps the greatest challenge is that of reconciling the *technological cycle* of industrial modernization with the corresponding *political cycle*; the former requires a long-term commitment by programs and firms while the latter operates on the shorter time horizon of two-to-four years. In the first part of 1994 administration, officials were confident that the federal government was assuming a long-term responsibility for industrial modernization (*ModComm* 1994). However, the subsequent shift in Congressional leadership and the possibility of a presidential changeover after the 1996 election has raised doubts about the federal role. It increasingly appears that the TRP will be a one-shot approach, giving a massive but temporary federal boost to public industrial modernization efforts. This was perhaps inevitable as the use of TRP as a funding vehicle for industrial modernization reflected short-term political compromises: first, in eluding budget deficit problems by using defense funds, and second, by cloaking the program with a national defense mission to avoid the appearance of "industrial policy" (still a politically sensitive concept in the U.S.). These compromises are now becoming unravelled. The proponents of industrial modernization will have to make their case in other ways if federal funding is to be continued for the many local programs spawned by NIST's TRP resources. It may be possible to persuade the new Congressional leadership to reauthorize large-scale civilian funding for NIST's MEP program on the grounds of program effectiveness, competitiveness, jobs, or state and industry support (rather than as technology policy, per se), but this is not a certain sale.

Even if federal, civilian-side funding is secured and NIST is given a budget for industrial modernization commensurate with its charter, the existing level of multi-agency fragmentation within the federal government will still make it hard to achieve coordination and consistency. In addition to the Defense Department, many other federal agencies – including Energy, NASA, Agriculture, the federal laboratories, the Economic Development Administration, and the Small Business Administration – have become involved in modernization and technology deployment initiatives. Similar problems exist at state levels, where interagency and interinstitutional competition is a frequent occurrence.

If federal leadership fades, the weight of responsibility for further development of U.S. industrial modernization efforts will shift to the states. Such swings are not unusual in U.S. economic and social policy. There is a strong tradition of state and local autonomy in designing programs to fit particular needs, and this localization is generally a desirable feature of industrial modernization programs. The TRP competition showed that state and local governments and regional business communities are extremely interested in industrial modernization and technology deployment. At the same time, without supplementary federal funding and a national framework to guide program development, distill best practices, and disseminate and transfer those practices around the country, state industrial modernization efforts will certainly be weakened.

5. The Japanese Situation: The Changing Position of SMEs

In Japan, even more than in the U.S., the small-firm manufacturing sector has grown in terms of employment and production over recent decades.[25] Japan's 850,000 small manufacturers employ three-quarters of the country's manufacturing workers and produce more than half of the manufacturing value added. These small manufacturers are diverse (Chalmers 1989). About half are workshops with three or fewer workers, while the remaining 432,000 small enterprises each employ between 4 and 299 people. Often, the smallest workshops are engaged in traditional Japanese crafts, while other small firms are labor-intensive operations, producing simple components or carrying out routine tasks for larger companies. Some are labor-only subcontractors, with no plant or equipment of their own, who send personnel to work inside the plants of other companies. But many small firms are involved directly in modern manufacturing using and developing advanced technologies.

In technological terms the gap between small and large firms is smaller in Japan than in the United States, as shown in Table 2; small Japanese companies are more likely to use new technologies and techniques than their U.S. counterparts. Almost 50 percent more Japanese SMEs than U.S. SMEs

[25]Japan's Small and Medium Enterprise Law defines a "small and medium enterprise" in manufacturing as one with 300 employees or less or with capital of ¥100 million or less. A "small-scale enterprise" in manufacturing is defined as one with 20 employees or less.

use numerical or computer-numerical control machine tools, and they use more than four times as many advanced machining centers and robots. Worker training – essential to the proper use of new technology – is also relatively strong in small Japanese firms.

Japan's hundreds of thousands of small, flexible, and technologically proficient manufacturers are sources of high-quality inputs and technological enhancements to larger companies. Contracting in Japan is typically organized in a pyramid fashion, with large manufacturers at the top supplied by smaller firms in multiple lower tiers. Long-term relationships between the tiers of smaller and bigger firms have given the smaller units the confidence to invest in new technology, workforce training, and ongoing product and process improvement (Dore 1986; Trevor and Christie 1988). But this "carrot" is usually accompanied by the "stick" of continuous improvement; most large companies require strenuous cost, quality, and delivery schedules, further driving smaller suppliers to modernize.

While many SMEs are closely tied to their larger customers, they also seek outside sources of support. Large firms may help their suppliers by sharing information, technology, and personnel – but not always; large companies may have little time to deal with the problems of smaller firms, small suppliers may be more specialized than their larger customers, or small firms may need special training or technological expertise. In such cases, small firms have to look beyond their larger customers for assistance. They may also seek outside help to deal with problems of adjustment. As large Japanese customers rationalize or internationalize production, many smaller firms are trying to develop new products to offset reductions in their traditional business lines. In other instances, large firms are themselves diversifying, compelling their smaller suppliers to shift into new technologies. Increasing competition from low-cost Asian suppliers, a highly valued yen, and the difficulty of attracting young, skilled workers (who prefer larger firms) are also stimulating small firms to invest in new labor-saving technologies and to upgrade working conditions.

The bursting of Japan's bubble economy of the late 1980s and the onset of a deep recession after 1991 has intensified the pressures on Japanese SMEs to rethink their technology and business linkages and strategies. Major companies, especially in the hard-hit automotive and consumer electronics industries, were forced to slash costs and curtail production during the recession, leading them to reduce or cut off orders from their smaller suppliers. This has occurred to such an extent as to call into question the future of Japan's tight parent-company – supplier relationships. While these links are so embedded that any fundamental changes will take a long time, the recession has convinced many small enterprises not to rely so heavily on a single major buyer. Cost-cutting drives have led to a drop in SME capital investment, which in 1992 was down 9 percent from the level of 1988. SME business closures are up, while start-ups are down. At the same time, the recession appears to have opened new possibilities for SMEs with innovative technologies to exploit niche markets which larger companies are disinclined to enter.

The current recession is thus likely to give an impetus to a trend that had been gathering momentum during the 1980s. An expanding segment of small Japanese manufacturers wants to develop, control, and sell their own products and technologies in domestic and international markets, without the constraints set by larger customers. One indication of this is the declining proportion of Japanese small manufacturers exclusively engaged in subcontracting, which dropped from two-thirds to just over one-half during the 1980s. The last decade saw an increase in R&D spending and personnel in many small firms, and an improvement in design capabilities. Geographical clusters of small, innovative firms have formed in Tokyo and other large cities, and in less-urbanized locations such as the Nagano Prefecture in central Japan (D. Friedman 1988). New horizontal and lateral relationships are being developed between and among small firms and in joint ventures with larger firms. A major aim of recent Japanese local and regional public technology policy is to reinforce and stimulate these emerging SME business development strategies.

6. Modernization Policies in Japan

Japan's policies toward small firms (those with 300 or fewer employees) have evolved significantly. In the years after World War II small firms were often viewed as a backward sector in Japan. As policymakers favored building up large firms, especially in heavy and mass-production industries, efforts were made to combine some small firms into bigger ones, and to shelter the rest from more efficient competition. While some protective measures still exist, the more recent thrust of Japanese small-firm policy has been to stimulate upgrading and modernization. Fostering innovative knowledge-intensive small firms is considered vital to Japan's shift toward high technology and more flexible production methods. Small-enterprise promotion and technology improvement are also seen as important regional economic development tools.

Modernization in Japan means not only strengthening technology, facilities, management, operations, and human resources in small firms themselves, but also improving entire sections of small enterprises, including intercompany and interindustry relationships. Japan's national small-enterprise laws and policies establish general mechanisms to provide finance, tax incentives, guidance, and assistance to individual companies and groups of small firms. There are also special measures to help small companies convert to new business lines and develop new products and technologies. The national government assists local programs that provide small firms with technology development support and guidance. Funds are made available for local technology centers, the clustering of companies for joint product development, marketing, training, improvements in design abilities, information dissemination, and research and entrepreneurship.

The principal central agency responsible for small firms in Japan is MITI's Small and Medium Enterprise Agency (SMEA), whose functions include overseeing small-industry guidance and technology development, subcontracting, enterprise promotion, and planning and research. Separate national councils for small and medium enterprise policymaking, stabilization, and modernization provide advice and review. Other agencies and bureaus contribute, including MITI's Machinery and Information Industries Bureau and the Agency of International Science and Technology (attached to MITI). An associated public organization, the Japan Small Business Corporation, provides guidance and financing for structural improvement and upgrading projects in small firms, training for enterprise personnel and local program staff, information and computing support, and business finance. The national government has also established three major financial institutions targeted at small and medium-sized firms, to complement private financing and promote specific modernization policy goals: the People's Finance Corporation, which extends funds to very small-scale firms; the Small Business Finance Corporation, which supplies longer-term funds to small and mid-sized firms, and the Shoko Chukin Bank, which finances small-firm cooperatives and industry organizations.[26]

[26]For discussion of Japanese policies for small firms, see Ministry of International Trade and Industry 1994 (and earlier issues of this annual publication).

A recent national policy encourages small firms with different specialties to work together to develop and commercialize new products.[27] This is carried out by establishing local technology plazas or meeting places, supporting mediators, sharing information, offering subsidies to business associations and fusion groups, and providing support for shared production and marketing. By 1994, about 2,500 SME fusion groups had been registered in Japan. Most of these were still at the stage of initial association and research, but some had moved to commercialize jointly developed new products (Shapira 1994a).

Prefectural and city offices of industry promotion operate at the local level, providing local services that include small-firm development and guidance. Equipment modernization loans and leasing systems for general, high-technology, and information-processing equipment are available for small firms, funded jointly by national and prefectural governments. Prefectures and cities also make additional funds and incentives available to local small firms for plant and equipment investments through tax relief, interest subsidies on private bank loans, and other allowances. Area business and industrial associations play an important role in small-industry modernization in Japan, a role supported by the public sector. Local chambers of commerce, industry federations, subcontractor promotion associations, and industry-specific structural improvement associations all receive public financial support.

An example of the integration of several of these programs is found in Ehime Prefecture, on the southern island of Shikoku, where a Towel Industry Resource Center is working to revitalize local small towel firms, by promoting new computer-aided design (CAD) and manufacturing methods, design consulting, training, and joint marketing. This center is a cooperative effort involving the local industry association, the prefectural and city governments, and MITI. Next door, programmers, CAD operators, and information specialists are trained in a new Computer College established with a foundation grant from the Ministry of Labor's Employment Promotion Corporation.[28]

The variety of programs in Japan, together with the fact that small-firm industrial modernization initiatives overlap with programs for general business development, regional development, and technology promotion, makes it

[27]Extra-Ordinary Law Concerning the Promotion of the Development of New Business Areas through Fusion of Knowledge of Small and Medium Enterprises in Different Industries ("Fusion" Law), 1988.

[28]Shikoku Towel Industry Association and Ehime Prefectural Local Industry Promotion Center, Imabari, Ehime Prefecture, interviews conducted February 23, 1993.

difficult to calculate precisely Japanese spending for small-firm technology assistance. Only a small portion of small-business assistance is counted in the regular central government tax budget; most central resources are provided from trust funds and other capital accounts. The U.S. Office of Technology Assessment estimates that more than $30 billion – or more than 5 percent of the national regular and capital budgets in Japan – goes to support small firms each year, including nonmanufacturers and loans but excluding prefectural and city contributions and spending on related regional development and technology programs (U.S. Congress 1990a).

6.1 Kohsetsushi Centers

The cornerstone of Japan's system for modernizing small firms is a nationwide network of local and prefectural, technology and testing centers. Known as Kohsetsushi Centers (an acronym for *koh*, public; *setsuritsu*, establishment; and *shikenjo*, testing laboratory), these are publicly sponsored institutions, with large engineering staffs, that serve as free or almost-free resources for manufacturers with 300 workers or less.[29] Japan began to establish these industrial research, experiment, and testing institutes at the turn of the twentieth century, based in part on the U.S. model of agricultural experiment stations and extension services. National and university institutes were the first to be founded, followed in the 1920s and 1930s by a number of local government centers to strengthen local industries. After World War II, additional local Kohsetsushi Centers were constituted. In recent years, a few new Centers have been added, while many older Kohsetsushi Centers have expanded or built new facilities.

Today, there is at least one Center in each of Japan's 47 prefectures, with 22 Centers in the Tokyo metropolitan region (Map 2). In total, there are more than 170 Centers, employing 6,800 people, including 5,200 engineers and technical personnel (Table 3). Prefectural and municipal governments administer the Centers and provide most of their budgets, the equivalent of U.S. $982 million in FY 1993. The central government typically provides 10 to 20 percent of the budget of each Center, with funds coming from MITI, the Japan Small Business Corporation, and the Japan Bicycle Development Association (which uses profits from gambling on bicycle racing to improve

[29]The discussion of Japan's Kohsetsushi Centers draws on field research in Japan in 1989, 1990, 1993 and 1994. The first part of this research is reported in greater detail in Shapira 1992; 1993b.

machinery and metalworking industries). Fee income from services to private firms is small.

Kohsetsushi staff spend up to half their time on research, mainly on applied projects focused toward local industries. Projects may be organized by the Center itself, or sponsored with local companies or universities. Small manufacturers often send one or two of their staff to work on Kohsetsushi research projects, providing opportunities for company personnel to gain research experience, develop new technical skills, and transfer technology back to their firms. The Centers run a variety of seminars and study meetings to disseminate information on research and new technologies to local firms, publish newsletters and research reports, and maintain technical libraries.

Conducting tests and examinations is another major Kohsetsushi activity. For nominal fees, Kohsetsushi laboratories will analyze materials and products, verify standards compliance, calibrate measuring instruments, and make sophisticated testing equipment available. These services are used extensively by Japan's small firms, with more than 680,000 tests and examinations conducted yearly. These services help small manufacturers to enhance quality, precision, and product development, and to resolve problems in materials and components.

To help small companies overcome technical difficulties and implement new technology, Kohsetsushi Centers provide advice and guidance services. For simple requests, enterprise managers call in by telephone or visit the Center; more complex problems are dealt with by field visits from Center staff to companies. Over 400,000 contacts are handled annually, consisting of 318,000 technological consultations, most of which take place at the Center, and over 21,000 instances where staff teams or advisers visit firms. About half of these visits are made through a program of technology advisers whereby Kohsetsushi Centers match company needs and requests with registered private manufacturing consultants. The advisers are initially reimbursed from local and central funds, allowing them to provide a first round of services at no cost to firms.

Training in new technologies for employees of local small manufacturers is provided through Kohsetsushi group and customized programs. Employees go to the Centers for classroom instruction and hands-on experience with advanced tools, computers, and software systems. Many Kohsetsushi Centers offer open laboratories, making their specialized equipment available for research, prototyping, and training. Kohsetsushi Centers also sponsor technology diffusion and network groups to encourage small firms to exchange information, share technology, and develop new products and markets. Each Center may sponsor several such groups, each comprising up to 30 local firms.

6.2 New Regional Technology Projects and Partnerships

Complementing the long-established Kohsetsushi Centers is a newer and still growing infrastructure of regional technology projects, industry resource centers, and technocenters. Concerned with lagging regional development outside of Japan's booming Tokyo-Nagoya-Osaka central core, MITI and other central government ministries have established a series of projects to promote the technological upgrading of existing industries and the development of new technologically based enterprises in other regions.

One of the most well-known of these initiatives is the Technopolis program which, in the 1980s, designated 26 areas to serve as nodes for high technology growth (Masser 1990). New technocenters have been built at many of these Technopolis sites to help introduce advanced technologies to local small firms. Regional technology development is also the aim of the Research Core program, which equips special facilities for promoting small-firm technology transfer, business incubation, and training. Sponsored by MITI, although funded mainly by local government and the private sector, ten Research Core locations have been chosen to date. Other regional technology projects championed by MITI and other ministries include the Key Facilities Concept which promotes facilities for information services and research in peripheral areas, and the New Media Community, which develops new information systems, in part for local-firm networking (Edgington 1995).

These regional technology initiatives are complemented by an expanding group of new, local-industry resource centers, city technocenters, and training institutes. For example, on the northern island of Hokkaido, the Muroran Technocenter is promoting the technological development of SMEs as part of a strategy to diversify the heavy industrial base of the local economy. The Technocenter assists area small manufacturers through training, consulting, marketing, networking, information distribution, research and development, and by making advanced production and information-processing machinery available for demonstration and local industry use. The center is supported by city, prefectural, local industry, and national government funds.[30]

[30]Interview and field visit, Muroran Technocenter, Muroran, Hokkaido, January 26, 1993.

Many of the new regional technology projects and industry centers are structured as *third sector* organizations. This usually involves a governing foundation comprised of public and private representatives, which allows more flexibility in activities and staffing, and enables resources to be leveraged from the private sector. Funding comes from private member companies, banks, utilities, and local and prefectural governments. The national government contributes equity capital and loans for third-sector organizations through an Industrial Infrastructure Improvement Foundation which is endowed with proceeds from the privatization of the telephone company, NTT.

Japan's third-sector approach borrows from the public–private partnership models developed in the United States. Policymakers have promoted the concept of the third sector in Japan to allow greater flexibility and autonomy than permitted in purely governmental operations. However, the desired effects have yet to be felt. Most third-sector organizations use seconded government personnel who employ management systems and methods little different from those found in the public sector. Moreover, during Japan's booming bubble-economy era, numerous third-sector organizations built (or were provided with) lavish showcase facilities.[31] In the current recession these buildings are costly to operate and hard to fill with revenue-generating activities. Further, private contributions are not easy to obtain, which has forced national and local governments to supply additional soft subsidies. Good young researchers are also said to be reluctant to join third-sector technology organizations because of the fear of instability and concerns about the centers' reputation. At the same time it must be noted that most third-sector organizations are at an early stage of development. Their primary sponsors show few signs of retreat and are likely to persist in providing support to help these organizations build capacity and become more effective.

7. Assessment of Japanese Approaches

The combination of Kohsetsushi Centers, regional and local technology projects, financial incentives, and other national and local policies provides an extensive support system for Japan's small firms. Most small firms have easy access to these assistance resources. The emphasis on examination,

[31]It is worth noting that in Japan's political and budgetary systems the construction of a capital facility occurs first, with funding for operations gradually being allocated afterwards. In this sense, third-sector centers are not unusual, although this observation helps to demonstrate that they are run under traditional public-sector procedures.

testing, and analysis in the Kohsetsushi Centers has been valuable in helping small firms meet high standards of quality, performance, and precision. The ready access to advice and guidance services, and advanced equipment, training, and information services assists firms in upgrading their operations, products, workforce, and strategies. And the Centers help companies establish collaborative local research and act as a bridge to national research laboratories and universities.

The Kohsetsushi Centers present an intriguing illustration of nationally standardized services, but with little direct national funding. Across the country, different Kohsetsushi Centers offer a range of remarkably standardized services and programs; they pursue the same policies and offer similar kinds of assistance to firms. Typically, the Centers allocate about one-half of staff time to research, with the balance divided between (a) examinations and analysis, and (b) technology advice and guidance. Additionally, most Centers sponsor technology diffusion groups or plazas, in conformance with MITI's emphasis on this strategy.

This harmonization of services is, of course, attributable to MITI, which has a great influence on the Kohsetsushi system even though its share of direct funding is small. There are several mechanisms through which MITI achieves this influence. First, the exercise of administrative guidance (*gyosei shido*) is a long-standing and prevalent practice in Japan. Although not unique to Japan, administrative guidance entails using extra-legal means of obtaining adherence to policies and practices which are deemed to be desirable by the government. Second, the linkages between the Kohsetsushi Centers and MITI's national research laboratories guide Kohsetsushi research approaches. Third, many elected and appointed government officials at the local level have links with MITI or are former MITI employees, facilitating influence through personal connections and personnel rotation. Fourth, MITI is able to carefully leverage the small amount of project funding it provides for new research projects, training programs, and technology assistance and diffusion activities, with funding from MITI-affiliated bodies such as the Japan Bicycle Development Association.

Japanese firms often turn to the Kohsetsushi Centers for aid in improving existing technologies and products. Companies report that local technology centers are helpful in this respect and that the free or nominal cost of assistance encourages them to use the public services. In most cases, Center staff are able to assist in these common problems, which usually do not require the latest technological expertise. Noting this, some Japanese observers criticize the weak quality of Kohsetsushi research and technology. Visits made to Centers indicate a measure of truth in this view. In some instances, laboratories are visibly decayed, underutilized, and full of old equipment; many projects are run by individual researchers who also have many other responsibilities; the average age of researchers is high; and the titles of some published papers are mundane (e.g., examining PC use in manufacturing enterprises). In other cases, Kohsetsushi research laboratories are well organized and equipped, and focused on ambitious research objectives. The general impression, however, is that most Kohsetsushi research is catch-up rather than pioneering.

Paradoxically, this weakness in research could be considered a strength of the Kohsetsushi system. Kohsetsushi researchers spend a lot of their time catching up on work done in national laboratories and other research centers. Some Kohsetsushi staff report that along with their other responsibilities for guidance and assistance, it is a constant challenge to improve their knowledge about developments in their fields. In this catch-up mode, Kohsetsushi staff are not at the leading edge of their fields. Rather, they are in an intermediate position where they are not too far ahead of their small-firm clients. In the past, this meant they could readily transfer knowledge to smaller firms that was useful and close to applications. However, there is an increasing concern that a growing number of Japanese SMEs have technological capabilities well ahead of those found in most Kohsetsushi Centers, leading to calls to upgrade their quality of research and technology.

In seeking to assist and foster these more innovative small firms, many local governments want to increase the role of local centers in innovative technologies by funding more future-oriented research. New third-sector institutions specializing in advanced technologies have also been established. However, while the physical infrastructure for these new technology centers is impressive, the "soft" infrastructure of personnel and operating procedures to support innovation is weak. Japan's rigid labor market makes it hard to attract talented young technologists to centers which have yet to establish a reputation, and professionals with industrial experience are unable to leave their current employers in mid-career. Additionally, despite their intent, most third-sector centers are managed by dispatched public-sector personnel, effectively introducing the very practices of inflexibility and risk aversion these centers were designed to overcome.

Indeed, staffing has emerged as a major issue for most of Japan's local technology centers. The Japanese have tended to use career personnel, generally with university engineering or science undergraduate degrees, to provide core services. These employees usually work their entire career with the sponsoring prefecture, often in a single center. This ensures stability and helps build long-term relationships with local firms, but staff skills can become outdated, and the low turnover limits opportunities to recruit young staff in new areas of technology. Kohsetsushi Centers are trying to address this question by hosting visiting researchers and increasing education and training for existing staff. But the Kohsetsushi Centers – along with the new third-sector technology centers – continue to find it difficult to attract the best young researchers and technical staff in areas of new technology.

In the future the policies of both central and local governments to actively promote technology upgrading in small and medium manufacturers will lead to additional emphasis on technology services, with commensurate increases in both resources and demand for providers at the local level. In some instances Kohsetsushi Centers are facing competition from the latest generation of regional technology programs. More frequently, Kohsetsushi Centers are working with new third-sector initiatives in regional technology partnerships. However, the growing variety of regional technology schemes in Japan requires improved coordination of resources, and raises critical questions of program relevance and effectiveness. The particular challenge is whether Japan's public and public–private modernization systems can be sufficiently flexible, targeted, innovative, and customer-driven to meet the needs of the small-manufacturers sector in a fast-changing technological and business environment.

In an acknowledgement of the challenges facing the Kohsetsushi system, the Small and Medium Enterprise Agency has established a Technology Policy Committee to provide advice about new legislation for SME technology promotion, and regarding the restructuring and future role of the Kohsetsushi Centers. This effort will contribute to a new national "vision for the 21st century for technology policies for small and medium sized businesses," set to be announced by 1995 (*Nikkan Kogyo Shimbun* 1993). It is likely that this vision will seek to shift the Kohsetsushi Centers away from testing and technology guidance, to focus more on advanced research in fundamental and applied industrial fields. This research will be related to regional needs and aim to nurture the technological strengths of SMEs. Preparing the ground for this new direction, a regional study committee has already recommended the improved integration of Kohsetsushi Centers in regional industrial policies, upgraded technological capabilities within the Centers, wider research collaboration with other institutions and companies, improved researcher training, and a more active role in working with local companies (Kinki

Bureau of Trade and Industry 1994). One official has suggested that the Kohsetsushi Center "in 10 years will probably have a completely different image than it has today" (*Tokyo JITA News* 1993).

8. Comparisons and Contrasts in U.S. and Japanese Approaches to Industrial Modernization

We have seen that both the U.S. and Japan are paying increased attention to SME industrial modernization. From small beginnings, the U.S. is now rapidly building an extensive yet decentralized national industrial modernization system. Japan, which has long had established programs, is now adding new elements to address the changing needs of its SMEs.

There are some interesting comparisons and contrasts to be drawn from U.S. and Japanese approaches (Table 4). Perhaps one of the most obvious concerns scale and geographical coverage. Experience in both Japan and the U.S. indicates that for modernization programs to make a real difference they need to have sufficient resources, linkages, and local leverage points to work with large numbers of firms. While Japan has long had a nationwide system of modernization service centers available to most SMEs, until recently the U.S. had a much smaller and patchier approach. But this picture is changing as the basic lesson that an appropriate scale of effort and an extensive field presence is necessary to match the size of the problem has been recognized in the U.S. Resources committed to industrial modernization are increasing by roughly an order of magnitude, and a majority of U.S. states now have or are committed to developing programs.

Program stability and time horizon are both related to the issue of scale. Experience in both countries suggests that modernization programs need to be stable and lasting to develop essential relationships of trust with firms and to deal with multiple and complex problems. In Japan, funding for Kohsetsushi Centers is regular and predictable, while funding of the new third-sector programs is perhaps a little less certain. Japanese program managers do complain that they do not get as much money as they would like, but they are in a very different position to their U.S. counterparts who often must expend inordinate efforts to generate funds to enable programs to continue.

In order to diversify income sources, many U.S. programs aggressively seek fee income from firms to cover costs, a strategy rarely seen in Japan where services tend to be offered free or at a nominal price. Generating fee income can be beneficial by giving market signals to programs, inducing them to provide the services that are most valued by the firms themselves – those being the services for which firms are willing to pay. However, the danger is that programs will focus too much on short-term individual project sales, thereby avoiding more systematic or strategic efforts where fee income is unlikely but which could have bigger long-term benefits and spillovers. There is also the possibility that programs will lose objectivity and the trust of clients if they are too aggressive in promoting their own income base. Indeed, companies often comment that they like industrial modernization programs because they do not push any particular technology or service, unlike equipment vendors or private consultants who have a strong motive to promote their own products and interests. Most practitioners, even in programs that generate revenues, agree that technology modernization programs need a stable and sufficient core of public support to avoid such difficulties and fully realize their potential.

Although the level of public funding has now increased, the key question for the United States is whether this effort can be sustained over the long term. The short-term government perspective, dominated by budgetary and electoral cycles, is in conflict with the essentially long-term nature of industrial modernization. Modernization is not a quick-fix jobs program and it fundamentally entails more than the one-time identification and adoption of new technology. Stability and patience are crucial – characteristics that are often difficult for policymakers to maintain in America's fluid, decentralized, and often fragmented federal–state political system. Despite the generally warm receptivity of state-level policymakers to the concept of upgrading the SME industrial base, these elected and appointed officials do not always fully understand the combinations of managerial, technological, and system-changing tasks that modernization requires. Meanwhile, changes in national political leadership are set to renew ideological disputes about the desirability of government intervention that may constrain the further development of modernization policies and programs and lead to a significant retrenchment in federal support. In contrast, Japan's centralized and bureaucratically led policy apparatus more readily adopts a long-run view, although the infighting between different ministries and the stifling of local program experimentation, innovation, and flexibility by centralized control are considerable weaknesses.

Personnel and staffing for industrial modernization is an important concern in both Japan and the U.S. Japan's Kohsetsushi engineers typically spend their entire careers with one Center. Under this staffing system, tight relationships are developed over time between Center staff and firms – a pattern which facilitates mutual confidence and effective technology transfer. The downside is that the technological and research skills of the staff can become obsolete, and the low rate of turnover makes it hard to bring in new people with expertise in emerging technology areas. This problem has gained attention as the Kohsetsushi Centers seek to advance the technology frontier of their research. Third-sector initiatives promise more staffing flexibility, but the restricted mobility of technological professionals in Japan limits their availability.

The very fluid labor market in the U.S. makes it is easier to hire – and to lose – competent staff with either general or specific expertise. While a few U.S. programs use academic faculty, the dominant trend is to hire industrially-experienced engineers, preferably with business management experience or training. Other types of professionals with expertise in training, general management, or marketing are also used. In addition to technical competence, program staff need to possess a range of capabilities, including: working with diverse firms; building confidence, enthusiasm, and support among a variety of constituencies; and involving other specialists, faculty, or consultants in order to meet particular needs. With the increased scale of U.S. modernization efforts, the number of personnel engaged in the system is growing dramatically, and complementary initiatives are being put into place to provide staff training, tool libraries, shared information resources, and other staff support. U.S. programs also pay more attention to assessment and evaluation – although not always done well – whereas formal evaluation is practically nonexistent in Japan.

The approach to service delivery presents both similarities and contrasts. In Japan, the range of information, technical, and assistance services provided to small firms is marked by two key characteristics: (1) remarkable standardization, with little variation from place to place; and (2) a strong engineering focus. Several of the more well-established U.S. programs offer engineering services akin to those provided in Japan, while a growing number of new MEP centers offer a common service core of assessment, assistance, referral, information, and training. However, as might be expected, the variety of approaches is far greater in the U.S. and there is a higher degree of program decentralization and diversity. Some U.S. programs are pursuing more comprehensive approaches, integrating engineering with management, training, organizational, and marketing assistance. Possibly due to differences in the internal capabilities and business environments of U.S. and Japanese small firms, several U.S. programs seek to go beyond the specific problem-solving approaches prevalent in Japan in order to stimulate their customers to pursue broader technological upgrade and business development paths. Moreover, U.S. programs exhibit far more in the way of experimentation and innovation than seen in Japanese efforts, for example in developing assessment tools, benchmarking measures, and telecommunication techniques. On the other hand, a feature emphasized in Japan but rarely seen in U.S. programs is the role of examination, testing, and analysis services. This reflects the high level of concern for quality, close tolerances, and accuracy shared by both small firms and their customers in Japan.

There is also an interesting asymmetry in the approach to technology. Modernization programs in both countries exhibit a pragmatic approach to technology. Experience has shown that much improvement can be obtained in many small and mid-sized manufacturers by utilizing off-the-shelf technologies rather than highly sophisticated and complex new technologies. However, the direction of movement is different. In the U.S., modernization programs have come to emphasize pragmatic technologies to counterbalance a research and innovation system that has been highly focused toward advanced technologies, including advanced defense technologies. In Japan, where there has been a higher diffusion rate of proven commercial process technologies, modernization programs are now stressing innovative research and new technology development. Japanese programs are aided by their stronger direct linkage with research, which occupies about one-half of Kohsetsushi staff effort. It remains to be seen whether Japan's local and regional technology centers, particularly the Kohsetsushi Centers, can implement new research approaches, organizational styles, and personnel systems to make them fully effective in new technology development and avoid unnecessary research duplication.

Finally, there are contrasts in terms of vision, commitment, coordination, and institutionalization. In the 1980s and early 1990s there was a lack of national leadership and consensus in the United States about the importance of manufacturing in general, and industrial modernization in particular. State efforts grew to fill this vacuum, establishing a laboratory of innovative, diverse, and decentralized approaches. Now the federal government has increased its policy, programmatic, and financial roles, based on a new commitment to strengthening the U.S. industrial base. This has raised critical questions of federal–state coordination, an issue which is endemic throughout most U.S. domestic policy areas. The U.S. is searching for ways to establish a nationwide industrial modernization partnership with the states and industry, while avoiding building federal programs on top of existing state programs. It is hoped that this partnership will lead to a comprehensive, long-term, and effective system which also respects local needs and variations. It is a difficult task, made even more complex by new uncertainties about the federal role in coming years. In Japan, the policy commitment to industrial modernization has been demonstrably stronger, and there is more robust political and financial support for a comprehensive system. But as the technological and business environment changes, Japan seeks to find ways to evolve its system into one that is more innovative, flexible, and decentralized. In its own way, this a challenge as exacting as any facing the United States.

Table 4: Structure and Organization of U.S. Industrial Modernization Programs and Japan's Kohsetsushi Centers

	United States: Federal and State Industrial Modernization Programs	**Japan: Kohsetsushi Centers**
Sponsors	States, nonprofits, universities & colleges, others	Prefectures and large cities
Primary users	Varies, but generally manufacturers with fewer than 500 employees	Small and medium enterprises, 300 or fewer employees
Services	Technology needs assessment; problem solving; technical advice and guidance; implementation assistance; brokerage and referral; training; information dissemination; demonstration	Research; technical advice and guidance; examination and analysis; information dissemination; training; open laboratories and equipment use; registered technical advisers; diffusion of technology groups
Standardization and scope of services	Not all states have programs; service mix varies by state; experimentation in service approach and delivery	Nationwide coverage; standardized services; almost no experimentation in services
Staffing	Flexible, with varied use of core staff and consultants; many staff have engineering and industrial experience, but varies by program	Lifetime staffing, little flexibility; almost all staff have engineering or technical qualifications

Table 4 (Continued).

	United States: Federal and State Industrial Modernization Programs	**Japan: Kohsetsushi Centers**
Research links	Few programs have own research, but several are within research institutions; many are linked with local universities; emerging links with federal labs	Centers maintain own research programs; links with national labs; some university links
Service delivery methods	Varies, from informal to formal assessments and written recommendations; up to 5 or more days of service; fee policies vary, from free to cost match to full fees for intensive services	Generally informal; large number of small interactions; some cooperative research projects with SMEs; no or nominal fees for service
Program evaluation	Varies, but NIST-supported programs use standard performance measures; external reviews by program sponsors, oversight agencies	No formal evaluation systems
National role	Federal funding match; competitive selection for federal funds; diffusion of best practices; demonstration projects; coordination and partnership within a decentralized, federal-state system	Some national funds; centralized guidance over local activities; determination of overall program direction; identification of key technology priorities. Promotion of new third-sector centers.
Program status	Expansion mode, but future federal role and funding uncertain	Mature but may soon be restructured; funding stable
Key Issues	Long-term stability and role of federal government; disseminating best practices; proving effectiveness; stimulating system changes	Lack of flexibility; staff expertise; promotion of innovation; level of technology; relationship to new regional third-sector initiatives

Map 1: Recent Federal–State Industrial Modernization Projects in the United States

▽	MTCs* [designated through 1993]
●	MTCs* [designated under TRP** in 1994]
O	Other manufacturing extension projects [designated under TRP** in 1994]
STEP	NIST State Technology Extension program, 1994
USNET	USNET (promotion of interfirm collaboration) state partner

*Manufacturing Technology Center
**Technology Reinvestment Project

Map 2: Kohsetsushi Centers and Regional Technology Projects in Japan

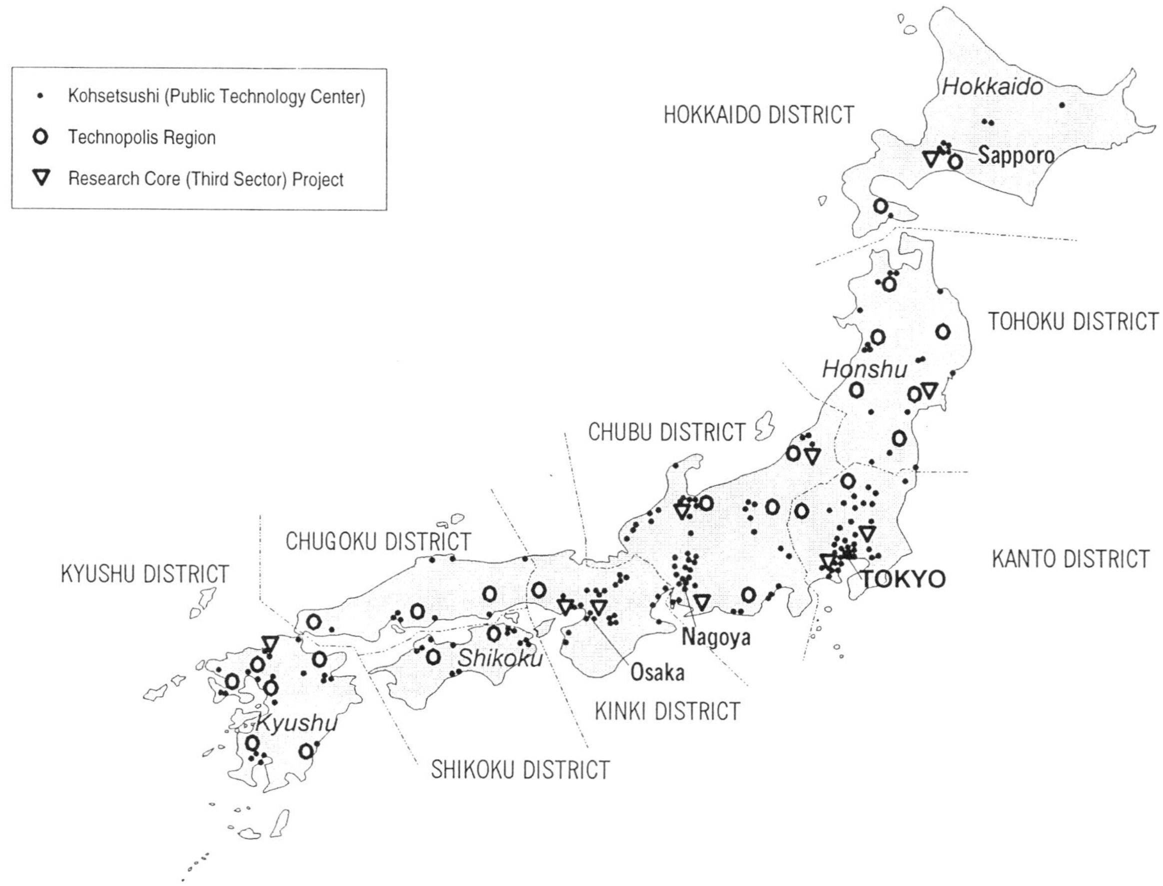

Abbreviations and Acronyms

ARPA	Advanced Research Projects Agency (Department of Defense)
EDI	Economic Development Institute (Georgia Institute of Technology)
GMEA	Georgia Manufacturing Extension Alliance
MEP	Manufacturing Extension Partnership
MOC	Manufacturing Outreach Centers
MRC	Manufacturers Resource Center
MTC	Manufacturing Technology Centers
NIST	National Institute of Standards and Technology
SME	small and Mid-Sized Manufacturing Enterprises
SMEA	Small and Medium Enterprise Agency (Japan, MITI)
STEP	State Technology Extension Program
TRP	Technology Reinvestment Project

References

Andrews, E. 1995. Congress and White House split on high tech. *New York Times*. 3 January, C19.

Berglund, D. and C. Coburn. 1995. *Partnerships: A Compendium of State and Federal Cooperative Technology Programs*. Columbus, OH: Battelle Press.

Best, M.H. 1990. *The New Competition: Institutions of Industrial Restructuring*. Cambridge, UK: Polity Press.

Bosworth, B. 1992. Learning to learn in networks. *Firm Connections* (May-June): 1,2.

Bosworth B. and S. Rosenfeld. 1993. *Significant Others: Exploring the Potential of Manufacturing Networks*. Regional Technology Strategies. Chapel Hill, NC.

Branscomb, L.M. 1993. The national technology debate. In *Empowering Technology: Implementing a US Strategy*, edited by L.M. Branscomb. Cambridge, MA: MIT Press.

Chalmers, N. 1989. *Industrial Relations in Japan: The Peripheral Workforce*. London and New York: Routledge.

Clifton, D., L.R. Edens, H.T. Johnson, and R.W. Springfield. 1989. Elements of an effective technology assistance policy to stimulate economic development. *Economic Development Quarterly* 3 (February): 52-57.

Clinton W.J. and A. Gore, Jr. 1992. *Manufacturing for the 21st Century: Turning Ideas Into Jobs.* National Campaign Headquarters. Little Rock, AR.

—. 1993. *Technology for America's Economic Growth: A New Direction to Build Economic Strength.* Executive Office of the President. Washington, D.C.

Combes, R.S. 1992. Origins of industrial extension: A historical case study. Masters Thesis, School of Public Policy, Georgia Institute of Technology. Atlanta, GA.

Connect, Inc. 1994. USNet electronic information exchange, networking directory forum. December.

Dertouzas, M, R.K. Lester, R.M. Solow, and the MIT Commission on Industrial Productivity. 1989. *Made in America: Regaining the Productive Edge.* Cambridge, MA: MIT Press.

Dore, Ronald. 1986. *Flexible Rigidities.* London: The Athlone Press.

—. 1987. *Taking Japan Seriously.* Stanford CA: Stanford University Press.

Economic Report of the President. 1994. Washington, D.C.: U.S. Government Printing Office.

Edgington, D.W. 1995. New strategies for technology development and information systems in Japanese cities and regions. In *Planning for Cities and Regions in Japan*, edited by P. Shapira, I. Masser, and D. Edgington. Liverpool, U.K.: Liverpool University Press.

Eisinger, P.K. 1988. *The Rise of the Entrepreneurial State: State and Local Economic Development Policy in the United States.* Madison, WI: University of Wisconsin Press.

Fogarty, M.S. and J.C. Lee. 1991. *A Manufacturing Industry Cluster Approach to Technology Deployment.* Center for Regional Economic Issues. Weatherhead School of Management, Case Western Reserve University. Cleveland, OH.

Freeman, C. 1987. *Technology Policy and Economic Performance: Lessons from Japan.* London: Pinter Publishers.

Friedman, D. 1988. *The Misunderstood Miracle: Industrial development and Political Change in Japan.* Ithaca: Cornell University Press.

Friedman, R.E. 1991. Networking comes to America. *The Entrepreneurial Economy Review* 9 (Spring): 3-12.

Giancola, D. 1991. Heat in Ohio: The difficult birth of an American network. *The Entrepreneurial Economy Review* 9 (Spring): 19-23.

Harrison, B. 1994. *Lean and Mean: The Changing Landscape of Corporate Power in the Age of Flexibility.* New York: Basic Books.

Industrial Technology Institute. 1987. *Technology and Applications.* Vol 1 of *Frostbelt Automation: The ITI Status Report on Great Lakes Manufacturing.* Ann Arbor, MI: Industrial Technology Institute.

Kelley, M. and H. Brooks. 1988. *The State of Computerized Manufacturing in the US Manufacturing.* Program on Technology, Public Policy, and Human Development, John F. Kennedy School of Government, Harvard University. Boston.

Kinki Bureau of Trade and Industry. 1994. *Reinforcement of Small and Medium Companies.* Report of the Committee on Kohsetsushi and Technology Reinforcement of Small and Medium Companies. Osaka.

Lichtenstein, G.A. 1992. *A Catalogue of US Manufacturing Networks.* National Institute of Standards and Technology, US Department of Commerce. Gaithersburg, MD.

Luria, D., R. Cole, A. Baum, et al. 1994. Fixing the manufacturing base: the allocation of industrial extension. *Journal of Policy Analysis and Management* 13 (Summer): 571-581.

Masser, I. 1990. Technology and regional development policy: A review of Japan's Technopolis programme. *Regional Studies* 24 (1): 41-53.

Ministry of International Trade and Industry, Small and Medium Enterprise Agency. 1994. *Small Business in Japan 1994, White Paper on Small and Medium Enterprises in Japan.* Tokyo.

ModComm. 1994. "Technology Policy is Here to Stay" says White House special assistant. Vol. 2 (Spring).

The Modernization Forum 1993. *MTC Project Development and Management.* Dearborn, MI.

——. 1994. *The Role and Importance of the Manufacturing Extension Partnership Centers.* Dearborn, MI.

National Coalition for Advanced Manufacturing. 1993. *Preparing Workers for the New Industrial Era: The Need for a Fundamental Shift in Federal Policy Toward Technical Education.* NACFAM. Washington, D.C.

——. 1994. *New Direction of the House Science Committee: NACFAM Alert.* Washington, D.C.

National Governors' Association. 1990. *Promoting Technological Excellence: The Role of State and Federal Extension Activities.* Washington, D.C.

——. 1991. *Increasing the Competitiveness of America's Manufacturers: A Review of State Industrial Extension Programs.* Washington, D.C.

Nikkan Kogyo Shimbun. 1993. MITI to formulate 'Technology Vision' for small, midsized enterprises. 29 June.

Osborne, D. 1988. *Laboratories of Democracy.* Boston, MA: Harvard Business School Press.

Performance Benchmarking Service. 1994. *The Bench Press* 1 (Fall).

Piore, M.J. and C. Sabel. 1984. *The Second Industrial Divide.* New York: Basic Books.

Pyke, F., G. Becattini, and W. Sengenberger. 1990. *Industrial Districts and Inter-Firm Co-operation in Italy.* International Labour Organization, International Institute for Labour Studies. Geneva.

Rephann, T. and P. Shapira. 1994. *New Technology Adoption in West Virginia: Implications for Manufacturing Modernization Policies.* Regional Research Institute, West Virginia University. Morgantown, WV.

Reilly, L. 1993. TRP filers ask for $8.4b. *Washington Technology*, 12 August.

Rosenfeld, S. 1992. *Manufacturing Networks and State Policy in North Carolina.* Southern Technology Council. Research Triangle Park, NC.

Sassen, S. 1989. New trends in the sociospatial organization of the New York City economy. In *Economic Restructuring and Political Response*, edited by R.A. Beauregard. Urban Affairs Annual Reviews (34): 69-113. Newbury Park, CA: Sage Publications.

Saxenian, A. 1994. *Regional Advantage: Culture and Competition in Silicon Valley and Route 128.* Cambridge, MA: Harvard University Press.

Scott, A.J. 1988. *Metropolis: From the Division of Labor to Urban Form.* Berkeley: University of California Press.

——. 1988. Flexible production systems and regional development. *International Journal of Urban and Regional Research* 12 (2): 171-185.

Sengenberger, W., G. Lovemen, and M.J. Piore. 1990. *The Re-Emergence of Small Enterprises.* International Labour Organization, International Institute for Labour Studies. Geneva.

Shapira, P. 1992. Modernizing small manufacturers in Japan: The role of local public technology centers. *Journal of Technology Transfer* (Winter).

—. 1993a. *Best Practices for Industrial Modernization.* School of Public Policy, Georgia Institute of Technology. Atlanta, GA.

—. 1993b. Helping manufacturers do better: Japan looks after the little guys. *IEEE Spectrum* 30 (September).

—. 1994a. Collaborative business exchange and technology fusion: The Japanese approach. *Firm Connections* (September/October): 10-12.

—. 1994b. *Technology Demonstration and Application Centers: Case Study Examples from the United States and Japan.* School of Public Policy, Georgia Institute of Technology, Atlanta, GA.

Shapira, P. and M. Geiger. 1990. Survey of technology use in West Virginia manufacturing. West Virginia University, Regional Research Institute. Research Paper 9001. Morgantown, WV.

Shapira, P., J.D. Roessner, and Richard Barke. 1995. New public infrastructures for small-firm industrial modernization in the USA. *Entrepreneurship and Regional Development* 7 (1).

Shapira, P. and J. Youtie. 1994. *USNet 1994* State baseline assessment: preliminary analysis. USNet Evaluation Paper 9402. Regional Technology Strategies. Chapel Hill, NC.

Shaw, B.T. 1987. *Strategies for Delivering Technology Assistance to Small Manufacturers: Lessons from Pennsylvania.* American Society of Mechanical Engineers. 87-WA-TS-9. New York.

Simons, G. 1991. The experience of the Northeast Manufacturing Technology Center. *Proceedings of the Second Meeting on Modernizing America's Industrial Base* (May). Detroit.

Sumiya, F. 1994. Small manufacturers face survival fight. *Nikkei Weekly*, 13 June, 1.

Technology Management Group. 1989. *How to Start and Operate an Effective Business and Technology Assistance Program.* Handbook Prepared for the US Department of Commerce, Economic Development Administration, Grant Number 99-06-07263. Philadelphia, PA.

Teitz, M.B. and P. Shapira. 1989. Growth and turbulence in the California economy. In *Deindustrialization and Regional Economic Transformation: The Experience of the United States*, edited by L. Rodwin. Winchester, MA: Unwin and Hyman.

Tokyo JITA News. 1993. Chugoku GIRI chief calls for restructuring of Kohsetsushi. 93FE08111 (June): 2-3.

Trevor, M. and I. Christie. 1988. *Manufacturers and Suppliers in Britain and Japan*. London: Policy Studies Institute, London.

U.S. Congress, Office of Technology Assessment. 1990a. *Making Things Better: Competing in Manufacturing*. OTA-ITE-443. Washington, D.C.: U.S. Government Printing Office.

——. 1990b. *Worker Training: Competing in the New International Economy*. OTA-ITE-457. Washington, D.C.: U.S. Government Printing Office.

U.S. Department of Commerce, Bureau of the Census. 1989. *Manufacturing Technology 1988*. Current Industrial Reports. Washington, D.C.

U.S. Department of Commerce, National Institute of Standards and Technology. 1990. *Technology-Based Economic Development: A Study of State and Federal Technical Extension Services*. Special Publication 786. Gaithersburg, MD.

——. 1992a. *Manufacturing Technology Centers: Broad Programmatic Issues*. Third Year Review Panel. Gaithersburg, MD.

——. 1992b. *Manufacturing Extension Partnership: A National Strategy for Manufacturing Excellence*. Gaithersburg, MD.

U.S. Department of Defense, Advanced Research Projects Agency. 1993 *Technology Reinvestment Project, Program Information Package*. Arlington, VA.

——. 1994. *Program Information Package for Technology Reinvestment Project, FY95 Competition*. Technology Reinvestment Project. Arlington, VA.

U.S. Department of Labor, 1993. *Employment and Earnings* 40 (August). Washington D.C.

U.S. Department of Labor, Bureau of Labor Statistics. 1993. *International Comparisons of Manufacturing Productivity and Unit Labor Costs, 1992*. USDL:93-348. 25 August.

U.S. General Accounting Office. 1991. *Technology Transfer: Federal Efforts to Enhance the Competitiveness of Small Manufacturers*. GAO/RCED-92-30. Washington, D.C.

Young, R.C., J.D. Francis, and C.H. Young. 1993. Innovation, high-technology use and flexibility in small manufacturing firms. *Growth and Change* 24:67-86.

Modernizing Manufacturing: Consultancy, Advisory, and Extension Services to Improve Small Business Performance

Graham Vickery*

Small firms are an essential part of the industrial fabric of modern economies. They provide a major part of production and employment and can be highly innovative. However small firms face a variety of major competitive challenges due to: structural change in industry; the widespread impacts of the introduction of information technology and other new technologies; the shift towards more R&D-intensive activities; and international competition in domestic and foreign markets. These challenges require a variety of strategic changes in products and services, in methods of production and supply of products and services, and in marketing and international operations. Overhaul of internal organization and upgrading of management and work-force skills and competencies are necessary responses to these changes.

To help small firms meet these challenges, many national, regional, and local governments have implemented management advisory services and technology assistance services (often termed *Industrial Extension Services*) to improve the performance of industrial firms. These services often provide introductory, or starter services to the large number of firms which have potential to improve performance. They usually operate through non-governmental or quasi-governmental institutions, private-sector consultants, or other business or sectoral associations to provide on-site, short-term technology and management assistance to individual small and medium-sized manufacturing enterprises. And in some cases they provide support to groups or consortia seeking common solutions to common challenges.

This chapter examines the use of these *Industrial Extension Services* to improve the performance of small business firms. We define and evaluate the concept, objectives, and design of technological advisory services. We then

*This chapter is a shortened version of more extensive work on the design and implementation of advisory services published by the OECD (1995). The views expressed in this chapter are not necessarily those of the OECD or of its member countries.

M. Teubal et al. (eds.), Technological Infrastructure Policy, 335–372.

use a variety of case studies to illustrate the key dimensions and practical consequences of program implementation.

1. Why Consultancy and Industrial Extension?

The small business firm sector has considerable untapped potential to increase its productivity. Because of small firms' economic importance, improvements in their performance would generate substantial benefits to the economy as a whole. As a consequence, many countries have introduced a broad range of special programs to assist small firms, addressing both the perceived widespread weaknesses in their performance and market failures in the demand for and supply of consultancy services and expertise to improve this performance.

1.1 The Importance of Small Firms

Small and medium manufacturing enterprises (SMEs) are an important part of the economic fabric of industrial countries. They:

- comprise over 90 percent of all manufacturing enterprises in most OECD countries;
- provide substantial employment; manufacturing establishments with fewer than 500 employees provide 60 to 80 percent of manufacturing employment in OECD countries and a substantially larger share of service employment, due to the smaller average size of service firms in the economies of OECD countries (see Table 1);[1]
- are significant sources of new jobs and employment growth; the share of employment in small and medium-sized establishments increased in the 1980s in many OECD countries – see Table 2 (OECD 1994; OECD 1985; Sengenberger, Loveman, and Piore 1990);
- generate a significant share of industrial output and value added (Table 2);

[1]OECD 1994 points out that small establishments contribute a high share of gross job gains *and* gross job losses, and that it seems important for the maintenance of healthy economic and employment performance to have a mix of small, medium and large enterprises with strong links between them. See also OECD 1985, Chapter 4.

Table 1: Employment Shares by Establishment Size: Total Economy and Manufacturing (percent)

Country	Year	Number of Employees			
		<20	20–99	100–499	500+
United States					
total economy	1990	26.1	29.3	24.5	20.1
manufacturing	1990	7.8	21.0	34.7	36.4
Japan					
total economy	1981	49.4	27.6	11.2	11.7
manufacturing	1990	23.7[1]	30.9	17.8[2]	27.6[3]
Canada					
manufacturing	1989	40.4		35.6	23.9
France					
total economy	1990[4]	28.8	20.4	16.4	34.4
Germany					
total economy	1990[4]	18.8	26.8	16.9	37.5
Italy					
total economy	1990	51.1[5]	22.4[6]	10.0	16.5
Norway					
total economy	1985	17.8	30.6	15.3[7]	36.4[7]
Switzerland					
total economy	1985	42.6	26.7	19.6	11.1
manufacturing	1985	19.6	26.9	30.6	23.0
United Kingdom					
total economy	1989	34.0	16.7	13.3	36.1
manufacturing	1989	11.2	23.3	35.6	30.3

[1] 4–19 employees.
[2] 100–299 employees.
[3] 300+ employees.
[4] Excluding self-employed.
[5] 0–9 employees.
[6] 10–99 employees.
[7] 100–199 employees and 200+ employees.

source: OECD, SME data base and national sources.

Table 2: Comparisons of Small and Medium-Sized Establishments in Manufacturing (percent of total)

Country	Year	Establishment	Employment	Value added
Japan (<100 employees)	1972	97.8	53.6	37.2
	1977	98.0	56.4	39.9
	1982	96.6	55.6	38.6
	1988	96.4	55.2	37.2
Japan (<300 employees)	1972	99.4	69.5	53.9
	1977	99.4	71.7	56.6
	1982	99.2	71.9	55.9
	1988	99.1	72.9	55.5
United Kingdom (<200 employees)	1975	96.3	21.9	18.2
	1977	96.6	22.5	18.5
	1982	95.5	35.3[1]	31.0[1]
	1988	96.9	39.2	31.7
United States (<250 employees)	1972	95.6	42.7	36.9
	1977	96.0	43.4	36.3
	1982	96.2	46.9	38.4
	1988	96.1	47.4	38.9

[1] Sample coverage changed.

Source: OECD and P. Shapira 1991.

- can be important for technological innovation, applications of new technology, and for providing industrial flexibility; they can also provide cost advantages due to lower wages and more flexible work organization than in larger establishments (Sengenberger, Loveman, and Piore 1990; OECD 1993);
- manufacture, supply, and increasingly develop and design the major share of components and subassemblies to larger firms in many industries; with increasing globalization, many small manufacturing firms are also becoming more international by exporting; by becoming partners with foreign firms in international trading, sourcing, and production networks; and by setting up their own production and service activities in foreign countries.

1.2 Weaknesses in the Small-Firm Sector

The performance of small firms is often undermined by inadequate and inefficiently used internal resources and external inputs of services (management, specialist skills, technological, marketing). There may also be market imperfections and slow market adjustments in the provision of these resources and services. As a result, substantial numbers of small firms neither perform well nor make the contribution that they could to the economy. Small firms have significantly lower levels of labor productivity than large firms – around 75-85 percent of levels in larger firms (Table 3). Furthermore, differences between small and large production units widened during the 1980s in some OECD countries (see Table 4 for data for the United Kingdom).

Table 3: Indices of Labor Productivity by Establishment Size: 1980s

Employees	France	Germany	Italy	United Kingdom
20–99	76.1	84.7	75.0	86.7
100–499	85.7	85.5	87.0	90.1
500–999	87.3	86.2	96.0	96.0
1,000+	100.0	100.0	100.0	100.0

Source: Adapted from W. Sengenberger 1990; UK Census of Production.

Table 4: Labor Productivity Indices: Manufacturing Establishments United Kingdom

Employees	1974/75	1979	1983	1988[1]
1–99	83.7	87.6	85.5	78.4
100–199	92.0	92.9	90.9	88.3
200–499	98.2	100.1	98.3	95.1
500–999	107.4	103.3	106.5	108.3
1,000–1,499	100.0	102.6	102.9	113.4
1,500+	108.9	107.7	114.7	128.4

[1] From 1987 the reporting unit is the "business".

Source: Calculated from CSO, UK Census of Production, net output per employee.

Table 5: Use of New Technology by Employment Size of Firms: 1988
(percent of total enterprises or establishments using the technology)

Type of technology	Australia		Canada (1989)		Japan		Sweden (1986)		United Kingdom (1987)		United States	
	10–199 establishments	200+ establishments	20–99 establishments	500+ establishments	<300 enterprises	300+ enterprises	5–19 enterprises	500+ enterprises	1–199 establishments	200+ establishments	20–499 establishments	500+ establishments
CAD/CAE	8.3	39.0	27.0	75.0	39.1	75.2	7.7	57.7	5.6	42.0	36.3	82.6
CN/CNC machine tools	11.8	34.8	21.0	65.0	57.4	79.4	n.a.	n.a.	6.2	33.0	39.6	69.8
Flexible manufacturing cells	0.9	5.1	7.0	31.0	39.4	67.4	n.a.	n.a.	n.a.	n.a.	9.1	35.9
Robots, pick-and-place equipment	1.5	14.7	3.0	30.0	22.6	62.2	n.a.	n.a.	1.4	15.0	5.5	43.3
Robots, assembly	0.3	4.3	n.a.	n.a.	8.3	41.1	n.a.	n.a.	0.5	9.0	3.9	35.0
Automated storage/ retrieval	1.3	11.1	1.0	11.0	10.9	44.9	1.6	26.7	n.a.	n.a.	1.9	24.4
Automated guided vehicles	0.5	2.7	2.0	12.0	n.a.	n.a.	n.a.	n.a.	n.a.	n.a.	0.8	13.1

Note: Information for Canada and the United States covers metalworking and machinery industries (equivalent to ISIC 38). Other countries cover all of manufacturing. Survey methods vary.

Source: Compiled from national surveys.

More telling, small firms are slower to adopt and assimilate new technologies, management strategies, and advanced manufacturing technologies (AMT); this is evident in all countries and across all manufacturing sectors (see Table 5). Differences in the use of AMT between smaller and larger firms are notable for all technologies, but they are most pronounced in the use of advanced and complex equipment such as that used in flexible manufacturing, or storage and handling systems, which require greater skills and resources to plan, install, and to run. Differences are less pronounced for smaller, well-established pieces of equipment such as CNC machine tools and computer-aided design equipment, which can be installed singly and where greater expertise has been built up for use in smaller firms.

Weaknesses in demand

Internal weaknesses of SMEs (especially smaller firms with 40–100 employees) lead to market failures on the demand side for external business and industrial services and expertise.[2] Among the weaknesses occurring most frequently are:

- Lack of information on applications of new technologies and management methods, especially those based on microelectronic and computer-controlled manufacturing technologies, to keep abreast of rapidly changing products and processes.
- Insufficient experience with and suspicion of outside sources of advice. SMEs often express doubts about the value of using consultants and external resources to determine the best management approaches and production technologies to remain competitive in an increasingly diversified and globalized economy.
- Lack of knowledge and experience absorbing and adapting outside advice so as to transform it into effective internal advantages.
- Limited internal management abilities to master advanced production techniques and employ strategic management methods even when they are available and have been transferred to the firm. Managers of many small enterprises are unable to make efficient use of internal and external resources to grow and to improve their long-term potential. This can be a crucial problem for firms with around 40 employees, where organizational and technological demands outstrip the capacity of a single general manager or engineer.

[2]For a discussion of general problems see: OECD 1989a; OECD 1989c; OECD 1989d; and OECD 1991. For problems specific to small firms, see: OECD 1989b, Part III and subsequent years; OECD 1990: and OECD 1993.

- Shortages of finance combined with the relatively high cost of introducing new technologies, as well as an unwillingness to accept external financing or equity to help implement these advances.

Weaknesses in supply

A range of failures also exists on the supply side for industrial services and expertise:

- Little attention is paid to small manufacturing enterprises by the consulting community. Large firms may already use consultancy on a regular basis to solve well-defined technical problems. Hence many different kinds of consultancy – in functional areas such as accountancy, management, engineering, software, human-resource development and training, and advanced manufacturing organization and development – have grown up to service large firms as consultancy and external services become a more common input into their operations. The small-firm market is usually fragmented, difficult to aggregate, and not worth pursuing because of the small size of contracts and limited payoffs (Sauviat 1991).
- Especially acute are shortages of comprehensive consultancy and advisory services for small firms which provide the range of external services, skills, and competencies necessary to: improve product design and quality; introduce strategic production and logistical management; manage complex organizational change; and raise skill levels. These are far more challenging tasks than the simple introduction of new production systems.
- There has also been a range of weaknesses in financial markets and in accounting methods to supply finance on appropriate terms to meet the cost and time obstacles inherent in changing production methods and firm organization, and moving towards more skill-intensive and knowledge-intensive development, production, and marketing methods.
- Until recently there has been relatively limited response to meet these special needs. Consulting, training, and technical assistance services have not been offered widely by industry and trade associations, commercial vendors, other firms, governments, and neutral third-party organizations concerned with improving the short- and long-term performance of SMEs. The services that have been provided – often only for purposes of generating fees – are better adapted to larger firms or more advanced technologies.

1.3 The Government Response

Government responses are changing rapidly. Many national, regional, and local governments have implemented management advisory services and technology assistance services (often termed *Industrial Extension Services*) to improve the performance of industrial firms. These services often provide introductory services to the large number of firms that have potential to improve performance. They usually operate through nongovernmental or parastatal institutions, business or sectoral associations, and private-sector consultants to provide on-site, short-term technology and management assistance to individual small and medium-sized manufacturing enterprises. In some cases they provide support to groups or consortia seeking common solutions to common challenges.

Some of the wide range of consultancy, advisory, and industrial extension services and programs recently in operation in OECD countries are listed in Tables 6 and 7. These tables provide a selective overview of some of the many recent initiatives introduced at national and subnational level in OECD countries. Programs vary widely in their institutional delivery, but they share many common features – notably, all aim at improving the long-term industrial performance of small firms by increasing the flow and effectiveness of external expert information and services.

Programs emphasize:

- advisory, consultancy, and management assistance,
- training and continuing education,
- the promotion of small-firm networking and cooperation, and
- technology diffusion.

In some countries and at the subnational level, assistance is organized into Industrial Extension Services providing a wide range of services, or referrals to services which the organizations do not provide directly. Many programs have evolved towards providing or subsidizing comprehensive advisory and consultancy services on strategic management techniques, human resource development, and the facilitation of long-term and ongoing organizational change. At the same time, extension services have moved away from providing advice and support exclusively to implement particular hardware solutions.

Table 6: Selected Consultancy, Advisory, and Extension Programs

Australia's National Industry Extension Service aims to increase SME competitiveness by using outside expertise to improve management functions.

Austria has a special program on computer-integrated manufacturing (Innovation and Technology Fund) designed to increase applications, funding up to 50 percent of costs of information collection, consultancy, and training.

Canada has a range of business guidance and counselling activities including technology outreach centres (TOP), technology advisory and information (IRAP), manufacturing assessment, and manufacturing technology applications program (AMTAP consultancy support).

Denmark's Technological Development Programme partly aimed at improving the use of advanced manufacturing technology by SMEs. This was followed by programs for network development to improve small firm performance.

Finland's industry policy has an important focus on improving the technological and human resources and industrial infrastructure which supports specialization and competitiveness of SMEs, building on Finland's long experience with regional policy.

France has an integrated program (ATOUT) to improve diffusion and use of microelectronics and new materials, and improve computerized information flows in SMEs.

Germany has a wide range of programs to support SMEs including a scheme to improve management in SMEs by subsidizing 50 percent of consultancy costs, more in the case of start-ups.

Ireland provides general support services to small firms, technology-related support (particularly the National Technology Audit Programme) and business development services and support.

Japan introduced a law in 1988 to promote the collective exploration of business problems by SMEs to develop joint solutions to these problems. This supplements the long-established "Kohsetsushi" local guidance network.

The **Netherlands** has programs to improve the performance of subcontractors and suppliers across a range of manufacturing industries, and promote quality and logistics management.

In **Norway** the Business Development using New Technology (BUNT) program was designed to improve the strategic orientation of small firms.

Table 6 (continued).

The **Portuguese** PEDIP program stengthened the infrastructure supplying consultancy, advisory, and mangement assistance to modernise established industries, and supported productivity, quality and design infrastructure projects. PEDIP2 will have an important focus on enterprise strategy to improve competitiveness, productivity, quality, and human-resource development.
In **Spain** the Technological Development Plan aimed to improve SME performance, and the Industrial Technology Action Plan fosters wider use of advanced technology, an improved technological infrastructure, and specialised training, with special efforts in less-favored regions.
Sweden supplemented its long-running applied research and demonstration programs in manufacturing technology with a subsidized network of technology consultants.
Switzerland has emphasised training and technology upgrading as part of computer-integrated manufacturing initiatives to improve regional diffusion of technology to SMEs, based on established and extensive private-public institutions and networks to promote applied research and technology diffusion.
Turkey is focusing on improving the infrastructure for SMEs, particularly the supply of technical consultants and management expertise, quality and marketing upgrading, and improving the training system.
The **United Kingdom** Enterprise Initiative Consultancy Scheme improved essential management and technical skills in a high proportion of SMEs and increased long-term use of external resources in management. It is being succeeded with a consultancy brokerage and diagnostic service.
The **United States** has many state-level technology diffusion and mangement assistance programs. Third-party services to improve performance are common features. Federal initiatives have expanded and coordinated manufacturing extension.

Source: Compilation drawn from OECD 1995, and OECD, *Industrial Policy in OECD Countries: Annual Review*, various years.

Table 7: Features of Selected Consultancy, Advisory, and Technology Extension Programs

	Australia NIES	Canada AMTAP	Denmark CIM strategy	Germany FUKMU	France ATOUT	Ireland Technology Audit	Japan Kohsetsushi	Norway BUNT	Portugal PEDIP 1,2,5,6	United Kingdom Consultancy Initiatives	United States NYS[1] ITES/IEP
Principal Service areas covered											
Business planning	x	x	x	x	x	x	x	x	x	x	x
Technology assistance	x	x	x	x	x	x	x	x	x	x	x
Design	x				x	x		x	x	x	
Quality	x	x				x	x	x	x	x	x
Marketing	x			x	x	x		x	x	x	x
Management/ Financial Systems	x	x	x	x	x	x		x	x	x	x
Training/ human resources	x	x			x	x	x	x			x
Assisted firm size, upper limit	SMEs	20–100 employees main targets	Mainly SMEs	Industry <DM10M turnover for general consultancy	<2000 employees	<Ir£3M turnover <60 employees larger can participate	<¥100M capital <300 employees	20–100 employees main targets	Mainly SMEs	<500 employees	<500 employees 100 employees main targets
Length of support	Variable	Variable, implementation support in some cases	Variable		Variable, implementation support in some cases	10–15 days	Usually short	20 days (2 phases)	Variable	5–15 days	Variable
Public-sector share of service costs (%) In parentheses ( ) follow-up implemetation share %	50, free diagnosis	Max. 75 (50)	75–50	60–50	50	80 (50)	100–50	75	50	50 66 in special areas 1993: 33 and 50	50 maximum
Number of firms/projects in program per year	2500	130 + group projects	50	24000		120	25000 guidance visits	100 + group projects	200 + group projects	11000 completed	200

[1] The New York State consultancy/extension programs are given as an example. There are increasing numbers of U.S. state-level initiatives supplemented by Federal manufacturing technology extension.

Source: Compiled from national sources, see table 6.

Despite the widespread introduction of consultancy programs, there has been substantial uncertainty about the most effective ways for governments to help strengthen management and technological resources in enterprises. Fortunately, some countries have begun extensive program evaluations. These help determine the strengths and shortcomings of programs and approaches; assist in making program delivery more effective; and clarify what can – and cannot – be achieved by the programs. This chapter summarizes the most important features of consultancy, advisory and industrial extension programs, and draws broad conclusions regarding their design and implementation.[3]

2. Aims and Design of Consultancy, Advisory, and Extension Services

OECD countries have introduced a broad range of special programs to assist small firms because of their economic importance, widespread weaknesses in their performance, and market failures in the demand for and supply of consultancy services and expertise to improve this performance. These programs have sometimes been introduced on a small-scale trial basis relative to the number of potential targets and beneficiaries of the services. Although the provision of small-business advisory services is sometimes modeled along the lines of agricultural extension services, nowhere – except perhaps in Japan – are they as extensive or ambitious as the dense network of agricultural extension services in these countries; this indicates the early and exploratory stage of small-business extension services.

Initial development of publicly backed consultancy and extension services has been aimed at improving the diffusion and use of advanced technologies, particularly those connected with microelectronics and computer-controlled equipment. Recently, services have been extended to a wider range of small firms and to more functional business areas; this is illustrated in Figure 1 by the breadth of the support and service networks shown. These services have also largely been introduced during a period when many of the traditional areas of industrial support, notably for industrial investment, have been declining (see Figure 2). The programs on which this chapter focuses aim to expand effective use of external resources and improve performance by increasing:

[3]Much of the chapter is based upon survey work by: National Academy of Engineering 1993; Competitiveness Policy Council 1993; Hanna, Guy, and Arnold 1994; OECD 1989a; and OECD 1991. References to recent evaluations of programs in individual countries are given in the separate bibliography.

Figure 1: The Complete Network of a Firm

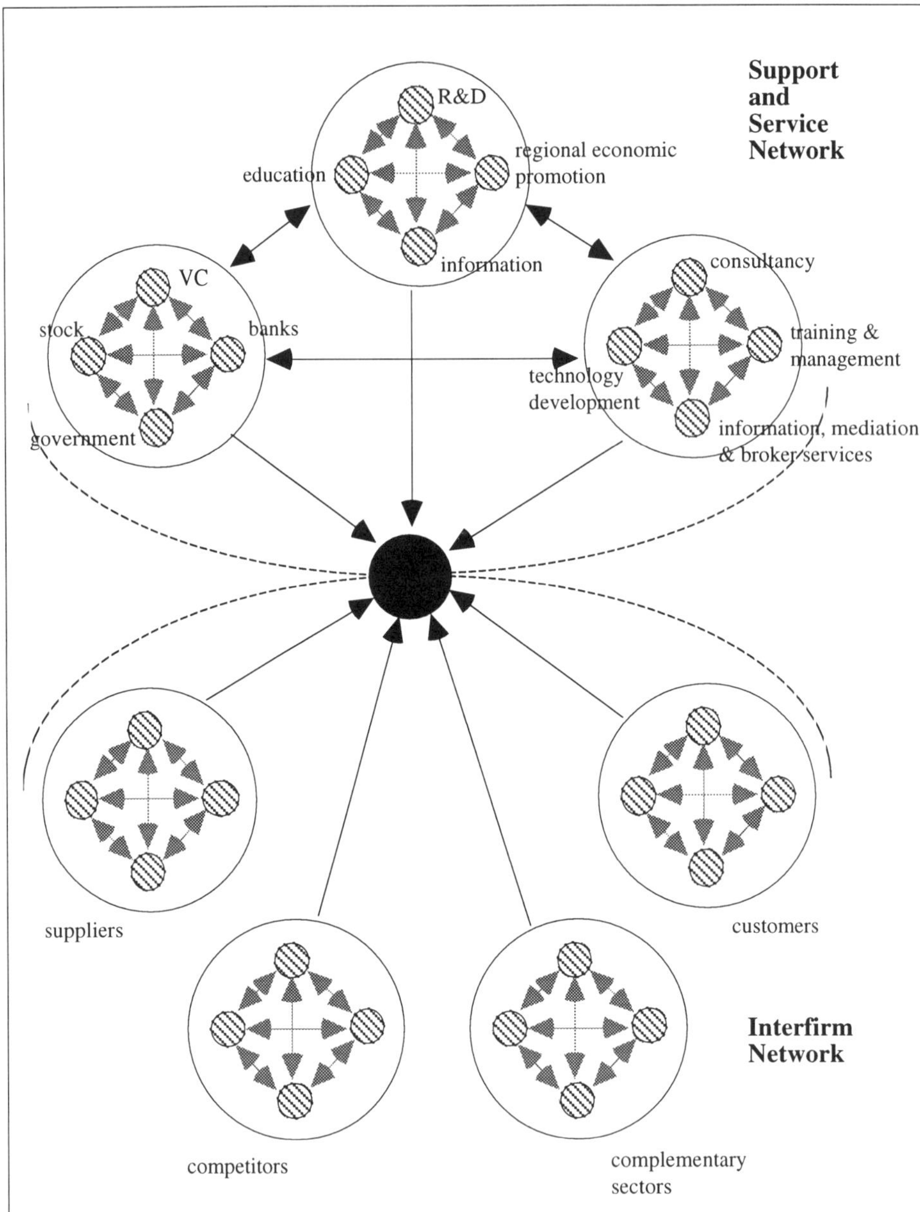

Source: Braunling 1994, p. 93.

Figure 2: Policy Objectives of Reported Industrial Support Expenditures, Net Cost to Government in Current $US Billion

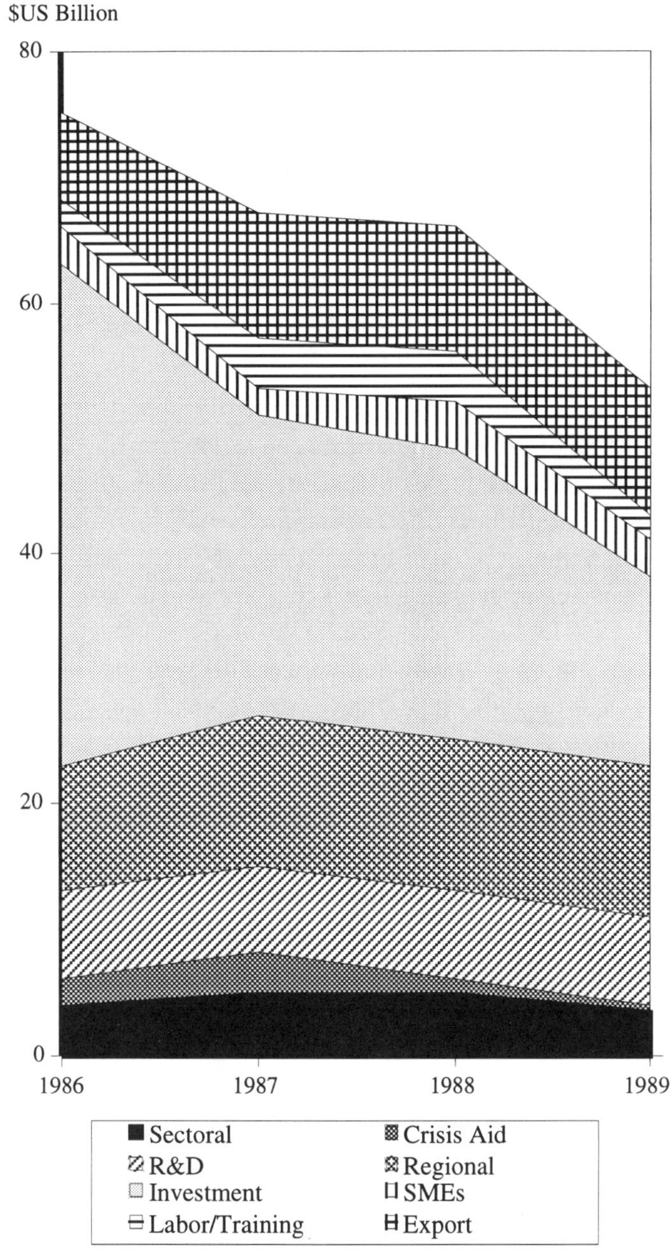

Source: Database for Industrial Support Programmes, OECD.

- awareness of the potential of new manufacturing and management methods, and of problems encountered when implementing strategically-oriented solutions;
- the effective use of resources outside of small firms, particularly through the use of consultancy and advisory services;
- market exploration and market development as part of strategic management approaches;
- the availability of training and the supply of skills and competencies, and by expanding firm-based training;
- the use and effectiveness of publicly backed market-driven intermediaries and enabling institutions which can effectively provide a range of different services, or supply information on them.

These areas were outlined as being particularly important in increasing the effective diffusion of IT-based advanced manufacturing technologies. They have been expanded and extended in many of the consultancy-based programs now in operation in OECD countries.

Typical programs provide a broad range of subsidized management assistance and technology services. They often provide initial services through field agents working for nongovernmental institutions or through private-sector consultants working on site with individual SMEs to provide short-term assistance. Services range from providing very short-term advice in response to specific enquiries; to a few days of consulting at no cost to the client; up to several weeks of assistance with around one-half of the cost paid by the government or delivery agency; and in some cases (but less commonly) assistance with implementation of production and business solutions.

Services provided are designed to:

- help managers strengthen their overall ability to maintain or enhance their competitiveness;
- point out problems and reasons why actions being considered should be modified, deferred, or discarded;
- raise the general level of awareness among managers of SMEs about:
    - turning to consultants and outside organizations for assistance;
    - employing advanced manufacturing technologies and other new technologies in their production activities and products;
    - reviewing management, training, design, marketing, and accounting practices and methods in the light of resource-enhancement strategies recommended by consultancy and advisory services;

- encourage managers to focus on longer-term strategies and problems than those encountered in day-to-day operations and crisis solving that consume management priorities and energies.

2.1 Clarifying Key Issues

Successful programs must address a broad range of questions in the planning stage to ensure that goals are clearly set, targets identified, and implementation methods articulated. Important questions to be addressed include:

- What problems are being addressed; must particular market failures be overcome; and are these short or long term?
- What should be expected of programs introduced to address identified problems?
- What types of clients are best served by consultancy and extension programs?
- Should services emphasize or be restricted to technology applications, or should they be more broadly conceived to include management assistance, product design, quality control, market development, training, and other services?
- How can governments ensure that services respond to market demands and client needs?
- What level of support should be provided for services and for how long?
- Who should provide the services – government agencies, intermediary organizations, universities and higher education institutions, or private consultants?
- How can governments obtain good returns on their investments in support programs?

Country studies and expert opinion suggest that there are common characteristics and indicators of success regarding program design and implementation, and that these are accessible to governments wishing to initiate or continue consultancy and industrial extension programs. These qualities are discussed in the following sections under two headings: program design and implementation.

3. Key Dimensions of Program Design

Having discussed the aims and range of small-business extension programs found in industrial countries, we next turn to the specific dimensions of these programs. Here we examine *additionality*, funding issues, investing in intangibles, provision of services through NGOs, cost sharing, and problems of coordination.

3.1 Additionality: Achieving Positive Returns from Public Expenditures

A common objective of publicly funded industrial support programs is to assure that such investments are complements to – rather than substitutes for – activities undertaken by the firms themselves. This concept of *additionality* means that public expenditures generate private-sector, firm-level activities that would not have occurred in the absence of public funding, and that these result in positive, extra returns. Experience suggests that the greatest potential for additionality is found in small manufacturing firms (40–100 employees) which are not engaged in high-technology and research-intensive areas of electronics, instrumentation, and machinery. These small firms may have most difficulty in obtaining additional technological and management resources to overcome internal weaknesses.

Particular care needs to be taken to target firms with economic promise, but which are otherwise unlikely to initiate improvements or alter strategies without intervention from consultancy or advisory services, that is, to assure that the intervention satisfies the additionality objective. One way of accomplishing this is to narrowly define target groups in order to clearly identify the firms' unmet requirements. Program providers should then be able to build expertise to serve the specific needs of this client group that are likely to achieve additional measurable impacts through service provision.

In Norway, for example, it was estimated that only 10 percent of SMEs have adequate management and technological resources. Firms with fewer than 200 employees typically have very little long-term planning capacity. Those with fewer than 40 employees often rely on one person for management and strategic decision making, and typically the manager has very little time for any longer-term planning at all. This group, particularly firms with 40–100 employees has become the prime target for the national consultancy program (Finne, Levin, and Nilssen 1993).

In the Netherlands, the program to improve subcontracting and outsourcing performance was run on an experimental basis with small groups of firms in eight different product markets to develop general tools which subsequently could be used more widely. The aim was to create improved subcontracting networks that would enhance overall innovative performance among small suppliers as well as increase flexibility, quality, and efficiency. The selected product areas and subcontractor participants were carefully targeted to maximize demonstration effects (Netherlands Ministry of Economic Affairs 1991).

There are, however, tradeoffs between promoting successful projects and maximizing additionality. Early success is important in programs to build the demonstration effect, and to accomplish this, programs may have to initially serve aggressive, technology-oriented, high-potential firms in order to demonstrate early success. Such firms will most probably self-select for programs and be effective early users of services. However, promoted projects are more likely to be those that these firms would have undertaken anyway, thus failing to satisfy the additionality objective.

Thus we see that the firms that are outward looking in their search for resources are also most prone to benefit quickly from extension programs; this is exactly the group that is most likely to implement modernization and improvement strategies successfully. Their strategies contrast with the group which is slower – or unable – to seek external assistance, but for which program assistance may be an important or even necessary factor for improving performance; this is the ultimate target group of firms for extension programs if additionality is to be realized. The important task is to ensure that this second group of firms – more difficult to reach initially – is able to participate fully and benefit from external services to improve internal performance.

3.2 Realizing and Measuring Expected Results

Consultancy and industrial extension programs are designed to overcome market failures in the supply and use of short-term technological services; but typically the level of funding provided is insufficient to reach more than a small proportion of those firms that could benefit from external services. While these services help improve the performance of many individual small firms, the realized benefits may appear modest, while measuring the effects is often difficult. There are several reasons for this:

- The levels of support typically provided are too low to bring about substantial short-term increases in national economic performance.

- The majority of SMEs are slow to seek out and utilize the extension services.
- The causal relationships between government support and specific final economic outcomes have not been fully established.
- It is difficult distinguish actions taken solely on the basis of program recommendations from those which would have taken place anyway or which are undertaken for other reasons.

Public investments to expand the supply of services to SMEs tend to be low compared with SME expenditures on fixed capital, training, and research. Unless public expenditures are extensive, effective outreach only affects a limited proportion of small firms; the consequent short-term impacts on national or even regional economic performance – output, employment, productivity, or exports – are unlikely to be substantial, while even long-term effects are difficult to distinguish.

Furthermore, industry take-up of the services provided is often slow, even with programs which are based on long institutional experience and established infrastructure, such as the Danish computer-integrated manufacturing strategy program (Exhibit 1) or the Georgia Tech Program in the United States (Exhibits 2 and 3). The most economically and technologically adept firms are usually the first to use programs. The large majority of middle-level firms are slower to use external services.

Results to justify program expenditures are generally slow to appear. This difficulty in demonstrating clear results in the short term has significant impacts on program funding. Policy makers need to be convinced of the relationship between program support and clearly established results, at least in the long run, to make the necessary commitments to adequate and stable funding of small-business extension services.

3.3 Funding Levels and Stability

In order to produce significant and lasting results, programs should probably be funded for a minimum of five years and at a level substantial enough to reach 10-20 percent of target firms. This should at least enable the provision of basic information on strategic business planning and generate measurable impact. Observations of programs in Australia and the United States suggest one full-time professional person may be needed for every 100 SMEs in the target group, to provide minimum levels of information and some advisory and assistance services.

Programs in Australia, Germany, Ireland, Japan, Norway, and the United Kingdom have probably been adequate to ensure an effective level of outreach. They at least cover priority target areas (Australia, Ireland, Norway), and in some cases are extensive and long-standing enough to provide broad coverage for a large proportion of small firms (Germany, Japan, the United Kingdom). For this kind of program, broad outreach is an essential feature to increase strategic capabilities and improve the mix of resources and sources of expertise used by a large number of small firms.

3.4 Investing in Intangibles

Successful programs should increase client appreciation of the importance of investment in nonphysical assets (strategic organizational change, training, skill upgrading, design, quality improvements, and marketing). Such intangible investment may be undersupplied in SMEs, because of the difficulty of financing such uncollateralized investments; their untried benefits may be underestimated by SMEs, and quality control of these services raises special problems. The increasing pressures of competition and industrial restructuring in a dynamic environment require that SMEs adopt a broad view of investment that includes purchase of services delivered through consultancy, advisory, and extension services. Increasingly, technology diffusion programs that once emphasized the one-way transfer of technical information are recognizing the complementary necessity of upgrading management skills in small recipient firms; this has led to a redefinition of goals to include continuing and improved education, training, knowledge, and information transfer in an increasingly competitive industrial environment.

3.5 User Involvement I: Use of Nongovernment Intermediary Organizations

Small-business technological extension services generally operate more effectively and efficiently when they are not provided directly by government departments and by civil servants. While funding support from the government is essential, industrial extension services are more effectively provided by nongovernment organizations that depend in part on income from the business community. These intermediary organizations have a range of advantages over direct government provision:

- They promote greater responsiveness to market demands and the needs of SMEs.

- Service delivery tends to be entrepreneurial rather than constrained to narrowly defined services that fit within the mission of a single agency.
- They encourage greater flexibility in personnel management, both in regards to direct employment and in the use of contract personnel.
- Private organizations and industry associations can deliver services with high quality, greater flexibility, and lower costs.

Nongovernmental delivery organizations can take a wide variety of forms including: third-party organizations such as industry associations, industry support institutions, spin-off groups from universities, and private consultants qualified by the government or third-party organizations. Industry and professional associations (as in the United Kingdom) or specialized government-backed technical institutions (as in Norway) can also play a lead role in ensuring the quality of services provided by private consultants as well as in administering day-to-day running of the service delivery. The most commonly used nongovernmental delivery bodies are:

- industry or sectoral associations or special service bodies established by industry or business groups;
- professional societies whose membership covers the skills base needed to provide services; and
- special groups or advisory services jointly established by government and the private sector.

The record of universities in service delivery is mixed. Economic development and technology transfer are often far removed from the mission and culture of higher-education institutions. Notable exceptions are to be found, such as the special missions given to modernization services in a number of U.S. universities (e.g., the industrial extension services at Georgia Tech (Exhibit 2) and the University of Maryland); these programs share the characteristic that they are carefully designed and separately funded services which operate outside the mainstream of university teaching and research.

In a number of countries, broad directions and guidelines for the delivery of comprehensive services are being developed by mixed industry – government – labor bodies. This ensures that the wide range of services delivered reflects not only concerns for improving technical business efficiency, but also takes into account the wider aims of human resource development and management strategies. An example of cooperative formulation of guidelines is the Australian NIES program (Exhibit 4). Policy guidance is jointly formulated by: the Australian Manufacturing Council (government, employers, unions);

the Small Business Council; other private-sector organizations; and a variety of governmental representatives from Commonwealth, State, and Territory industry departments, as well as the Australian trade promotion body.

3.6 User Involvement II: Sharing Costs with Clients

Effective service provision is demand driven and client oriented. When clients bear a substantial proportion of the cost of services they tend to both value them more and ensure that they remain relevant. Joint funding, with firms providing a reasonable share of costs, has four main advantages. Cost sharing:

- ensures that services will be demand driven;
- provides an efficient mechanism for assessing market demand;
- generates user-based incentives to improve the management and efficiency of service delivery; and
- leads to greater use of full-price external services in the future when the supply infrastructure works effectively.

In most countries with extensive programs, the public share is around 50 percent of consultancy costs. In order to promote their use by SMEs, most programs provide initial exploratory consultancy and advice at low or no cost to define possible areas of weakness and suggest possible follow-up. In countries where extensive longer-term implementation and development activities continue to be subsidized, the level of support is usually reduced to around 25 percent. It is unreasonable for programs to be totally self-supporting, with clients paying the entire cost, as this can distort overall aims and efficacy of service delivery. Cost sharing should ensure that services provided are considered worthwhile by the client, with the public sector paying a maximum of one-half of the costs.

3.7 Coordinating a Broad Range of Services

Experience with programs has shown that more than technological solutions are needed to boost small-firm performance. Many programs originated with the objective of providing technology to overcome specific problems implementing new information-technology and production-technology applications. Subsequently, more strategic approaches were adopted where technological functions are only one of a set of functions needing improvement. World-class manufacturing and services, quality and design, strategic management approaches, and advanced human resource management practices are some of the elements of these programs. There has also been an evolution away from many of the traditional approaches to small-firm policy, including the tendency to focus on finance problems.

SMEs which turn to programs for technological assistance usually need a combination of technological, management, organizational, and training services rather than entirely new applications. This has been confirmed by the experience of service providers and their small-firm clients in many countries. This argues for the provision of as broad a range of managerial and technical improvement services as possible, while assuring service quality and timeliness in delivery. Such programs ideally include training and human resource management, along with advising capacities for organizational improvements, quality control, design, business systems, and production management and marketing.

In several consultancy and extension programs there is no special technology focus. In the United Kingdom's Consultancy Initiative, for example, a general initiative covering manufacturing and service systems also covered design and logistics and general reorganization of production involved in adopting new manufacturing and service systems. New programs in Portugal, Spain, and Turkey support building the broad technological infrastructure including advice and consultancy services, and are giving considerable emphasis to training and skill formation.

Technology-oriented programs in Austria and Switzerland on computer-integrated manufacturing also placed extensive emphasis on training and extension services as well as on developing new manufacturing applications for small firms. French programs for the diffusion of microelectronics and new materials to small firms place considerable emphasis on improving management functions and training. There is a separate program focus on increasing SME management and organizational capabilities and information flow by linking different computerized business functions more effectively.

4. Program Implementation

A number of key characteristics are associated with successful implementation of services. Table 8 lists these characteristics generally, while Exhibits 3 and 5 identify the success factors associated with two subnational services in the United States. These characteristics are common across a range of OECD countries, and they are discussed below.

Table 8: Characteristics Associated with Successful Implementation of Advisory, Consultancy, and Extension Programs

Detailed planning processes, including a strategic audit of the rationale, objectives, needs, resources, and methods to design the program.
Adequate and stable funding over a reasonably extended period of time (3-5 years).
Simple program administration and implementation.
Publicity and efforts to build demand within the target group for extension services, particularly for lagging firms.
Considerable freedom for firms in selecting consultants.
Qualification and training of persons providing one-on-one services to individual clients.
Integrated monitoring, assessment, and evaluation that provides quick feedback to program managers 1-2 years after program initiation.

Source: Compiled from expert opinions and case studies.

4.1 Explicit Targeting of Services and Clients at the Planning Stage

Careful, explicit planning characterizes successful programs. Three planning activities that are typically established at the outset of most programs are: identification of objectives, establishment of monitoring methods, and determination of an evaluation procedure. But planning broad-range small-business services also requires:

- definition of target groups for the supply of services;
- determination of new services to provide, as well as how to merge existing services with new activities;
- deciding how to use consultants from the private sector in conjunction with resources from industry associations, other nonprofit organizations, and government, in order to provide an efficiently functioning consultancy service.

The planning process should produce a clear, unambiguous statement of program objectives and clients to be served. For example, the Danish Council of Technology had the overall responsibility for detailed planning of the comprehensive Danish Technology Development Program. The Council established advisory planning groups with broad representation from government agencies and from the private sector, to assist in a planning and implementation phase that lasted for over one year. The PEDIP program in Portugal paid considerable attention to building the consultancy and service infrastructure in an industrializing country, and included clear goals for the technological infrastructure agencies as part of the program.

These and similar examples indicate that preset goals should include some kind of quantifiable economic targets for small-business performance (productivity, employment, profitability – even if not all can be attributed to the services); achievements in product development and process improvements; improvements in the technical characteristics of production, as well as information required for administrative accountability.

It is advantageous for policy development if program objectives are linked with policy monitoring procedures, and midterm and final evaluations are built into them, particularly for programs with fixed time limits. This supports and encourages the early revision of programs where necessary, and helps to build up program expertise in determining targets, impacts, and additionality.

4.2 Building Demand for Services

It is very important to publicize programs to increase awareness of the importance of the strategic functional activities that program services support, and to build demand for these services. In the United Kingdom's Consultancy Initiatives, professional associations were active along with the Department of Trade and Industry (DTI) in promoting the programs, as well as running them as contractors on behalf of the DTI (see Exhibit 6). The independent assessment of the Danish Technology Development Program, including computer-integrated manufacturing consultancy and demonstration (Exhibit 1), concluded:

> Activities aimed at creating publicity about the development being promoted by the program are important in their own right. As for example, demonstration activities, conferences, exchanges of know-how, meetings and publishing regular news letters. The impact of such activities on the dispositions of industry may be equal to or even bigger than the impact of direct support... (National Agency of Industry and Trade 1990, Summary).

In Australia, the National Industry Extension Service initiated a nationwide publicity and information drive to raise awareness of the new scheme, and to ensure that its comprehensive coverage and decentralized state-level delivery were understood. Methods included: a series of meetings and conferences on design, enterprise workshops, export and business awards, an information and awareness program, a national bimonthly information magazine, and a computerized information system. These all contributed to an increased awareness of the problems that the program was designed to address and the means to overcome these problems.

This kind of initial coverage approaches that of the long-established *Kohsetsushi* network of decentralized applied research, technology transfer, and guidance centers in Japan. The *Kohsetsushi* network annually deals with over 3 enquiries or contacts per 100 manufacturing employees per year, equivalent to 2 enquiries for every 3 small and medium-sized manufacturing firms, or to the participation of an estimated 70,000 small Japanese firms (1 in every 10) in horizontal networks encouraged by the 1988 "Fusion Law". By comparison, in Australia there are 1.5 enquiries or contacts per 100 manufacturing employees, equivalent to around 1 enquiry for every 2 manufacturing SMEs. Other countries, for example Finland, have attempted to build on the extensive outreach experience with regional policy to improve the financial and technological capabilities and resources of widely dispersed small firms.

One major challenge facing programs is to encourage small firms to use consultants and external resources on a regular basis. Many small firms are distrustful of government programs and suspect that they will face a substantial amount of administrative and bureaucratic work, often beset by delays, in order to get assistance. Programs can dispel these concerns by having simple qualification and application processes. These can be particularly effective if they are delivered through decentralized contractors or managers who are already part of the technological and business infrastructure and have experience in providing business services to small firms.

4.3 Quality Assurance I: Qualification and Training of Service Providers

The impacts of industrial extension activities depend ultimately on the effectiveness of service providers and consultants. Proficiency in three areas is particularly crucial (as is the avoidance of important pitfalls):

- The ability to independently diagnose the basic strategic problems and evaluate the needs of small firms (rather than accepting the views of the firm management).

- The capacity to develop appropriate solutions to the specific problems identified, and to propose practical methods of implementation (rather than providing prepackaged and predetermined solutions).
- The persuasive talent to convince management that solutions are appropriate and beneficial for the enterprise (rather than presenting the recommendations as an *expert report*).

Service providers who have prior business experience technical and managerial backgrounds, and interpersonal skills are most likely to be effective consultants.

Special training of consultants may be required to ensure that an adequate pool of external sources of expertise of high quality exists, and that potential program deliverers understand program objectives. Norway developed a training and qualification program for consultants wishing to provide industrial consultancy and strategic advisory services to SMEs. In the Norway's *Business Development using New Technology* program, 130 experienced consultants (often, but not always, from consultant groups) were selected from a large number of applicants. They were trained to provide a pool of expertise in strategic analysis and process consulting. Not all consultants who undertook training subsequently provided services, as firms had flexibility within the program to choose consultants. This training package has been adopted by other European countries and by the European Commission in a pilot program to improve consultancy and advisory infrastructures.

4.4 Quality Assurance II: Client Selection of Consultants

To ensure competition in service delivery and a good match between client needs and consultant skills, some programs encourage a wide choice of consultants and advisory services. The freedom of clients to select consultants from a qualified list and the latitude to define the terms of reference most significant to them are valuable features of a number of programs, including the UK Consultancy Initiatives, national programs in Australia, Canada (Exhibit 7), and Norway, and subnational programs in the United States. Freedom of choice may result in higher participation in programs, as consultants work with firms to encourage participation. It may also increase competition among service providers for clients, resulting in improved service quality as well as contributing to general awareness of the value of using external resources.

4.5 Integrated Monitoring, Assessment, and Evaluation

Monitoring and evaluation must be built into programs at the planning stage (Shapira 1993; Shapira, Youtie, and Roessner 1994). Identification of the variables to be measured and methods of collecting firm-level data need to be established before the first services are delivered. This requires unambiguous statements of program objectives and of the characteristics of firm performance to be evaluated. On the other hand, the evaluation should not focus exclusively on short-term, output-oriented measures of performance such as increased production, employment, or profitability. They must also consider changes in nonphysical investments such as R&D, software, organizational change, and training, all of which are vital for long-term viability of small firms.

Responsibility for independent assessment of performance must be decided at an early stage. Ideally there should be a midterm evaluation 1-2 years after program initiation as well as regular collection of monitoring data to enable program adjustment during operation.

The Denmark National Agency of Industry and Trade designed and conducted an exemplary evaluation of the Danish Technological Development Program. The evaluation pointed out that:

> In order to ensure adaptation and/or improvement of the program (once it becomes active), it is necessary (to have) a carefully planned, comprehensive and systematic registration and assessment of relevant data of experience in the program administration. Furthermore, a quick and externally executed evaluation at a relatively early stage of the program (after 1-2 years of experience) would be appropriate (to provide for possible adjustment of the program). (National Agency of Industry and Trade 1990, Summary)

From the outset, the United Kingdom Department of Trade and Industry established procedures to monitor and evaluate the Consultancy Initiatives and allocated around one percent of program funds for this purpose. DTI retained a reputable external firm with a good track record to conduct detailed evaluations, including extensive use of structured face-to-face interviews with managers from firms served by the Consultancy Initiatives. The objectives of the evaluation were to:

- assess impacts of the initiatives on business performance;
- measure changes in attitudes of firms towards using consultants; and
- identify ways to improve program delivery and administration during its lifetime.

5. Lessons from Programs

Industrial Extension Services are intended to solve immediate strategic problems and help build internal capabilities. They are intended to overcome *market failures* in the demand for – and supply of – external services. On the demand side, they demonstrate the value of using external resources and expertise to overcome internal managerial and technological shortcomings. Lack of information, uncertainties, and inertia translate into weak demand for new manufacturing technologies, and management and organizational improvements which are potentially very valuable to small firms. On the supply side there may also be market failures due to problems developing and diffusing production methods and management techniques which are adapted and deliverable to a dispersed population of small firms. Advisory service programs usually have the general aim of expanding the infrastructure of technological and business support services, particularly for firms with fewer than 100 employees.

Important features of successful programs to deliver external services to small firms include:

- Careful attention to rationale and objectives to ensure *additionality* – government funding should not simply displace private funding, but induce positive, incremental changes in firm behavior.
- Broad outreach and awareness to ensure that a wide range of firms use services.
- Supply and coordination of a wide range of services covering the strategic (rather than narrowly technical) needs of the small-business sector.
- An emphasis on often-neglected investment in nonphysical assets, on building capabilities and upgrading managerial and technical skills within the firm, rather than providing it with information.
- Responsiveness to specific local circumstances of the firm and its environment – through, for instance, decentralized local delivery mechanisms – in order to ensure that particular preplanned technological solutions are not overemphasized.
- Simple organization and delivery of services.
- Delivery of services through third-party intermediaries, under the cooperative direction of government and business sectors.
- Ensuring the quality of services delivered – through consultant training and accreditation, and through cost sharing – so that firms value the services provided.

- Incorporation of regular monitoring and early evaluation in initial program design to ensure feedback to program providers, adjustment of service delivery, and better design of future programs.

Evaluations of existing programs establish that they are cost-effective ways of improving firm performance, particularly for firms in the 40-100 employee size group. Firms which participate in these programs usually demonstrate better-than-average output growth, exports, productivity, and profitability. They adopt more strategic approaches to problem solving and are generally more innovative. They maintained better-than-average performance during the recent recession, although evaluations of future performance estimated before the recession may have been too optimistic on positive impacts on firm performance.

Total government investment in these programs is usually small, and subsidy per service-provided is low. To be effective, firms have to devote considerable resources (3-5 times) over and above the subsidy element. Outreach efforts have been extensive, but identification of target firms has been a problem. Participants usually self-select, which means that firms with the greatest potential for success tend to receive assistance. These firms are usually not in the top 10 percent of dynamic high-tech performers, but are usually in the upper-middle group facing new competitive pressures.

Programs to promote networks of small firms have been developed. Although these networks may have high organization and transactions costs, if they are carefully structured they can help overcome the problem of outreach. When comprised of firms in close geographic proximity facing similar problems, these networks can be cost-effective means of identifying, targeting, and providing services to firms with the greatest potential to benefit from improved inputs of strategic management services and technology.

Government programs to build the consultancy and advisory infrastructure cannot substitute for market provision of these services. However, they can extend the supply of these services and expand demand within the small-business sector, with positive effects on both. To do this successfully, programs must be flexible and have clear objectives in addressing particular market failures. Finally, effective programs work within a commercial framework, offering a wide choice of consultants, intermediaries, services, and terms of reference.

Exhibit 1: The Danish Technological Development Program: Computer-Integrated Manufacturing

In 1987, the Danish Government adopted the Computer-Integrated Manufacturing (CIM) Strategy Programme to help firms of all sizes apply CIM technologies. Consulting firms and technological service institutes were qualified for the work and the program supported development of analytic tools, software and training to help consultants and firm management with the CIM strategy.

The program contained three sequential phases:

Phase 1: A free day-long introduction to CIM;

Phase 2: An evaluation to determine whether CIM would be relevant to the enterprise. Slightly over half the costs covered by TDP.

Phase 3: Strategy development to build management resources and the knowledge base to implement a CIM system. Grants from the Danish Industrial Research and Development Fund covered 25 percent of the total costs to a maximum of DKr 200 000.

Source: Compiled from national reports.

Exhibit 2: Georgia Institute of Technology Industrial Extension Service United States

The Industrial Extension Service was established in 1960 at one of the best technical universities in the United States. Using a network of regional offices and field staff, the service provides manufacturers and local communities with information and technical assistance on new technologies, management and marketing assistance, and access to problem-solving engineering skills at the Georgia Institute of Technology.

A firm interested in securing services usually contacts a regional office. An initial site visit to the firm collects information about the business, including production methods, management structure, and financial conditions. Staff diagnose problems, provide recommendations, or tap the expertise elsewhere at Georgia Tech or from consultants.

Staff provide each client with a few days of service free of charge. Clients pay for the additional costs of extended consultancies.

Source: Compiled from national reports.

Exhibit 3: Factors for Success of the Georgia Institute of Technology Industrial Extension Service, United States

Provide timely and quality services.

Employ a highly qualified professional staff with relevant technical expertise and a broad understanding of the small manufacturers' environment.

Be independent in the sense that it has no stake in the outcome of the advice other than success of the firm.

Be affiliated with a trusted institution and establish credibility with the local business community.

Maintain personal contact with the firm's owner or manager.

Supplement technology assistance with overall business assistance (human resources, information systems, marketing, finance). Technology changes affect other aspects of operating and managing the small firm.

Provide follow-up. Implementing new management and technology systems in small firms may require considerable time.

Source: Adapted from: D.S. Clifton (1988), "Georgia Tech Research Institute: 30 Years of Outreach", mimeo, Georgia Institute of Technology, Atlanta, GA, USA.

Exhibit 4: The Australian National Industry Extension Service

The National Industry Extension Services (NIES) was established in 1986, as a joint initiative between the Commonwealth, State and Territory governments to improve the performance of small and medium-sized manufacturing and service companies.

NIES aims to assist small firms attain international competitiveness by using outside expertise to improve management functions. The long-term aim is to develop the private sector infrastructure of consultancy services to support SME development.

NIES provides information and referral services and access to a network of private sector consultants, and subsidizes enterprise improvement projects. Services are delivered largely at the state level either by state government departments or by special agencies. They cover strategic and financial planning, design, quality, human resources, networking, and new market development.

Information and advice is free. NIES pays 50 percent of consultancy costs with limits depending on the programme.

Source: Compiled from national reports.

Exhibit 5: Important Factors for Success The NorthEast Manufacturing Technology Center, United States

If there is no learning on the part of the firm, there is no technology transfer. Training and technical support are integral parts of the implementation of the new technology.

The small firm usually does not have the skilled personnel needed to take advantage of advanced technology. Direct technical support is not only required during implementation but must be available to the firm after the project is completed.

Consultants generally do not provide technology transfer services to the small firm. Commercial services are often too expensive for the firm and require commitments of personnel on the part of the consultant that could be used to better advantage on other larger clients.

Small firms need a neutral party to aid them in evaluating alternatives. The majority of small firms receive information on new technologies from sales people who are trying to sell them a particular kind and brand of technology (usually hardware), and do not provide a range of options and integrated solutions covering a wide range of services.

The small firm needs a series of projects timed to match the ability of the firm to absorb each change. The overall impact may take well over a year.

The NorthEast Manufacturing Technology Center was one of the first of seven centers set up on a regional basis in the United States, financed initially through the National Institute of Standards and Technology. They are designed to improve the manufacturing performance of small and medium-sized firms through the application of integrated modernization strategies involving advanced manufacturing technologies.

Source: Adapted from G.R. Simons, "The Experience of the NorthEast Manufacturing Technology Center", mimeo, Rensselaer Polytechnic Institute, Troy, NY, USA.

Exhibit 6: The United Kingdom Consultancy Initiatives

In 1988, the United Kingdom Department of Trade and Industry (DTI) launched six related consultancy initiatives as a comprehensive, self-help package of advice and guidance to British business, aimed at improving the performance of small and medium-sized enterprises with less than 500 employees. These initiatives were managed in a closely coordinated fashion by third-party service deliverers and all involved the use of subsidized private sector consultants to help SMEs. Service firms and manufacturing firms participated in the initiatives.

The Consultancy Initiatives enabled SMEs to use consultants for: business planning including developing business objectives; product design and design management; financial and information systems; manufacturing and services systems and the introduction of modern production and handling methods; marketing; quality management and establishment of total quality approaches to operations.

DTI paid half of the costs (and two-thirds in Assisted and Urban Programme Areas) of five to fifteen days of consultant assistance, with average DTI contribution around £3,400. Subsidy was reduced to one-third and one-half respectively in 1993. The client firm selected consultants from a roster of individuals and firms qualified by the service deliverers.

Source: Compiled from national reports.

Exhibit 7: The Canadian Advanced Manufacturing Technology Application Program

AMTAP, established in 1989, provided financial assistance to help SMEs hire outside expertise to investigate the feasibility of upgrading manufacturing through application of advanced manufacturing technologies (such as CAD/CAM, robotics, flexible manufacturing and total quality control). The program contributed to costs of consultant evaluations of the commercial and technical feasibility of AMT applications to substantially upgrade manufacturing processes, review long-term strategy, define areas of improvement, prepare an implementation plan, and in some cases provide implementation assistance.

Up to 75 percent of consultant costs up to C$25,000 were covered. Requests were not funded if they involved routine application of new equipment or methods, were already under way, or were supported by third parties. Consultancies concentrated equally on implementation of AMT technology, production planning and related management practices and training and human resources.

Source: Compiled from national reports.

References

Braunling, G. 1994. The innovation infrastructure: A systems approach and role of the European Community (in Hebrew). In *Modernization and Growth in the Small Business Sector: Lessons for Public Policy*, edited by M. Justman. Jerusalem: Industrial Development Policy Group, The Jerusalem Institute for Israel Studies.

Competitiveness Policy Council. 1993. *Forging the Future: Policy for American Manufacturing*. Washington, D.C.

Hanna, N, K. Guy, and E. Arnold. 1994. Information technology diffusion: Experience of industrial countries and lessons for developing countries. World Bank. Mimeo.

National Academy of Engineering. 1993. *Mastering a New Role. Shaping Technology Policy for National Economic Performance.* Washington, D.C.

National Agency of Industry and Trade. 1990. Summary of *Evaluation of the Danish Technological Development Program*.

OECD. 1985. Employment outlook. Chap. 4 in *Employment, Small and Large Firms: Where Have the Jobs Come From?* Paris.

—. 1989a. *Government Policies and the Diffusion of Microelectronics*. Paris.

—. 1989b. Part III of *Industrial Policy in OECD Countries, Annual Review 1989* (and subsequent *Annual Reviews*). Paris.

—. 1989c. *Labour Market Flexibility: Trends in Enterprises*. Paris.

—. 1989d. *New Technologies in the 1990s: A Socio-Economic Strategy*. Paris.

—. 1990. Seminar on problems financing SMEs. Mimeo. Paris.

—. 1991. *Managing Manpower for Advanced Manufacturing Technology*. Paris.

---. 1993. *Small and Medium-sized Enterprises: Technology and Competitiveness*. Paris.

—. 1994. Employment outlook. Chap. 3 in *Job Gains and Job Losses in Firms*. Paris.

—. 1995. *Boosting Businesses: Advisory services*. Paris.

Sauviat, C. 1991. Conditions d'émergence et caractéristiques du développement d'un marché de l'expertise et du conseil en France. Document de travail 91.01. IRES.

Sengenberger, W., G. Loveman and M.J. Piore, eds. 1990. *The Re-emergence of Small Enterprises: Industrial Restructuring in Industrialised Countries.* Geneva: International Institute for Labour Studies.

Shapira, P. 1993. Best practices for industrial modernization. Georgia Institute of Technology. Mimeo.

Shapira, P., J. Youtie, and J. D. Roessner. 1994. Evaluating industrial modernization programs: Issues, practices, and strategies. Georgia Institute of Technology. Mimeo.

Country-Specific Evaluations

Australia

Bureau of Industry Economics. 1992. The National Industry Extension Service (NIES) and industry development. Occasional Paper No. 9.

Bureau of Industry Economics. 1993. R&D, innovation and competitiveness – An evaluation of the research and development tax concession. Research Report 50, AGPS, Canberra.

Price Waterhouse. 1992. A*n Evaluation of the National Industry Extension Service.* Vol. 1 and 2.

Canada

Mudde, M. and T. Colven. 1991. General conclusions drawn from the first 122 evaluations of AMTAP consultants' final reports. ISTC. Mimeo.

Mudde, M. 1993. ISTC program evaluation study: Advanced Manufacturing Technology Application Program. (AMTAP). Mimeo.

Denmark

National Agency of Industry and Trade (NAIT). 1990. *Evaluation of the Danish Technological Development Program.*

Ireland

Fitzpatrick Associates and EOLAS Science and Technology Evaluation Unit. 1992. Review of the Technology Audit Program.

EOLAS. 1992. Inventory of the Irish National Consultancy System: Euro-Bunt Project. Mimeo.

Japan

Shapira, P. 1991. Japan's Kohsetsushi Program of Regional public examination and technology centers for upgrading small and mid-size manufacturing firms. Mimeo.

Vickery, G. and Blair, L. 1989. *Government Policies and the Diffusion of Microelectronics*. OECD.

Netherlands

Ministry of Economic Affairs. 1991. *Over het Programma Toeleveren & Uitbesteden Voortgangsrapportage.*

Ministry of Economic Affairs. 1991. *Programma Kwaliteit en Logistiek Voortgangsrapportage.*

Norway

Finne, H., M. Levin, and T. Nilssen. 1993. *Evaluation of the BUNT Program – Summary report*. SITEF IFIM. Trondheim.

Portugal

Ministry of Industry and Energy. 1993. Relatório de Exeção PEDIP 1988-92.

Sweden

Arnold, E. and K. Guy. 1992. *Evaluation of the IT4 Program: Final report*. Technopolis and SPRU. Brighton, UK.

Arnold, E. and J. Bessant. 1993. *NUTEK evaluation of the AMT Action Areas*. Technopolis and Brighton University. Brighton, UK.

Eriksson, A. H. and A. Nilsson. 1989. Renewal funds – implementation, evaluation, distribution. The Swedish Center for Working Life.

United Kingdom

Segal Quince Wicksteed. 1989, 1991, 1994. *Evaluation of the Consultancy Initiatives*.

United States

Shapira, P. 1993. Best practices for industrial modernization. Georgia Institute of Technology. Mimeo.

Shapira, P., J. Youtie, and J. D. Roessner. 1994. Evaluating industrial modernization programs: Issues, practices, and strategies. Georgia Institute of Technology. Mimeo.

Economics of Science, Technology and Innovation

1. A. Phillips, A.P. Phillips and T.R. Phillips: *Biz Jets*. Technology and Market Structure in the Corporate Jet Aircraft Industry. 1994 ISBN 0-7923-2660-1
2. M.P. Feldman: *The Geography of Innovation*. 1994 ISBN 0-7923-2698-9
3. C. Antonelli: *The Economics of Localized Technological Change and Industrial Dynamics*. 1995 ISBN 0-7923-2910-4
4. G. Becher and S. Kuhlmann (eds.): *Evaluation of Technology Policy Programmes in Germany*. 1995 ISBN 0-7923-3115-X
5. B. Carlsson (ed.): *Technological Systems and Economic Performance: The Case of Factory Automation*. 1995 ISBN 0-7923-3512-0
6. G.E. Flueckiger: *Control, Information, and Technological Change*. 1995 ISBN 0-7923-3667-4
7. M. Teubal, D. Foray, M. Justman and E. Zuscovitch (eds.): *Technological Infrastructure Policy*. An International Perspective. 1996 ISBN 0-7923-3835-9
8. G. Eliasson: *Firm Objectives, Controls and Organization*. The Use of Information and the Transfer of Knowledge within the Firm. 1996 ISBN 0-7923-3870-7

KLUWER ACADEMIC PUBLISHERS – DORDRECHT / BOSTON / LONDON